Multi-Method (Geo-) Thermochronology and Trace Elements Tracing Magmatism, Mineralization and Tectonic Evolution

Multi-Method (Geo-) Thermochronology and Trace Elements Tracing Magmatism, Mineralization and Tectonic Evolution

Guest Editors

Fan Yang
Cun Zhang
Jian Chang

MDPI

Basel • Beijing • Wuhan • Barcelona • Belgrade • Novi Sad • Cluj • Manchester

Guest Editors

Fan Yang
School of Earth Sciences
Lanzhou University
Lanzhou
China

Cun Zhang
School of Materials Science
and Engineering
Qilu University of Technology
Jinan
China

Jian Chang
National Key Laboratory
of Petroleum Resources
and Engineering
China University of Petroleum
Beijing
China

Editorial Office
MDPI AG
Grosspeteranlage 5
4052 Basel, Switzerland

This is a reprint of the Special Issue, published open access by the journal *Minerals* (ISSN 2075-163X), freely accessible at: www.mdpi.com/journal/minerals/special_issues/9K5W6A8N9F.

For citation purposes, cite each article independently as indicated on the article page online and using the guide below:

Lastname, A.A.; Lastname, B.B. Article Title. *Journal Name* **Year**, *Volume Number*, Page Range.

ISBN 978-3-7258-3406-8 (Hbk)
ISBN 978-3-7258-3405-1 (PDF)
https://doi.org/10.3390/books978-3-7258-3405-1

Contents

minerals

Editorial

Editorial for the Special Issue "Multi-Method (Geo-) Thermochronology and Trace Elements Tracing Magmatism, Mineralization and Tectonic Evolution"

Fan Yang [1,*], Cun Zhang [2] and Jian Chang [3]

[1] Key Laboratory of Mineral Resources in Western China (Gansu Province), School of Earth Sciences, Lanzhou University, Lanzhou 730000, China
[2] School of Materials Science and Engineering, Qilu University of Technology (Shandong Academy of Sciences), Jinan 250353, China; gemzhangc@qlu.edu.cn
[3] National Key Laboratory of Petroleum Resources and Engineering, China University of Petroleum, Beijing 102249, China; changjian@cup.edu.cn
* Correspondence: fanyang@lzu.edu.cn

check for updates

Received: 6 February 2025
Accepted: 10 February 2025
Published: 11 February 2025

Citation: Yang, F.; Zhang, C.; Chang, J. Editorial for the Special Issue "Multi-Method (Geo-) Thermochronology and Trace Elements Tracing Magmatism, Mineralization and Tectonic Evolution". *Minerals* **2025**, *15*, 169. https://doi.org/10.3390/min15020169

With the rapid development of analytical techniques, especially the in situ Lu-Hf, Rb-Sr, U-Pb, fission-track and (U-Th)/He dating of garnet, zircon, apatite and other accessory minerals, several important geological issues have been successfully resolved or re-determined in the past decade [1–4]. Additionally, machine learning approaches help us determine the genetic types of different mineral deposits [5]. Among these, precise temporal evolution and duration of magmatism and mineralized processes, as well as uplift and exhumation history of mineral deposits and basins within orogenic belts and cratons, can assist the exploration of mineral and petroleum resources [6–9]. Further, in situ mineral-scale trace element concentrations could also be simultaneously determined using LA-ICP-MS in situ accessory mineral dating; this would provide new perspectives on the formation and evolution of major geological objects via integration with corresponding ages [1,2]. Although early studies focused on most scientific issues in magmatism, mineralization and tectonic evolution within different geological objects, several aspects of these investigations have not shown agreement.

Thus, this Special Issue, "Multi-Method (Geo-) Thermochronology and Trace Elements Tracing Magmatism, Mineralization and Tectonic Evolution", highlights state-of-the-art analytical techniques for exploring accurate formation and tectonic mechanisms, as well as new scientific issues related to geological objects (e.g., intrusions, mineral deposits, basins). In total, seventeen manuscripts were submitted for consideration for the Special Issue, and all of them were subject to the rigorous peer review process of *Minerals*. Eventually, ten articles with high qualities contributed by 62 different researchers from about 37 different institutions were accepted for publication and inclusion in this Special Issue.

The contributions are listed in Table 1; most (1, 2, 3, 4, 6, 7, 8, 10) relate to the field of petrology, mineralogy and mineral deposits. Contribution 5 focuses on sedimentology and tectonics, and contribution 9 involves regional tectonics. All submitted articles strictly adhere to the requirement and scope of the Special Issue. The authors employ advanced methodologies including in situ trace element geochemistry, stable isotope geochemistry (e.g., He, Ar, S), multi-technique and high-precision geochronology (e.g., zircon U-Pb, Rb-Sr dating) and thermochronology (e.g., apatite fission-track, (U-Th)/He), as well as simulation and machine learning methods, to conduct comprehensive studies on different rocks, minerals and regional terranes.

Table 1. Detailed analysis of the ten contributions published in this Special Issue.

Contribution N#	Title	Research Area	Author
1	Petrogenesis of the Laoshan Suite in the Jiaodong Peninsula (Eastern China): An Oxidized Low Ba–Sr A1-Type Granite	Petrology; mineral deposits	Li et al.
2	Genesis of the Daliuhang Gold Deposit, Jiaodong Peninsula, Eastern China: Constraints from H-O-S-Pb-He-Ar Isotopes, and Geochronology	Mineral deposits	Tian et al.
3	Mineralogy and Geochemical Characteristics of Scheelite Deposit at Xuebaoding in Pingwu, Sichuan Province	Mineralogy	Cao et al.
4	Mineralogical Characteristics and Genetic Types of Pyrite with Different Occurrence: Constraints from Spectroscopy, Geochemistry and δ^{34}S Stable Isotopes	Mineralogy	Ma et al.
5	Present-Day Geothermal Regime and Thermal Evolution of the Fukang Sag in the Junggar Basin, Northwest China	Sedimentology; Tectonics	Guo et al.
6	Occurrence of State of Gold in Crude Oil and Its Economic Significance	Mineral deposits; Petroleum geology	Ni et al.
7	In Situ Geochemical and Rb–Sr Dating Analysis of Sulfides from a Gold Deposit Offshore of Northern Sanshandao, Jiaodong Peninsula, North China: Implications for Gold Mineralization	Mineralogy; Mineral deposits	Tian et al.
8	Interpretable Machine Learning for Geochemical Anomaly Delineation in the Yuanbo Nang District, Gansu Province, China	Mineral deposits	Zhang et al.
9	Thermochronological Constraints on the Tectonic History of the Arabian–Nubian Shield's Northern Tip, Sinai, Egypt	Tectonics	Mansour et al.
10	Detrital Zircon Dating, Deformation Stages, and Tectonics of the Pane Chaung Formation and Surrounding Units in the Western Indo-Burma Range, Southeast Asia	Petrology; Tectonics	Zhang et al.

These investigations elucidate magmatic activities, mineralization processes and tectonic evolution within and around orogenic belts and cratons. Contributions 1, 2, 6, 7 and 8 investigate gold enrichments and mineralization. Through extensive experimental research on the affinity of gold with rock bodies including granite and associated Au-hosting minerals such as pyrite and crude oil, these articles offer unique insights into the tectonic environments of gold mineralization and precipitation. Other articles, such as contributions 3 and 4, focus on the mineralogy and formation process of scheelite and pyrite. The remaining two studies (9, 10) focus on a specific area to reveal the issues of tectonic evolution for different geological objects. The details of all contributions are provided below.

Contribution 1 conducts comprehensive analyses of whole-rock major and trace elements, along with LA–ICP–MS zircon U–Pb geochronology and geochemistry, to explore the genetic link between gold mineralization and Mesozoic granites in the Jiaodong Peninsula, Eastern China. The study concludes that oxidized adakitic magmas likely played a significant role in gold enrichment and mineralization. In contrast, the Laoshan A-type granite, characterized by high oxygen fugacity, appears to inhibit gold formation, primarily due to its distal location from the ore-controlling fault zones.

Contribution 2 systematically investigates the genesis of the Daliuhang gold deposit (Jiaodong Peninsula, Eastern China) through multi-isotopic (H-O-S-Pb-He-Ar) geochemistry and geochronology. The results demonstrate that mantle-derived gold-bearing fluids enriched in volatiles migrated along tectonic weak zones. Fluid phase separation at granite-hosted fault sites facilitated gold extraction from granodiorite, ultimately forming economically significant mineralization.

Contribution 3 characterizes gem-grade scheelite mineralization in the Xuebaoding deposit (Pingwu, Western China) through mineralogical and geochemical analyses. The study establishes a genetic association with A-type granites, constraining scheelite formation to the Early Jurassic under low-sodium conditions.

Contribution 4 employs advanced analytical methods to differentiate sedimentary versus hydrothermal pyrite based on crystallinity, trace element signatures and sulfur isotopes. Sedimentary pyrite exhibits sulfur sourced from granite–sedimentary interactions, while hydrothermal varieties show mixed mantle-derived sulfur with seawater sulfate contributions.

Contribution 5 reconstructs the thermal evolution of the Fukang Sag (Junggar Basin, NW China) using thermochronology (apatite (U-Th)/He, fission-track) and thermal modeling. Findings indicate Late Triassic maturation of Carboniferous–Permian source rocks in the eastern sag, highlighting implications for hydrocarbon generation.

Contribution 6 investigates gold occurrence mechanisms in crude oil from the Linnan Depression (Bohai Bay Basin, Eastern China). Gold enrichment is attributed to complexation with organic acids/thiols, establishing crude oil as a potential transport medium for precious metals under specific geological conditions.

Contribution 7 (by the same team as Contribution 2) employs in situ sulfide geochemistry and Rb-Sr dating to constrain gold mineralization processes in the Jiaodong Peninsula, Eastern China. The results propose a mixing model between deep magmatic–hydrothermal fluids and seawater as the critical precipitation mechanisms.

Contribution 8 pioneers interpretable machine learning applications for geochemical anomaly detection in the Yuanbo Nang District, NW China, with an aim to extract richer insights from the ML modeling of a geochemical exploration dataset. The work elucidates the impact of different trace elements (e.g., As, Sb) on gold mineralization. The findings hold substantial guiding value for future mineral exploration and geological surveys.

Contribution 9 reconstructs the tectonic history of the Arabian–Nubian Shield's northern margin (Sinai, Egypt) through thermochronology, identifying four cooling phases linked to (1) Neoproterozoic post-accretion erosion, (2) Devonian–Carboniferous Variscan tectonism, (3) Cretaceous Gondwana breakup and (4) Oligocene–Miocene Gulf of Suez rifting.

Contribution 10 utilizes detrital zircon geochronology to decipher deformation stages in the Indo-Burma Range's Pane Chaung Formation. The results reveal dual subduction dynamics during Gondwana–microcontinent interactions in SE Asia, and the characteristic of orogenic doublet subduction systems in the Indo-Myanmar orogenic belts.

In the Special Issue, most researchers employed traditional approaches to conduct investigations into the genetic mechanisms and tectonic evolution of certain minerals, rocks and other geological objects. However, all the contributions ultimately failed to contemplate establishing tectonic model diagrams to clearly interpret the fluid migration, source characteristics and enrichment processes of different elements. Quantitative evaluation in this area is required in the future. In the Special Issue, five contributions, accounting for 50% of total articles, focus on gold deposits, highlighting that formation mechanisms and tectonic evolution models of gold deposits remain significant research hotspots and challenges in contemporary academia. Despite extensive studies investigating the precipitation mechanisms of gold, most research explores its pre-mineralization, formation processes and age information from associated Au-hosting minerals. In addition, studies demonstrating temporal evolution, tectonic exhumation and preservation potential are still lacking in research on the post-mineralization assessment of gold deposits [10].

In summary, the Special Issue confirms that a growing number of researchers are working on various fields of mineralogy, petrology, mineral deposits and tectonics and sedimentology. Certainly, future research directions ought to be further pursued in terms of new production areas, new methods and interdisciplinary intersections. Hopefully, this series of articles will contribute to a better understanding of scientific issues of magmatism, mineralization and tectonic evolution in different geological objects.

Author Contributions: Conceptualisation, F.Y. and C.Z.; methodology, F.Y. and C.Z.; writing—original draft preparation, all authors; writing—review and editing, F.Y., C.Z. and J.C. All authors have read and agreed to the published version of the manuscript.

Data Availability Statement: The original contributions presented in the study are included in the article material; further inquiries can be directed to the corresponding authors.

Acknowledgments: The Guest Editors sincerely thank all authors, reviewers and editorial staff of *Minerals* for their efforts in publishing this Special Issue. This Special Issue was sponsored by the National Natural Science Foundation of China (42202077).

Conflicts of Interest: The authors declare that they have no known competing financial interests or personal relationships that could have appeared to influence the work reported in this paper.

References

1. Zhang, L.J.; Zhou, T.F. Minerals in-situ LA-ICPMS trace elements study and the applications in ore deposit genesis and exploration. *Acta Petrol. Sin.* **2017**, *33*, 3437–3452, (In Chinese with English Abstract).
2. Deng, J.; Qiu, K.F.; Wang, Q.F.; Goldfarb, R.; Yang, L.Q.; Zi, J.W.; Geng, J.Z.; Ma, Y. In situ dating of hydrothermal monazite and implications for the geodynamic controls on ore formation in the Jiaodong gold province, eastern China. *Eco Geol.* **2020**, *115*, 671–685. [CrossRef]
3. Ketcham, R.A.; Carter, A.; Donelick, R.A.; Barbarand, J.; Hurford, A.J. Improved modeling of fssion-track annealing in apatite. *Am. Mineral.* **2007**, *92*, 799–810. [CrossRef]
4. Flowers, R.M.; Ketcham, R.A.; Shuster, D.L.; Farley, K.A. Apatite (U–Th)/He thermochronometry using a radiation damage accumulation and annealing model. *Geochim. Cosmochim. Acta* **2009**, *73*, 2347–2365. [CrossRef]
5. Qiu, K.F.; Zhou, T.; Chew, D.; Hou, Z.L.; Müller, A.; Yu, H.C.; Lee, R.G.; Chen, H.; Deng, J. Apatite trace element composition as an indicator of ore deposit types: A machine learning approach. *Am. Mineral.* **2024**, *109*, 303–314. [CrossRef]
6. Alam, M.; Li, S.R.; Santosh, M.; Shah, A.; Yuan, M.W.; Khan, H.; Qureshi, J.A.; Zeng, Y.J. Morphological, thermoelectrical, geochemical and isotopic anatomy of auriferous pyrite from the Bagrote valley placer deposits, North Pakistan: Implications for ore genesis and gold exploration. *Ore Geol. Rev.* **2019**, *112*, 103008. [CrossRef]
7. Poulin, R.S.; Kontak, D.J.; McDonald, A.; McClenaghan, M.B. Assessing scheelite as an ore-deposit discriminator using its trace-element and REE chemistry. *Can. Mineral.* **2018**, *56*, 265–302. [CrossRef]
8. Chen, Y.J.; Santosh, M. Triassic tectonics and mineral systems in the Qinling Orogen, central China. *Geol. J.* **2014**, *49*, 338–358. [CrossRef]
9. Dong, Y.P.; Zhang, G.W.; Neubauer, F.; Liu, X.M.; Genser, J.; Hauzenberger, C. Tectonic evolution of the Qinling orogen, China: Review and synthesis. *J. Asian Earth Sci.* **2011**, *41*, 213–237. [CrossRef]
10. Yang, F.; Wen, Y.; Jepson, G.; Santosh, M.; Wu, L.; Shen, X.; Ali, H. Prolonged exhumation and preservation of the Yuku molybdenum ore field, East Qinling, China: Constraints from medium-to low-temperature thermochronology. *Ore Geol. Rev.* **2024**, *167*, 105973. [CrossRef]

minerals

MDPI

Article

Detrital Zircon Dating, Deformation Stages, and Tectonics of the Pane Chaung Formation and Surrounding Units in the Western Indo-Burma Range, Southeast Asia

Ji'en Zhang [1,*], Wenjiao Xiao [1,2,*], John Wakabayashi [3], Fulong Cai [4] and Kyaing Sein [5]

1 State Key Laboratory of Lithospheric and Environmental Coevolution, Institute of Geology and Geophysics, Chinese Academy of Sciences, Beijing 100029, China
2 Xinjiang Research Center for Mineral Resources, Xinjiang Institute of Ecology and Geography, Chinese Academy of Sciences, Urumqi 830011, China
3 Department of Earth and Environmental Sciences, California State University, Fresno, CA 93740, USA; jwakabayashi@csufresno.edu
4 Key Laboratory of Continental Collision and Plateau Uplift, Institute of Tibetan Plateau Research, Chinese Academy of Sciences, Beijing 100101, China; flcai@itpcas.ac.cn
5 Myanmar Geosciences Society, Yangon 11052, Myanmar; kyaingsein@gmail.com
* Correspondence: zhangjien@mail.iggcas.ac.cn (J.Z.); wj-xiao@mail.iggcas.ac.cn (W.X.)

Abstract: The Indo-Burma Range (IBR), as one of the youngest accreted units in the Eastern Neotethys, plays a crucial role in understanding the interactive relationships between the Gondwana supercontinent and its rifted microcontinents in SE Asia. However, its basement nature and tectonic evolution remain debated. Here, we conducted a comprehensive structural analysis across six sections within the IBR and correlated Late Triassic flysch units between the Western IBR (Pane Chaung Formation) and the Tethyan Himalaya. Within the Mindat section, the eastern segment of the Pane Chaung Formation unit displays top-to-east vergent overturned folds, indicating eastward backthrusting, in contrast to the prevailing top-to-west vergence structures in Kalemyo, Natchaung, Magwe and the western segment of the Mindat flysch unit. By reconstruction of this backthrust sheet, a megathrust separates the Pane Chaung Formation unit in the footwall to the west from schist units in the hanging wall to the east. The Pane Chaung Formation unit in the Western IBR and its counterparts in the Tethyan Himalaya share common characteristics, including herringbone cross-beddings, Carnian–Norian *Halobia* fossils, and dominant detrital zircons of 220–280, 500–620, 900–1000, and 1100–1140 Ma. Alongside the Paleozoic strata and Precambrian one-stage model ages of Mesozoic dikes, as evidenced by $\varepsilon_{Nd}(t)$ (−13.4 to −0.1) and $\varepsilon_{Hf}(t)$ (−24.2 to −0.1) in the Tethyan Himalaya, these facts suggest that the major tectonic units of the Western IBR–Tethyan Himalaya are the result of the amalgamation of a microcontinent with the West Burma Block. The transition from OIB to E-MORB and N-MORB, the rapid deepening of sedimentary waters, and the presence of the 155–152 Ma Indian ocean crust collectively indicate that the microcontinent rifted from the host East Gondwana as a fragment of the Argoland archipelago in the Late Jurassic. This identification sheds light on the orogenic processes of the doublet subduction zones in the Indo-Myanmar orogenic belt.

Keywords: Indo-Burma Range; Myanmar; Argoland; Neotethys Ocean; microcontinent; Triassic

check for updates

Citation: Zhang, J.; Xiao, W.; Wakabayashi, J.; Cai, F.; Sein, K. Detrital Zircon Dating, Deformation Stages, and Tectonics of the Pane Chaung Formation and Surrounding Units in the Western Indo-Burma Range, Southeast Asia. *Minerals* **2024**, *14*, 1290. https://doi.org/10.3390/min14121290

Academic Editor: Jean-Michel Lafon

Received: 4 November 2024
Revised: 10 December 2024
Accepted: 12 December 2024
Published: 19 December 2024

1. Introduction

The interactions between the Eastern Gondwana and Laurasia supercontinents formed the collisional orogen in Tibet and the accretionary orogen in SE Asia (Figure 1a) [1–4]. Bridging these two orogenic belts results in the Indo-Myanmar orogenic belt of the Eastern Neotethys (Figure 1a). There, the Indo-Burma Range (IBR), recognized as one of the youngest accreted units [5–7], acts as a crucial locus for deciphering the complex interplay between accretion and collision processes. The IBR is located in Myanmar and Nagaland,

India (Figure 1b) [5–7]. Numerous geologists have conducted extensive investigations in this region [5,6,8–13]. However, the nature and tectonic affinities of its major units remain debated: a microcontinent [1,10], an accretionary complex [13], or a composite of a microcontinent and accretionary complex [5,14]. The basic evidence and/or arguments for the nature and tectonic affinities of its major units include the following.

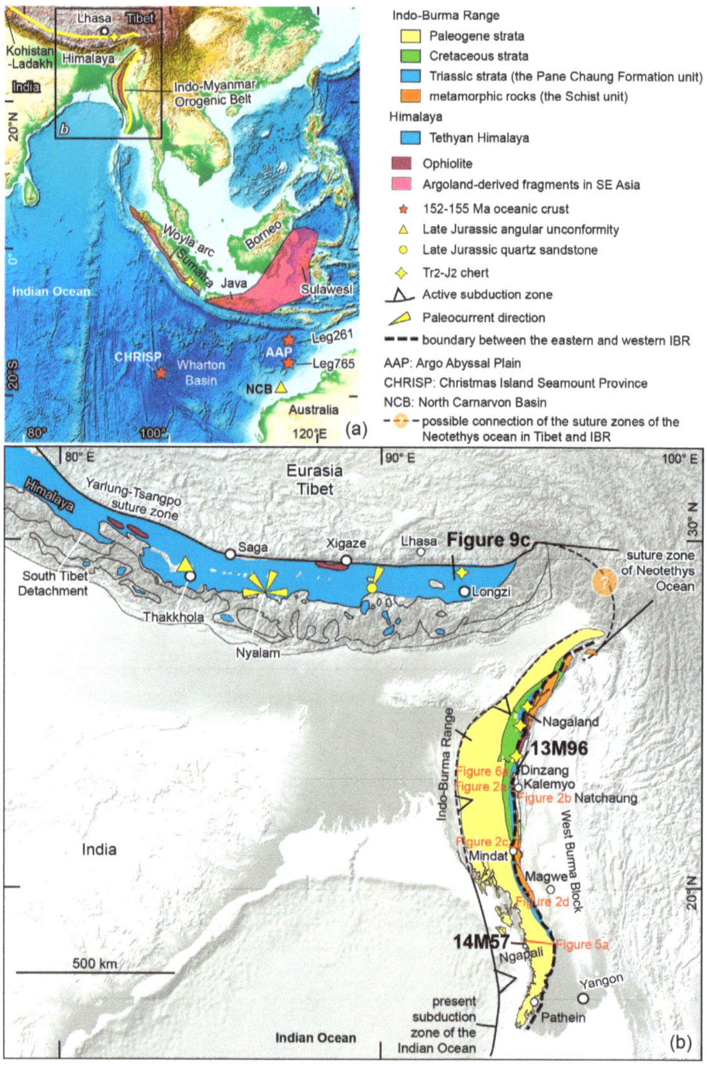

Figure 1. (**a**) A schematic map of SE Asia, Southern Tibet, NW Australia, and the NE Indian Ocean, illustrating the Late Jurassic oceanic crust and the Argoland-derived fragments in SE Asia [1,15,16]. The base map is sourced from the National Oceanic and Atmospheric Administration (2012). (**b**) Tectonic units within the Indo-Burma Range and Tethyan Himalaya. Ages of cherts in ophiolitic mélanges, and interbedded siltstones and cherts are from [7,8,17,18]. Paleocurrent directions, inferred from cross-bedding of Late Jurassic and Triassic sediments, are based on [19–21]. The Late Jurassic angular unconformity in NW Australia is documented in [22], and in Southern Tibet, is from [23]. Locations of samples 13M96 and 14M57 are marked.

(1) Based on the presence of the Kanpetlet Schists near Mindat (Myanmar) and the Naga Metamorphics (Nagaland, India), which comprise schist and gneisses yielding detrital zircon U-Pb age peaks at 550 and 950 Ma, the basement of the IBR has been postulated as a microcontinent (Figure 1b) [8–10]. The discovery of Triassic-aged detrital zircons (239 Ma) within the Kanpetlet Schists, however, has challenged this interpretation [12,13,24].

(2) The accretionary complex model is supported by the presence of ophiolites, blueschists, and eclogites in Nagaland (Figure 1b) [8,25,26], as well as the occurrence of MOR- and OIB-type metabasalts, metacherts, and metasediments in the Kanpetlet Schists in the IBR (Myanmar) (Figure 2) [7,13,27]. Zhang, Xiao [13] proposed that the entire IBR could be an accretionary complex, but it fails to explain the dominance of Precambrian detrital zircons in the Naga Metamorphics [8].

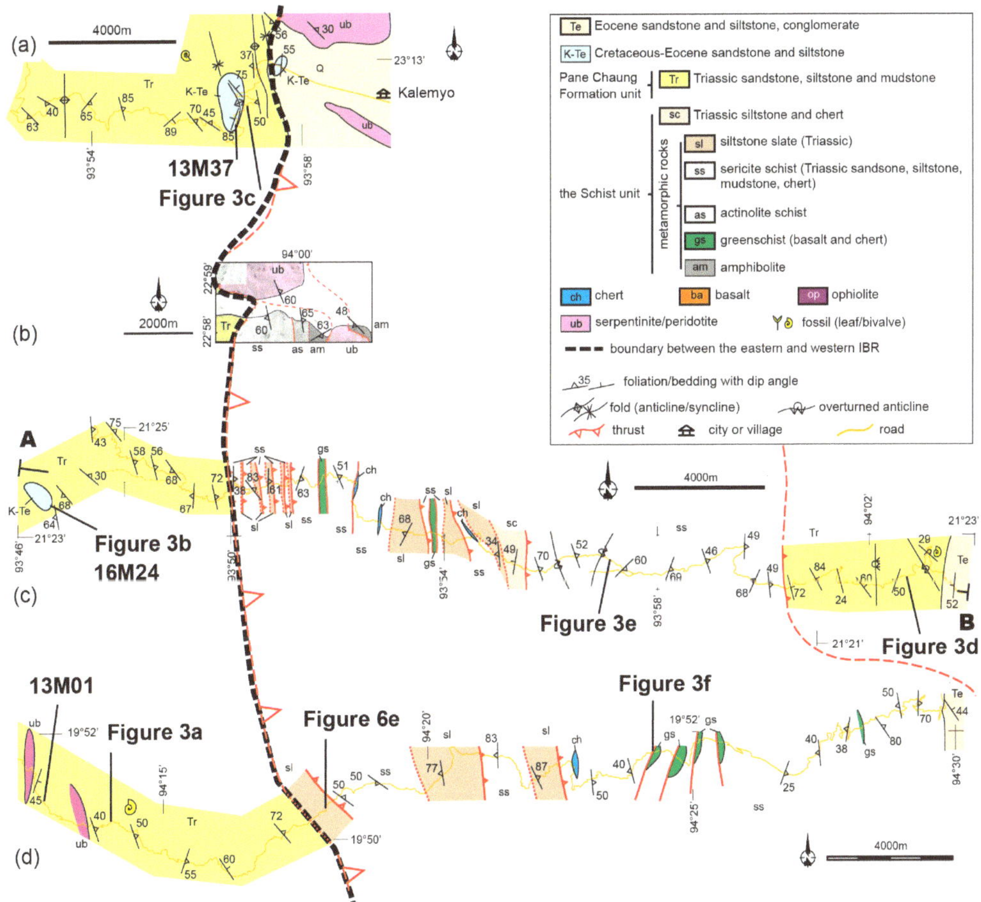

Figure 2. Detailed geological and structural maps of (**a**) the Kalemyo section, (**b**) the Natchaung section, (**c**) the Mindat section, and (**d**) the Magwe section. A megathrust delineates the Late Triassic Pane Chaung Formation units in the footwall to the west from the schist unit in the hanging wall to the east, except in the Mindat section, where an Eastern Pane Chaung Formation unit is present, which was supposed as a thrust sheet backthrust from the west segment. Locations of cross-sections A–B are marked. Samples 13M01, 13M37, 16M24 are located.

Zhang, Chung [24], and Advokaat and van Hinsbergen [15] proposed that an Argoland archipelago was developed, based on similarities of the ages and ε_{Hf} (*t*) of detrital zircons with those from the North Carnarvon Basin. Acharyya [5] proposed that certain rock units in the Eastern IBR represent thrust sheets originating from the West Burma Block, which comprise Precambrian and Paleozoic metamorphic rocks (Figure 1b) [7,28]. The widespread faults within the Naga Metamorphics [8,10] adds complexity to this interpretation. In the Western IBR, Late Triassic flysch, such as the Pane Chaung Formation, and Cenozoic sediments blanket the region; there is no exposure of Paleozoic and Precambrian rocks [7,9]. Therefore, the aforementioned controversial models and facts demonstrate the inadequate constraint on the nature and affinities of the major units in the IBR [6].

In this paper, based on our geological mapping and structural analysis of the Late Triassic Pane Chaung Formation and schist units within the IBR, and the correlation between the IBR and the Tethyan Himalaya (Figure 1b), we aim to delineate the nature and affinities of the major units in the Western IBR.

2. Regional Geology

The Indo-Myanmar orogenic belt is located between the Eastern Gondwana and Laurasia supercontinents (Figure 1a) [1–4]. Its westmost segment, the IBR, is a tectonic unit separating the Neotethys Ocean to the east from the Indian Ocean to the west (Figure 1a), and documents the most recent subduction–accretion events of the Neotethys Ocean [5–7]. Spanning a N-S length of 1300 km, the IBR widens to a maximum of 300 km along the ophiolitic belt, the suture zone of the Neotethys Ocean in the east to the offshore of the Indian Ocean in the west (Figure 1). Within the suture zone of the Neotethys Ocean, some Middle–Late Jurassic cherts occur in ophiolitic mélange (Figure 1b) [7,8,18,29]. Seismic tomography and waveform modeling conducted in the middle segment of the IBR in Myanmar have unveiled doublet subducting zones: one is along the eastern margin of the IBR to the east, and one is the product of the subducting Indian oceanic slab to the west [30].

The Western IBR consists of the Late Triassic Pane Chaung Formation unit, which is the oldest unmetamorphosed sediment, containing fossils of *Halobia* and *Daonella*, and is predominantly covered by Paleogene sediments [9,13]. The Eastern IBR is characterized by numerous small ophiolitic fragments and a schist unit, which consists of schists, and blocks of eclogite and blueschist in Nagaland [6–9]. One schist unit, the low-grade Kanpetlet Schist, was prominently featured in the core that was surrounded by the Late Triassic Pane Chaung Formation in the Mindat section [6,13,31]. The IBR underwent multi-episode deformation, including the convergent deformation and Cenozoic deformation, likely involving significant strain partitioning [6].

Cretaceous and Paleocene sediments are limited to the Eastern IBR (Figure 1b) [7,9, 32,33]. The Cretaceous rocks mainly consist of limestones containing Albian–Cenomanian ammonites, corals, and planktonic foraminifera, *Globotruncana* and *Ammonoides* [9,34]. Paleocene sandstone overlays the western segment of the Late Triassic flysch, 29 km west of the Mindat town (Figures 1b and 2c), displaying steep to subvertical bedding (Figure 3b). The youngest group of detrital zircons from sample 16M24 yielded a U-Pb age of 63.6 Ma (Figure 4c).

Eocene sediments are widespread throughout the Western IBR (Figure 1b), and locally across the entire IBR, as seen in the Nagpali section (Figure 5a). In Myanmar, sandstone–siltstone–shale successions locally rest with angular unconformity on Late Triassic and Cretaceous sediments [35], some being interlayered with Eocene limestones (Figure 1b). These formations are rich in leaf fossils (Figure 5a), including tens of cm-sized *Musa basjoo* Siebold leaf fossils found atop a mountain peak 26 km west of Kalemyo [7,12], indicative of a tidal flat environment [9]. In Nagaland (India), the strata contain several cm to meter-sized blocks of Cretaceous limestone, ophiolitic, and metamorphic rocks, and develop ripple marks and cross-beddings [5,36]. Two samples (13M37 from the Ngapali offshore and 14M57 from west of Pathein, Figure 1b) yielded the youngest groups of detrital zircons of U-Pb ages of 47.8–45.4 Ma (Figure 4d,e), aligning with the Eocene ages indicated by fossil evidence.

Figure 3. Photographs illustrate lithologies in the IBR. (**a**) Late Triassic Pane Chaung Formation flysch in the Magwe section, (**b**) Eocene sandstone in the Mindat section, (**c**) sole marks at the base of the Pane Chaung Formation sandstone in the Kalemyo section, (**d**) cross-bedding in the Late Triassic sandstone in the Mindat section, the coin is for scale, (**e**) sericite schist (metasiltstone) and quartzite (metachert) in the Mindat section, and (**f**) metabasalt and metachert in the Magwe section.

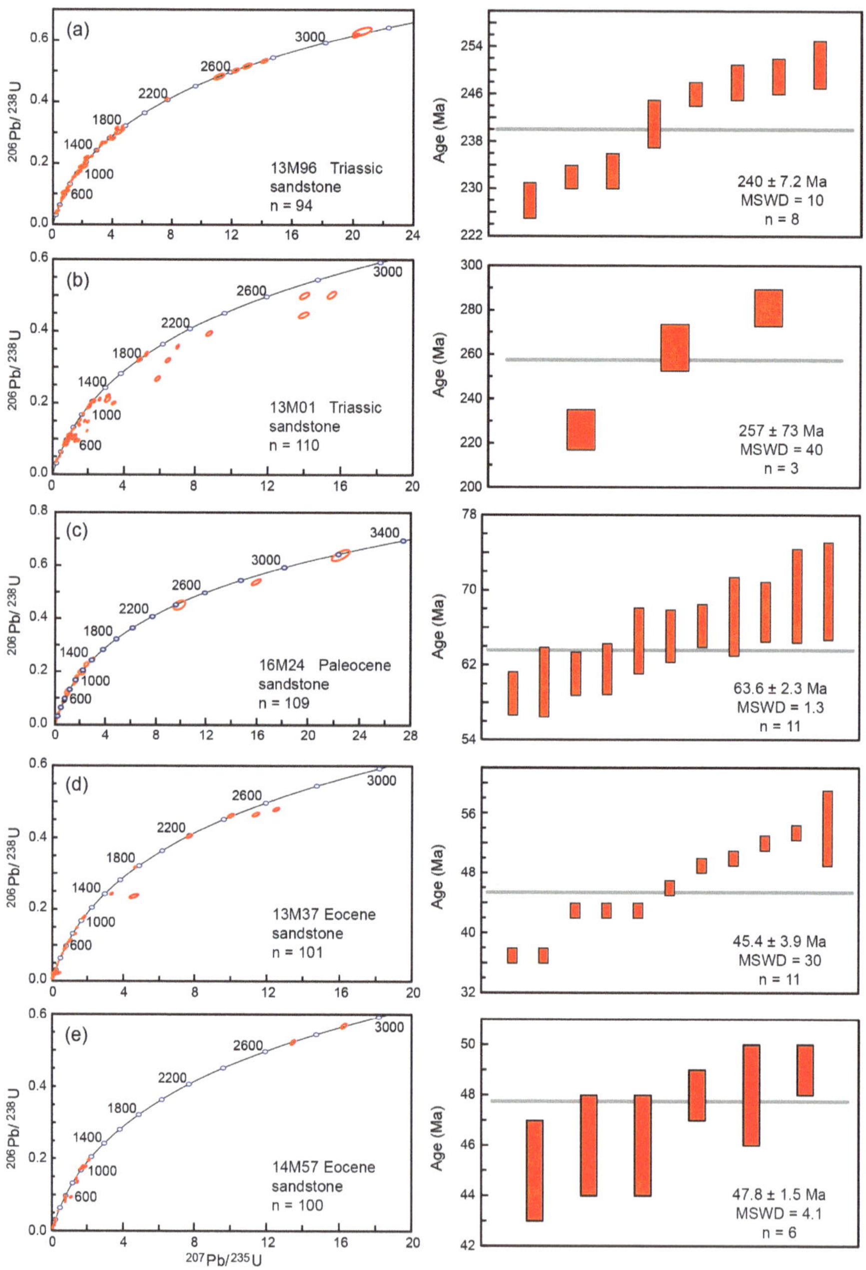

Figure 4. (**a**,**b**) Detrital zircon ages of sandstones 13M96 and 13M01 from the Pane Chaung Formation in the Kalemyo and Magwe sections, respectively, indicate that their youngest groups date back to the Triassic period. (**c**) Detrital zircons of sandstone 16M24 from the Mindat section show the youngest age group as Paleocene, and (**d**,**e**) detrital zircons of Eocene sandstones 13M37 and 14M57 from the Kalemyo and Ngapali sections, respectively. The right columns are the ages of $^{206}Pb/^{238}U$.

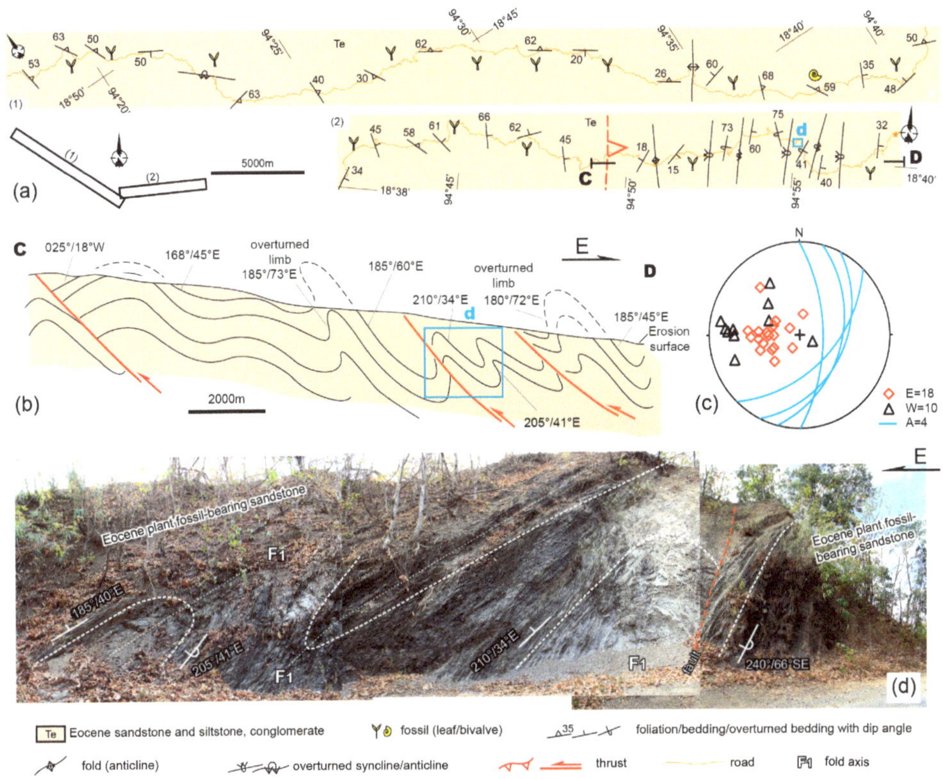

Figure 5. (**a**) Lithological and structural map of the Ngapali section illustrates the Eocene strata traversing the entire IBR from the eastern margin of the IBR to the offshore of the Indian Ocean. (**b**) Cross-section C–D depicts overturned folds with westward vergence developing in the eastern segment of this section. (**c**) Stereonet diagrams (lower hemisphere) display the attitudes of the east (E) and west (W) limbs of anticlines and their axial surfaces (A). (**d**) Westward-verging overturned folds in the Eastern Ngapali section.

3. Analytic Procedure and Results of Detrital Zircon Dating

3.1. Analytic Procedure for LA-Q ICP-MS U-Pb Dating of Detrital Zircon

More than 100 randomly selected zircons were selected per sample, separated with heavy liquid and magnetic techniques, mounted on adhesive tape, then enclosed in epoxy resin, and polished to about half of their diameter. After being photographed under reflected and transmitted light, the samples were photographed by CL imaging with a SX51 Electron Probe Microanalyzer at the Institute of Geology and Geophysics, Chinese Academy of Sciences in Beijing (IGGCAS). Then, an Agilent 7500a quadruple (Q)-ICPMS with a 193 nm excimer ArF laser-ablation system at IGGCAS was used for the determination of the zircon U-Pb age. Detailed analytical procedures can be found in [37], and only an outline is given here.

Before analysis, the sample surface was cleaned with ethanol to eliminate possible contamination. Every 5 zircons were analyzed with a standard Harvard zircon 91500 and National Institute of Standards & Technology standard glass NIST SRM 610 (determining trace elements in zircons). Each spot analysis consisted of approximately 30 s background acquisition and 40 s sample data acquisition. During experiments, the ablation material was sent to the Q-ICPMS for U-Pb analyses. $^{207}Pb/^{206}Pb$, $^{206}Pb/^{238}U$, $^{207}U/^{235}U$, $^{208}Pb/^{232}Th$ ratios were corrected by using zircon 91500 as an external standard. The fractionation correction and results were calculated using GLITTER 4.0. Common Pb was corrected according to the method proposed by [38]. The weighted mean U-Pb ages and Concordia plots were processed using ISOPLOT 4.15.

3.2. Sampling and Dating Results

We collected two Triassic samples (13M01, 13M96), one Paleocene sample (16M24), and two Eocene samples (13M37, 14M57) of sandstones for detrital zircon dating from four E-W-trending sections near the Kalemyo, Natchaung, Mindat, and Magwe areas (Figures 1b and 2). The data are presented in Table S1.

Samples 13M01 and 13M96 are sandstones of flysch (Figure 3a) from the Pane Chaung Formation unit at the Magwe (Figure 2d) and Kalemyo (Figure 1b) sections, respectively. Their detrital zircons have angular to subrounded shapes. Th/U ratios of 13M01 and 13M96 are 0.03–3.57 and 0.10–1.79, respectively (Table S1). U-Pb ages of 110 detrital zircons from 13M01 ranged from 226 to 3012 Ma, with the youngest group at 257 ± 73 Ma (MSWD = 40) (3 grains) (Figure 4b). Due to the large data error, we constrain the depositional age of the strata using the ages of the *Halobia* sp. and other samples as Triassic. U-Pb ages of 94 detrital zircons from 13M01 ranged from 228 to 3100 Ma, with the youngest group at 240 ± 7.2 Ma (MSWD = 10) (8 grains) (Figure 4a).

Sample 16M24 is a Paleocene sandstone from the Mindat section (Figures 2c and 3b). Its detrital zircons have angular to subrounded shapes. U-Pb ages of 109 detrital zircons range from 59 to 3215 Ma (a grain with 50.6 Ma is excluded), with the youngest group at 63.6 ± 2.3 Ma (MSWD = 1.3) (11 grains) (Figure 4c).

Samples 13M37 and 14M57 are Eocene sandstones at the Kalemyo section (Figure 2a) and Nagpali (Figure 1b), respectively. Their detrital zircons have angular to subrounded shapes. Th/U ratios of 13M37 and 14M57 are 0.03–3.45 and 0.03–2.00, respectively (Table S1). U-Pb ages of 101 detrital zircons from 13M37 ranged from 37 to 2741 Ma, with the youngest group at 45.4 ± 3.9 Ma (MSWD = 30) (11 grains) (Figure 4d). U-Pb ages of 100 detrital zircons from 14M57 ranged from 42 to 2889 Ma, with the youngest group at 47.8 ± 1.5 Ma (MSWD = 4.1) (6 grains) (Figure 4e).

4. Field Geology and Lithologies

4.1. Late Triassic Pane Chaung Formation Unit

The Late Triassic Pane Chaung Formation unit on the geological map of the Mindat-Saw Area, Myanmar [39] is primarily concentrated in the Western IBR (Figure 1b). Within the Mindat section, this rock unit is found on both the eastern and western sides of the Kanpetlet schist, the core of the Mindat Anticline (Figure 2c) [33]. In the Kalemyo, Natchaung and Magwe sections, the Pane Chaung Formation is situated west of the schist (Figure 2a,b,d) [7,14].

This unit consists of unmetamorphosed (calcareous) sandstone, siltstone, and mudstone (Figure 3a); they widespread develop sole marks (Figure 3c) and occasional herringbone cross-beddings (Figure 3d). The sandstones contain subrounded to subangular grains of feldspar, monocrystalline quartz, as well as sedimentary, volcanic, and metamorphic lithic grains, which were plotted within the recycled orogenic provenance field on the QtFL diagram [12]. Late Triassic fossils of *Halobia* and *Daonella* have been discovered [9,10,12,13]. The youngest groups of detrital zircons in sandstones (samples 13M96 and 13M01) from the Magwe and Kalemyo regions are 257-240 Ma (Figure 4a,b), consistent with 238 Ma from [11], 228 Ma from [12], and 256 Ma from [24]. These findings provide robust constraints on the Late Triassic age of this rock unit.

4.2. Schist Unit

The schist unit is mainly located in the Eastern IBR (Figures 1b and 2a,b,d) [33], and comprises sericite schist, phyllite, and thin-bedded quartzite (Figure 3e), occasionally associated with greenschist-facies basalt and chert (Figure 3f). This unit acts as the matrix for blocks of eclogite and blueschist in Nagaland (India) [25,26], and 119–114 Ma amphibolite in IBR (Myanmar) [13,40]. Along its eastern fringe, ophiolites comprise ultramafic rocks, mafic rocks, and cherts containing Middle Jurassic–Eocene radiolarian fossils [7,8,18,29].

Sericite schist interlayers with thin-bedded quartzite (Figure 3e), whose protoliths are siltstone and chert, indicative of deep-marine sedimentations [13]. Metabasalts in the Kalemyo, Mindat, and Magwe sections are characterized by MOR- and OIB-type features [7], consistent with those in Nagaland [27,41]. The coexistence of metabasalt and chert (Figure 3f) suggests that they are components of OPS (oceanic plate stratigraphy), remnants of oceanic crust [42]. Finding fossils in these metamorphosed rocks is challenging. The youngest groups of detrital zircons from sericite schists (samples 13M22 and 13M147) from the Mindat and Magwe sections yield Triassic ages of 239–228 Ma.

5. Structures

The structures within the IBR are complex, including folds, imbricated thrusts, and backthrusts. In the Late Triassic flysch units, structures display diverse polarities in different segments, including top-to-west and top-to-east vergent structures, with the latter likely linked to backthrusting.

5.1. Structures with Top-to-West Vergence in the Pane Chaung Formation Unit

Structures with top-to-west sense within the Late Triassic Pane Chaung Formation unit are observed in the Kalemyo and Magwe sections (Figure 2a,d), the Dinzang section (Figure 6a,b), and the western segment of the Triassic flysch unit of the Mindat section (Figures 2c and 7a). These structures consist of thrusts and fault-propagation overturned folds (Figure 6e), indicating that the thrusts are moving westward in polarity.

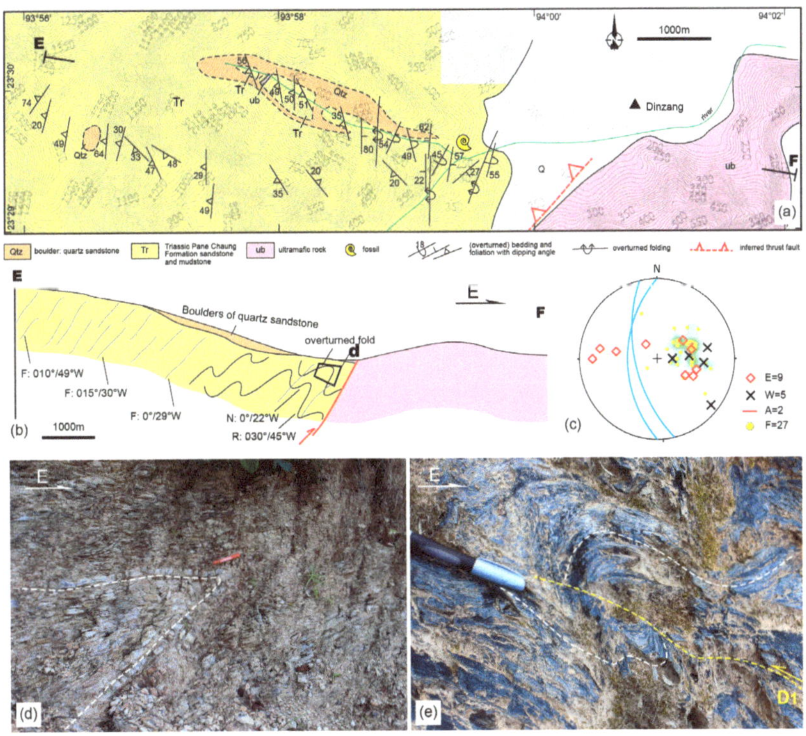

Figure 6. (**a**) Geological map and (**b**) cross-section of the Dinzang section illustrate the Late Triassic Pane Chaung Formation unit backthrusting over serpentinite. (**c**) Stereonet diagrams (lower hemisphere) display

the attitudes of the east (E) and west (W) limbs of anticlines, their axial surfaces (A), and tectonic foliation (F). (**d**) In the Dinzang section, an overturned anticline displays top-to-east vergence. (**e**) Thrust-propagation folds with top-to-west polarities in the Magwe section imply westward movement of the thrust. Its location is marked in Figure 2d.

5.2. Structures with Top-to-East Vergence in the Pane Chaung Formation Unit

We provide a detailed description of the top-to-east polarity backthrust-related structures, although they may also be a possibility as a result of a late chaotic gravitational faulting and sliding. These structures are situated west of the Dinzang Valley (Figure 6a,b) and in the eastern segment of the Triassic flysch unit of the Mindat section (Figure 7a). In the Dinzang section, overturned folds (Figure 6d) display normal west limbs of anticlines dipping at 22–84°; while some normal east limbs dip at 27–80° and others are overturned, dipping westward at 44–55° (Figure 6c). Their axial surfaces dip westward at 48–58° (Figure 6c), indicating that the fold-related thrusts transport eastward (Figure 6b).

The eastern segment of the Triassic Pane Chaung Formation unit in the Mindat section dominates the structures with top-to-east polarities (Figure 7), characterized by widespread superimposed overturned folds (Figure 7b). An outcrop 6.5 km east of the Mindat town presents a refolded overturned anticline (or recumbent fold) (Figure 8a). We define the first stage of folding as F_1 and the second stage as F_2. The east limb of F_1 has been overturned, as evidenced by the cross-beddings (Figure 8b), while the west limb remains normal. The axial surfaces of F_1 predominantly dip westward at 31–60° (Figure 8c). The overturned fold (F_1) has been refolded into an upright fold (F_2) with steeply dipping axial surfaces (Figure 8a). The normal west limbs of F_1 refold to develop normal west and east limbs of F_2 dipping at 43–58° and 32–65°, respectively. The overturned east limbs of F_1 refold to fold F_2, whose limbs dip westwards at 24–85° and eastwards at 15–85° (Figure 8c). The thrust responsible for generating these overturned folds are inferred to move eastward (Figure 7b).

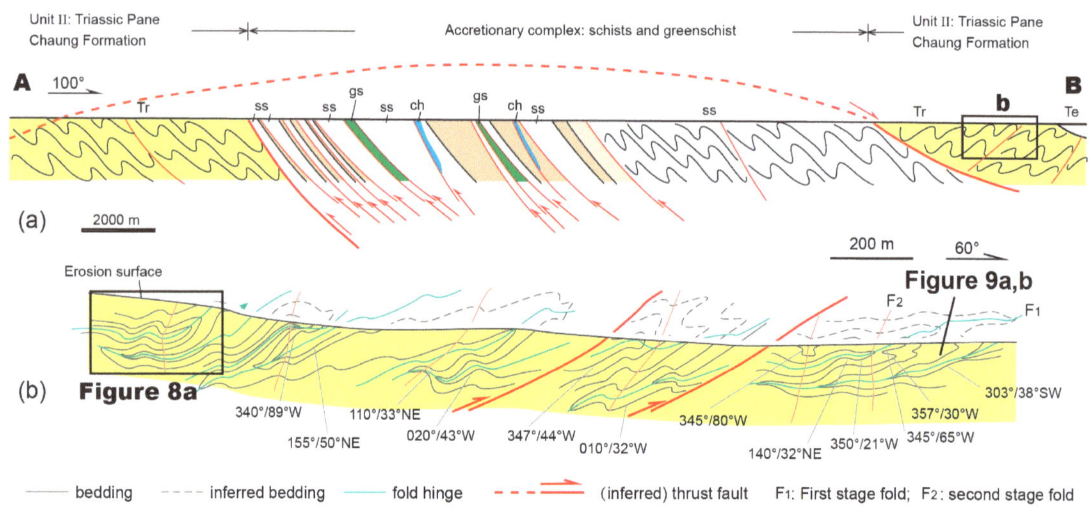

Figure 7. (**a**) Cross-section A-B of the Mindat section shows the schist unit in the core, flanked by two Late Triassic Pane Chaung Formation units to the east and west. The western unit displays a top-to-west polarity, while the eastern unit exhibits top-to-east vergence, as evidenced by (**b**) superimposed overturned folds. The first stage folds (F_1) have top-to-east vergence, whereas the subsequent folds (F_2) feature upright axial surfaces. For abbreviations of rocks, see Figure 2.

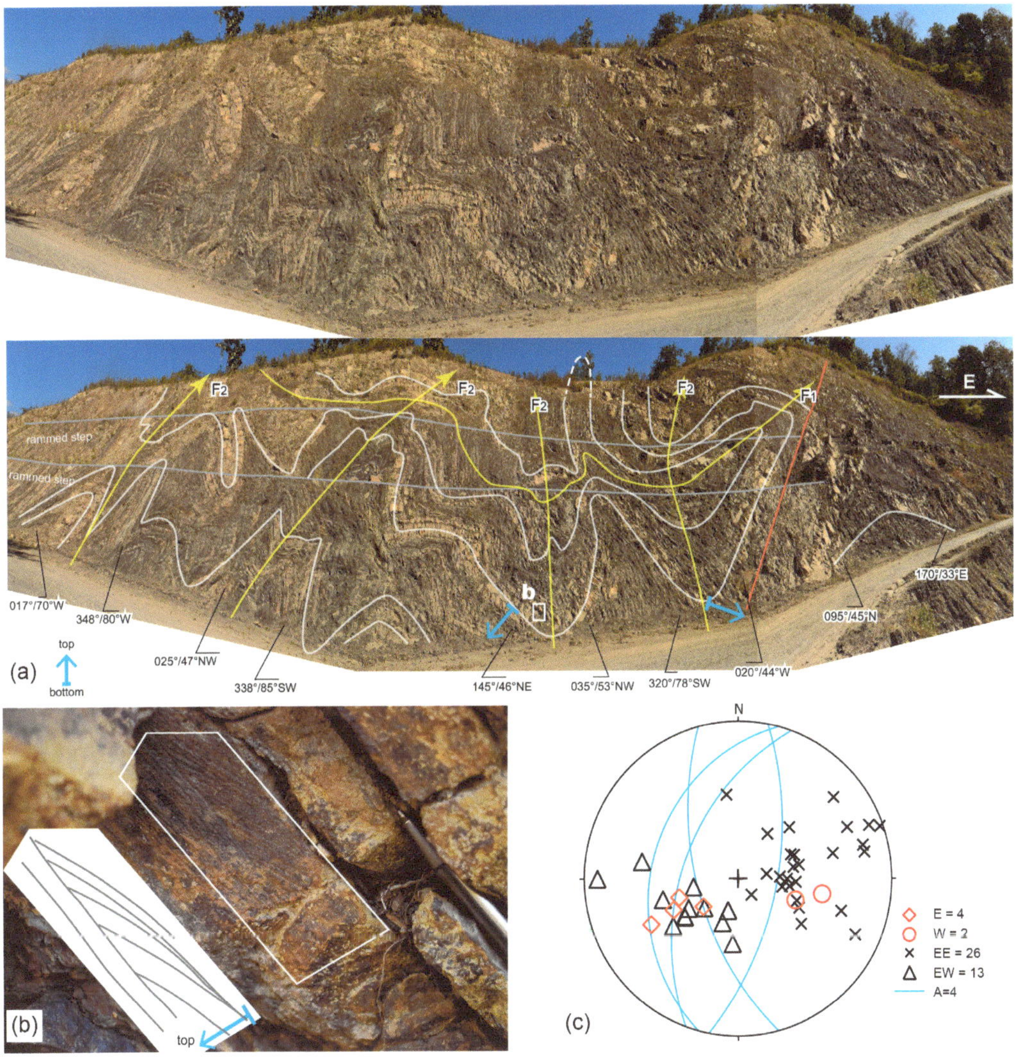

Figure 8. (**a**) An outcrop 6.5 km east of the Mindat town displays superimposed overturned folds, as evidenced by the overturned cross-bedding (**b**). (**c**) A stereonet diagram (lower hemisphere) displays the attitudes of the east (E) and west (W) limbs and axial surfaces (A) of F_1 anticlines, and the east (EE) and west (EW) limbs of the antiformal folds (F_2) that have refolded from the overturned east limbs of the first stage folds (F_1).

5.3. Imbricate Thrusts in Schist Unit

In the Mindat and Magwe sections, the repeated decreases in metamorphic grade from greenschist, to sericite schist and slate, demonstrate the development of imbricate thrusts (Figures 2c,d and 7a). These structures are characteristic of accretionary prisms.

5.4. Overturned Folds with Top-to-West Polarity in the Eocene Sediments

The Eocene strata have undergone extensive deformation, resulting in overturned folds (Figure 5b). In a particular outcrop (Figure 5d), these overturned folds display normal east limbs dipping at 15–68°, while both normal and overturned west limbs dip at 18° and 41-83°, respectively (Figure 5c). Their axial surfaces dip eastward at 37–61° (Figure 5c).

6. Deformation Stages

The stages of deformation are constrained by subduction ages, subduction directions, structural polarities, and affinities of tectonic units. The sequence between stage 2 and stage 3 is inferred logically without direct evidence because it is difficult to find in the field.

6.1. Stage 1 (D_1): Top-to-West Thrusting in the Schist and the Pane Chaung Formation Units

Blocks of 119–114 Ma amphibolite within the schist matrix constrain the initiation of imbricated thrusts in the Eastern IBR to at least the Early Cretaceous age during the subduction of the Neotethys Ocean [1,5–9,13,14,25,40]. The eastward subduction (present coordination) and subsequent closure of the Neotethys Ocean also contributed to the formation of thrusts with top-to-west polarities in the Late Triassic Pane Chaung Formation unit (Figure 7a).

6.2. Stage 2 (D_2): Top-to-East Backthrusting in the Pane Chaung Formation Unit

The top-to-east vergent overturned folds in the Dinzang valley (Figure 6a,b) and the eastern segment of the Triassic flysch unit in the Mindat section (Figure 7) suggest that the fold-related faults move eastward, contrasting with the top-to-west thrusting (D_1). The eastern Pane Chaung Formation unit in the Mindat section is situated in the hanging wall, different from the dominant features as the Pane Chaung Formation unit in the footwall and schist unit in the hanging wall in the Kalemyo, Natchaung, and Magwe sections, as well as the western one in the Mindat section (Figure 7a). We propose that the top-to-east thrusting represents backthrusting as the second deformation (D_2), following the top-to-west transportation.

6.3. Stage 3 (D_3): Top-to-West Thrusting in the Eocene Strata

If top-to-east displacement was acting on the Eocene strata, we would observe thrusts and associated folds with top-to-east polarities. However, the entire Eocene Ngapali section displays only top-to-west vergent structures (Figure 5a,b), lacking top-to-east movement. The top-to-west vergent structures in the Eocene strata (D_3) would develop subsequent to the top-to-east backthrusting (D_2).

7. Discussion

7.1. Tectonic Affinities and Nature of the Pane Chaung Formation Unit

In the Western IBR, the Late Triassic Pane Chaung Formation unit represents the oldest unmetamorphosed strata (Figure 2), extensively covered by Cretaceous–Eocene sediments (Figures 1b and 5a). Its tectonic affinities and nature remain elusive.

The Late Triassic Pane Chaung Formation unit in the Western IBR and the Langjiexue and Nieru Group units in the Tethyan Himalaya are correlated (Figure 1b). The Pane Chaung Formation in the Mindat area of Myanmar (Figure 9a) displays similar herringbone cross-beddings to those in the Langjiexue Group of the Tethyan Himalaya (Figure 9c). Additionally, both formations contain Carnian–Norian *Halobia* fossils (Figure 9b,d) [43], and detrital zircons dominating at 220–280, 500–620, 900–1000, and 1100–1140 Ma (Figure 9e) [11–13,18,24,44–52]. These facts collectively suggest a geographical contiguity of the Late Triassic flysch unit between the IBR and the Tethyan Himalaya [12,15,18].

In the Tethyan Himalaya, Mesozoic igneous dikes are widespread and intrude into the Triassic Langjiexue Group and Jurassic–Cretaceous strata (Figure 1b). Some Mid-Late Triassic and Early Cretaceous dikes display negative values of ε_{Nd} (t) (−13.4 to −0.1) and ε_{Hf} (t) (−24.2 to −0.1) (Figure 10a,b) [50,53–55], demonstrating a basement with Precambrian one-stage model ages (Figure 10c). Ordovician–Permian limestones and clastics in the Tethyan Himalaya (Figure 11) [56] further suggest a Paleozoic or older basement. These combined facts imply that the Tethyan Himalaya constituted a Precambrian continental sliver, on which the Pane Chaung Formation unit was deposited in the Western IBR (Figure 12).

The Triassic flysch units in the Tethyan Himalaya and Western IBR have paleocurrent directions spanning NW-, NE-, and E-W orientations, suggesting potential provenance from the south, east, or west (Figures 1b and 11) ([19,21] and this study). Detrital magmatic zircons of 200–400 Ma have $\varepsilon_{Hf}(t)$ values of −10 to +14, which correlate with those in the North Carnarvon Basin [57], indicating a strong affinity with NW Australia [18,45,52].

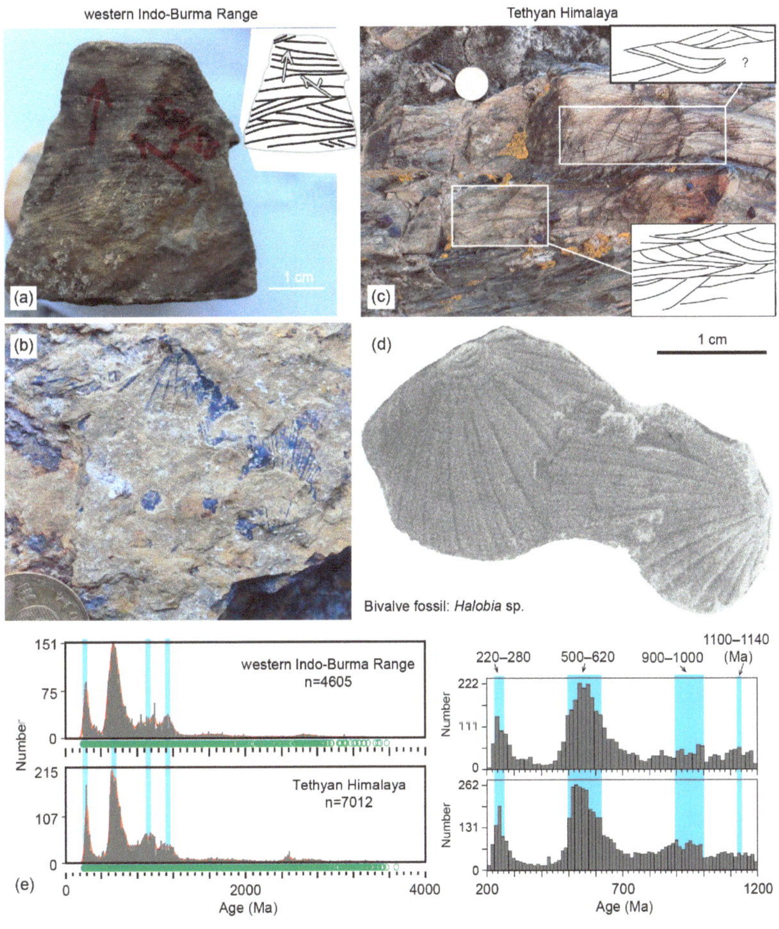

Figure 9. The correlation of the Pane Chaung Formation unit in the Western IBR with the Triassic flysch in the Tethyan Himalaya reveals their similarities of (**a**,**c**) herringbone cross-beddings, which are interpreted in inserted figures, (**b**,**d**) Carnian–Norian *Halobia* fossils in the Mindat area of Myanmar and the Langjiexue Group in the Tethyan Himalaya [43], and (**e**) detrital zircons with age peaks at 220–280, 500–620, 900–1000, and 1100–1140 Ma. Compiled detrital zircon data are sourced from the Pane Chaung Formation unit in the IBR [7,11–13,18,24] and in the Tethyan Himalaya [21,44–52,58,59].

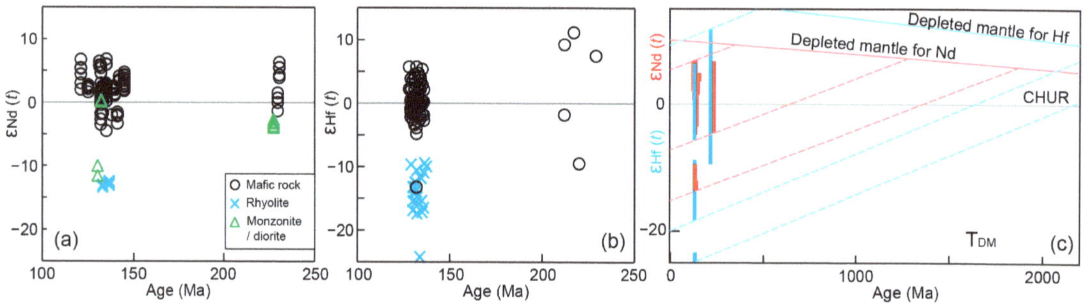

Figure 10. Compiled (**a**) ε_{Nd} (t) and (**b**) ε_{Hf} (t) isotopic data from Mesozoic mafic, rhyolite, and dioritic dikes in the Tethyan Himalaya demonstrate that some of them display negative values, indicative of Precambrian first-stage model ages (**c**). Data are sourced from [50,53–55].

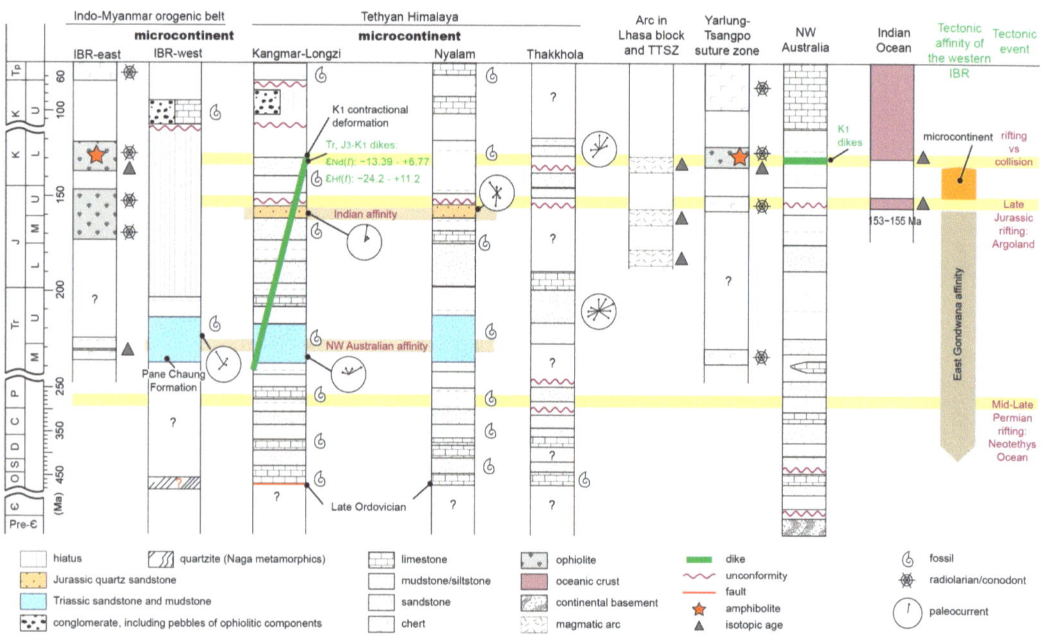

Figure 11. Comparison among the IBR, Tethyan Himalaya, NW Australia, Yarlung-Tsangpo suture zone, and the Indian Ocean. Comparable Late Triassic flysch units in the Western IBR (Pane Chaung Formation) and Tethyan Himalaya suggests that these regions shared tectonic processes. The Paleozoic strata and Precambrian one-stage model ages in the Tethyan Himalaya suggest that both the IBR and Tethyan Himalaya share a microcontinent basement. Evidence of an East Gondwana-affinity prior to the Late Jurassic period, followed by subsequent rifting and drifting from the host supercontinent during the Late Jurassic, demonstrates that the microcontinents in the Western IBR and Tethyan Himalaya are fragments of the Argoland archipelago. The stratigraphy column of the Tethyan Himalaya is from [56]. TTSZ: Trans-Tethyan subduction zone.

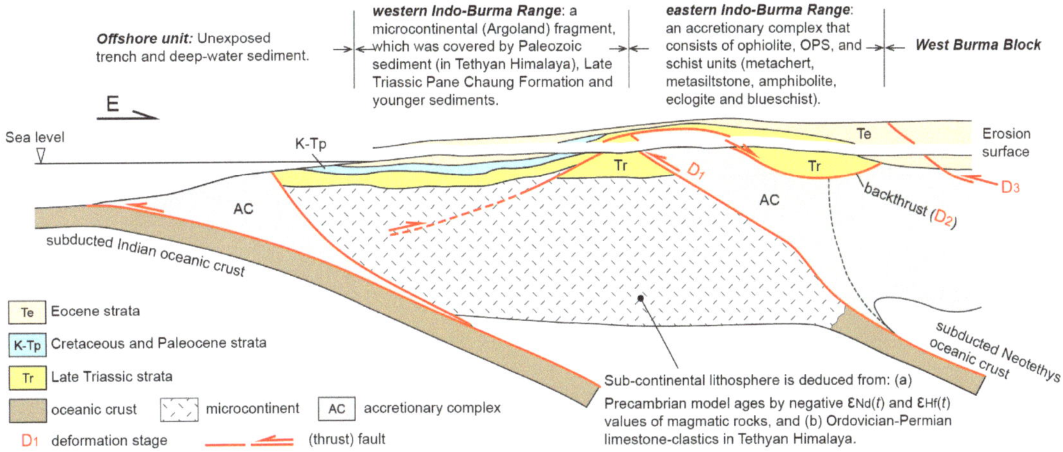

Figure 12. Tectonic model of the Indo-Burma Range illustrates an Argoland microcontinent in its western region, which was covered by the Late Triassic Pane Chaung Formation unit and Paleozoic sediments. The Eastern IBR features an accretionary complex comprised of ophiolite, OPS, schist, and high-pressure rocks. A backthrust displaced some Pane Chaung Formation units eastward as a thrust sheet above the accretionary complex. This Argoland microcontinent acts as a barrier, separating the Neotethys Ocean to the east from the Indian Ocean to the west. The subduction of these oceans has generated double subduction zones within the Indo-Myanmar orogenic belt.

7.2. Accretionary Complex for the Schist Unit in the Eastern IBR

The schists in the Eastern IBR comprise greenschist-facies MOR- and OIB-type basalt and chert (Figure 3f), sericite schist, phyllite, and thin-bedded quartzite as matrix (Figure 3e). They host blocks of eclogite, blueschist, and amphibolite [5–9,13,25–27,40]. These rocks together represent remnants of oceanic crust and overlying sediments. The D_1 imbricate thrusts in the schist unit (Figures 2 and 7a) are characteristic structures of accretionary prisms. The rock assemblages and structures collectively indicate an accretionary complex in the Eastern IBR, which off-scraped from the subducting Neotethys Ocean ([5,13] and references therein).

7.3. Paleogeographic Environments of the Pane Chaung Formation and Schist Units

The Neotethys Ocean suture zone is situated along the Yarlung-Tsangpo belt and the eastern margin of the IBR (Figure 1b). Within this suture zone, some chert blocks in ophiolitic mélanges in Southern Tibet (China) and South Sumatra (Indonesia) contain radiolarian fossils dating back to the Middle Triassic period (Figures 1 and 11) [17,60,61], indicating the existence of an open Neotethys Ocean in SE Asia during that time. Middle–Late Jurassic cherts along the IBR and Nagaland [7,8,18,29], together with MOR- and OIB-type basalts [7,27,41], suggest the ongoing development of the Neotethys Ocean in this region. Subduction-related magmatism and metamorphism commenced during the Early-Middle Jurassic epochs in Southern Tibet and Myanmar [7,26,62]. These lines of evidence imply that the Neotethys Ocean spreads without subduction during the Middle Triassic-Early Jurassic periods. The age coincides with the depositional age of the Pane Chaung Formation unit, the oldest unmetamorphosed cap-bed of the Argoland microcontinent in the Western IBR (Figure 12). This suggests that this unit was deposited along the passive continental margin of East Gondwana (Figure 13a,b) [18]. The protoliths of the Triassic sericite schist, phyllite, and thin-bedded quartzite in the accretionary complex (Figure 12) would originate from abyssal sediments of the passive continental margin, rather than a trench-slope basin (Figure 13a,b).

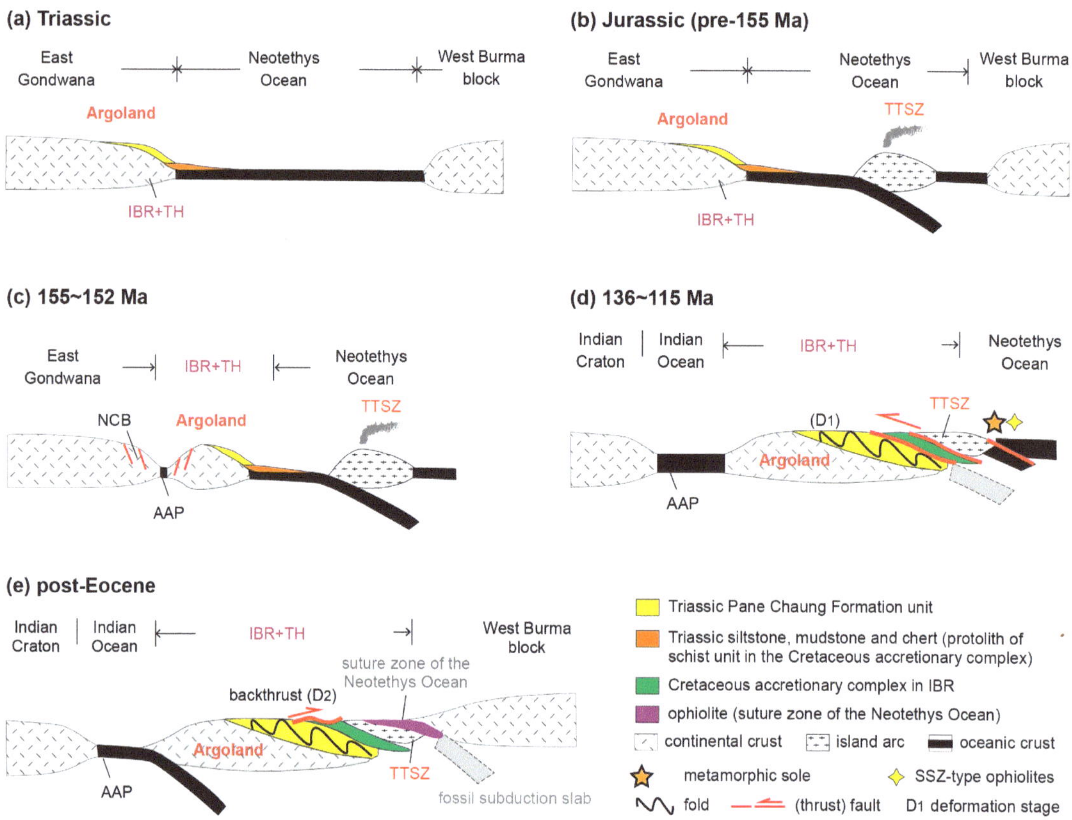

Figure 13. Tectonic evolution of the Indo-Myanmar orogenic belt and East Gondwana: (**a**) During the Triassic, the Neotethys Ocean separated the West Burma block (part of the Sibumasu continent) from East Gondwana, along whose passive continental margin the Pane Chaung Formation unit was deposited. (**b**) In the Jurassic, subduction of the Neotethys Ocean initiated an intra-oceanic arc of the TTSZ. (**c**) By the Late Jurassic, the Argoland microcontinent began to rift from East Gondwana, and (**d**) in the Early Cretaceous, the microcontinent collided with the TTSZ, resulting in folding in the Tethyan Himalaya and the development of metamorphic soles and SSZ-type ophiolites in the IBR and the Yarlung-Tsangpo belt. (**e**) Following the closure of the Neotethys Ocean, the Argoland microcontinent amalgamated with the West Burma block, leaving a fossil subducting slab in the mantle. This, together with the subducting Indian oceanic slab, formed double subduction zones in the Indo-Myanmar orogenic belt. This closure and collision also triggered backthrusting, displacing the Pane Chaung Formation unit eastward as the eastern segment in the Mindat section. AAP: Argo Abyssal Plain; IBR: Indo-Burma Range; NCB: North Carnarvon Basin; TH: Tethyan Himalaya; TTSZ: Trans-Tethyan subduction zone.

7.4. Paleo-Subduction Zones in the Indo-Myanmar Orogenic Belt

Late Jurassic quartz sandstones display NNE paleocurrent directions (Figures 1b and 11) and contain 95% quartzite pebbles with detrital zircon peaks at 550 and 950 Ma sourced from the Indian craton [20]. These features collectively suggest that the Tethyan Himalaya-Western IBR was geologically attached to East Gondwana during the Triassic to Late Jurassic periods (Figure 13a,b).

Early Cretaceous contractional deformation in the Tethyan Himalaya [63] suggests that this microcontinent resided in a compressional environment, rather than along the rift margin of East Gondwana (Figure 13d). This implies that this microcontinent had already drifted away from its host East Gondwana by that time (Figures 11 and 13c) [63].

Magmatic activity transitioning from 153 Ma OIB to 149–147 Ma E-MORB and 145 Ma N-MORB [63–66], coupled with the Late Jurassic rapid deepening of the sedimentary waters [67], indicates a contemporaneous Late Jurassic extension (Figure 13c). This extension aligns temporally with the Tithonian extension along the northern margin of the Indian craton (Figures 1b and 11) [23] and the presence of 152–155 Ma Indian ocean at the Argo Abyssal Plain and CHRISP (Figures 1a and 11) [16,68,69]. Moreover, they coincide with the Callovian escarpment, characterized by 1200 m sediments offshore of NW Australia due to faulting (Figure 11) [22], which was associated with the Late Jurassic break-up of East Gondwana [15,16,68]. This rift activity resulted in the formation of Argoland archipelagos from NW Australia that dispersed to SE Asia, including the Western IBR and Tethyan Himalaya (Figures 11, 12 and 13a–c) [15,24,63,70].

The in-situ constraints on the Argoland microcontinent in the Western IBR indicate that certain Precambrian and Paleozoic metamorphic rocks in Nagaland (Figure 11) [8] are probable representatives of its basement, rather than being thrust sheets originating from the West Burma Block [5]. Our study supports the interpretation of a fragment of the Argoland archipelago within the Indo-Maynmar orogenic belt [15,18,24], and further constrains its basement as a microcontinent (Figure 12).

The Argoland microcontinent, situated in the Western IBR–Tethyan Himalaya, rifted from East Gondwana in the Late Jurassic (Figure 13b,c) [15,63]. This rifting separated the Neotethys Ocean to the north/east (present coordination) and the Indian Ocean to the south (Figures 1a, 12 and 13c). In the Early Cretaceous, the Argoland microcontinent collided with an intra-oceanic arc of the Trans-Tethyan subduction zone, which developed in the Neotethys Ocean during the Jurassic–Cretaceous [71], resulting in contractional deformation in the Triassic flysch units in the Tethyan Himalaya and IBR (D_1 deformation) (Figure 13d) [63]. The final closure of the Neotethys Ocean at Paleogene led to the amalgamation of the IBR and West Burma block [35,72] and preserved a fossil subduction slab in the deep mantle. This, along with the current subducting Indian oceanic slab, formed double east-dipping subduction zones in the Indo-Myanmar orogenic belt (Figures 12 and 13e), which have been locally imaged by upper mantle seismic tomography and waveform modeling in Myanmar, revealing both the western modern subduction zone and the eastern remnant of the Neotethys oceanic slab [30]. Following the closure of the Neotethys Ocean, thrust displaced the Pane Chaung Formation unit backward over the schist unit in the accretionary complex in the IBR (Figures 12 and 13e).

Notably, Naing, Robinson [18] also explained the presence of double subduction slabs; however, our model has some differences: (1) In their model, the eastern subduction slab was located between the West Myanmar Terrane and Sibumasu Terrane, while we interpret it as the fossil slab of the Neotethys Ocean between the West Burma Block and the IBR (Figure 13e). (2) Our model considers the West Burma Block as part of the Sibumasu continent, whereas their model treats them as two independent terranes. (3) We adopt an intra-oceanic arc of the Trans-Tethyan subduction zone in the Neotethys Ocean, following [71], which can be correlated with the intra-oceanic Zedang arc in Tibet ([63] and references therein). In contrast, the model of Naing, Robinson [18] did not develop an intra-oceanic arc.

The recognition of the nature and tectonic affinities of the major units in the Western IBR enhances our understanding of the orogenic processes involving double subduction zones. It also proposed that archipelago architecture was developed there in the late Mesozoic that prompts a reevaluation of previous tectonic models of the Indo-Myanmar orogenic belt and further supports the multiple subduction model in East Tethys [5–8,73,74].

8. Conclusions

Our structural analysis and geochronological data, integrated with published data from the IBR and the Tethyan Himalaya, lead us to draw the following conclusions.

(1) In the Mindat section, two segments of the Late Triassic Pane Chaung Formation unit were investigated and further identified with detrital zircon dating. The Schist unit was confirmed to be an accretionary complex.
(2) The eastern segment displays top-to-east vergent overturned folds, which result from eastward backthrusting, in contrast to the prevailing structures with top-to-west vergence in other sections in the IBR. Restoration of this thrust sheet unveils a megathrust delineating the Pane Chaung Formation unit in the footwall to the west from schists in the hanging wall to the east.
(3) The correlation of Late Triassic flysch units between the Western IBR and the Tethyan Himalaya, coupled with the Paleozoic strata and Precambrian one-stage model ages of Mesozoic dikes in the Tethyan Himalaya, suggests that the major units of the Western IBR involved an emplaced microcontinent, possibly from a paleogeography of an archipelago along East Gondwana.
(4) The microcontinent rifted from its host East Gondwana during the Late Jurassic period. This is supported by the transition from OIB to E-MORB and N-MORB, the rapid deepening of sedimentary waters, and the existence of the 155–152 Ma Indian Ocean. Collectively, these facts indicate a fragment of the Argoland archipelago in the IBR, which is compatible with doublet subduction zones in Eastern Tethys.

Supplementary Materials: The following supporting information can be downloaded at: https://www.mdpi.com/article/10.3390/min14121290/s1. Table S1: Ages of detrital zircons in this study.

Author Contributions: Conceptualization, J.Z. and W.X.; methodology, J.Z. and W.X.; validation, J.Z., W.X. and F.C.; formal analysis, J.Z., W.X., J.W. and F.C.; investigation, J.Z., F.C. and K.S.; data curation, J.Z., W.X., J.W., F.C. and K.S.; writing—original draft preparation, J.Z., W.X., J.W. and F.C.; writing—review and editing, J.Z., W.X. and J.W. All authors have read and agreed to the published version of the manuscript.

Funding: This research was funded by the National Natural Science Foundation of China, grant number 42488201 and 42172247.

Data Availability Statement: Age data are presented as a Supplementary File.

Acknowledgments: We thank Fuyuan Wu, Lin Ding, Chunming Han, Yi Chen, and Shun Guo for the useful discussions on the geological problems of Myanmar. Win Swe, the past President of the Myanmar Geosciences Society, is thanked for the introduction to Myanmar geology. S.K. Acharyya from Jadavpur University (India), C.K. Morley from Chiang Mai University (Thailand), and E.L. Advokaat from Utrecht University (Netherlands) are thanked for their suggestions on the regional geology of the IBR.

Conflicts of Interest: The authors declare no conflicts of interest.

References

1. Metcalfe, I. Gondwanaland dispersion, Asia accretion and evolution of eastern Tethys. *Aust. J. Earth Sci.* **1996**, *43*, 605–623. [CrossRef]
2. Pan, G.T.; Chen, Z.L.; Li, X.Z.; Yan, Y.J.; Xu, X.; Xu, S.Q.; Jiang, X.S.; Wu, Y.L.; Luo, J.N.; Zhu, T.X.; et al. *Geological-Tectonic Evolution in the Eastern Tethys*; Geological Publishing House: Beijing, China, 1997; pp. 1–218, (In Chinese with English abstract).
3. Şengör, A.M.C. Plate tectonics and orogenic research after 25 years: A Tethys perspective. *Earth-Sci. Rev.* **1990**, *27*, 1–201.
4. Yin, A.; Harrison, T.M. Geologic evolution of the Himalayan-Tibetan orogen. *Annu. Rev. Earth Planet. Sci.* **2000**, *28*, 211–280. [CrossRef]
5. Acharyya, S.K. Indo-Burma Range: A belt of accreted microcontients, ophiolites and Mesozoic-Paleogene flyschoid sediments. *Int. J. Earth Sci.* **2015**, *104*, 1235–1251. [CrossRef]
6. Morley, C.K.; Naing, T.T.; Searle, M.P.; Robinson, S.A. Structural and tectonic development of the Indo-Burma ranges. *Earth-Sci. Rev.* **2020**, *200*, 102992. [CrossRef]
7. Zhang, J.; Xiao, W.J.; Windley, B.F.; Wakabayashi, J.; Cai, F.; Sein, K.; Wu, H.; Naing, S. Multiple alternating forearc- and backarc-ward migration of magmatism in the Indo-Myanmar Orogenic Belt since the Jurassic: Documentation of the orogenic architecture of eastern Neotethys in SE Asia. *Earth-Sci. Rev.* **2018**, *185*, 704–731. [CrossRef]

8. Aitchison, J.C.; Ao, A.; Bhowmik, S.; Clarke, G.L.; Ireland, T.R.; Kachovich, S.; Lokho, K.; Stojanovic, D.; Roeder, T.; Truscott, N.; et al. Tectonic Evolution of the Western Margin of the Burma Microplate Based on New Fossil and Radiometric Age Constraints. *Tectonics* **2019**, *38*, 1718–1741. [CrossRef]
9. Bannert, D.; Lyen, A.S.; Htay, T. *The Geology of the Indoburman Ranges in Myanmar*; Geologisches Jahrbuch Reihe B Regionale Geologie Ausland: Hannover, Germany, 2011; pp. 1–101.
10. Brunnschweiler, R.O. On the geology of the Indoburman Ranges. *J. Geol. Soc. Aust.* **1966**, *13*, 137–194. [CrossRef]
11. Sevastjanova, I.R.; Hall, M.; Rittner, S.M.T.L.; Paw, T.T.; Naing, D.H.; Alderton, G. Comfort, Myanmar and Asia united, Australia left behind long ago. *Gondwana Res.* **2016**, *32*, 24–40. [CrossRef]
12. Yao, W.L.; Ding, F.L.; Cai, H.Q.; Wang, Q.; Xu, T.; Za, W. Origin and tectonic evolution of upper Triassic Turbidites in the Indo-Burma Ranges, West Myanmar. *Tectonophysics* **2017**, *721*, 90–105. [CrossRef]
13. Zhang, J.e.; Xiao, W.; Windley, B.F.; Cai, F.; Sein, K.; Naing, S. Early Cretaceous wedge extrusion in the Indo-Burma Range accretionary complex: Implications for the Mesozoic subduction of Neotethys in SE Asia. *Int. J. Earth Sci.* **2017**, *106*, 1391–1408. [CrossRef]
14. Mitchell, A.H.G.; Hlaing, T.; Htay, N. The Chin Hills segment of the Indo-Burma Ranges: Not a simple accretioary wedge. In *Indo-Myanmar Ranges in the Tectonic Framework of Himalaya and Southeast Asia*; Soibam, I., Ed.; Geological Society of India Golden Jubilee: Bangalore, India, 2010; Volume 75, pp. 3–24.
15. Advokaat, E.L.; van Hinsbergen, D.J.J. Finding Argoland: Reconstructing a microcontinental archipelago from the SE Asian accretionary orogen. *Gondwana Res.* **2024**, *128*, 161–263.
16. Zahirovic, S.; Matthews, K.J.; Flament, N.; Müller, R.D.; Hill, K.C.; Seton, M.; Gurnis, M. Tectonic evolution and deep mantle structure of the eastern Tethys since the latest Jurassic. *Earth-Sci. Rev.* **2016**, *162*, 293–337. [CrossRef]
17. Chen, D.S.; Luo, H.; Wang, X.H.; Xu, B.; Matsuoka, A. Late Anisian radiolarian assemblages from the Yarlung-Tsangpo Suture Zone in the Jinlu area, Zedong, southern Tibet: Implications for the evolution of Neotethys. *Island Arc* **2019**, *28*, e12302. [CrossRef]
18. Naing, T.T.; Robinson, S.A.; Searle, M.P.; Morley, C.K.; Millar, I.; Green, O.R.; Bown, P.R.; Danelian, T.; Petrizzo, M.R.; Henderson, G.M. Age, depositional history and tectonics of the Indo-Myanmar Ranges, Myanmar. *J. Geol. Soc.* **2023**, *180*, 2022–2091. [CrossRef]
19. Li, X.H.; Zeng, Q.G.; Wang, C.S. Palaeocurrent data: Evidence for the source of the Langjiexue Group in southern Tibet. *Geol. Rev.* **2003**, *49*, 132–137, (In Chinese with English abstract).
20. Pan, W.Y. The Sedimentary Features and Paleogeographic Implication of the Late Jurassic Quartz Sandstones in Tethys Himalaya of Southern Tibet. Master's Thesis, China University of Geosciences, Beijing, China, 2018; pp. 1–66, (In Chinese with English abstract).
21. Zhang, B.S.; Wei, Y.S.; Garzanti, E.; Wang, C.S.; Chen, X.; Pan, W.Y.; Liu, Q.S. Sedimentologic and stratigraphic constraints on the orientation of the Late Triassic northern Indian passive continental margin. *Palaeogeogr. Palaeoclimatol. Palaeoecol.* **2019**, *533*, 109234. [CrossRef]
22. Gradstein, F.M. Leg 122–123, northwestern Australian margin; A stratigraphic and paleogeographic summary. *Proc. Ocean. Drill. Program Sci. Results* **1992**, *123*, 801–816.
23. Garzanti, E. Stratigraphy and sedimentary history of the Nepal Tethys Himalaya passive margin. *J. Asian Earth Sci.* **1999**, *17*, 805–827. [CrossRef]
24. Zhang, X.R.; Chung, S.L.; Tang, J.T.; Maulana, A.; Mawaleda, M.; Oo, T.; Tien, C.Y.; Lee, H.Y. Tracing Argoland in eastern Tethys and implications for India-Asia convergence. *GSA Bull.* **2021**, *133*, 1712–1722. [CrossRef]
25. Ao, A.; Bhowmik, S.K. Cold subduction of the Neotethys: The metamorphic record from finely banded lawsonite and epidote blueschists and associated metabasalts of the Nagaland Ophiolite Complex, India. *J. Metamorph. Geol.* **2014**, *32*, 829–860. [CrossRef]
26. Rajkakati, M.; Bhowmik, S.K.; Ao, A.; Ireland, T.R.; Avila, J.; Clarke, G.L.; Bhandari, A.; Aitchison, J.C. Thermal history of Early Jurassic eclogite facies metamorphism in the Nagaland Ophiolite Complex, NE India: New insights into pre-Cretaceous subduction channel tectonics within the Neo-Tethys. *Lithos* **2019**, *346–347*, 105166. [CrossRef]
27. Khogenkumar, S.; Singh, A.K.; Singh, R.K.B.; Khanna, P.P.; Singh, N.I.; Singh, W.I. Coexistence of MORB and OIB-type mafic volcanics in the Manipur Ophiolite Complex, Indo-Myanmar Orogenic Belt, northeast India: Implication for heterogeneous mantle source at the spreading zone. *J. Asian Earth Sci.* **2016**, *116*, 42–58. [CrossRef]
28. Yang, J.S.; Xu, Z.Q.; Duan, X.D.; Li, J.; Xiong, F.H.; Liu, Z.; Cai, Z.H.; Li, H.Q. Discovery of a Jurassic SSZ ophiolite in the Myitkyina region of Myanmar. *Acta Petrol. Sin.* **2012**, *28*, 1710–1730, (In Chinese with English abstract).
29. Baxter, A.T.; Aitchison, J.C.; Zyabrev, S.V.; Ali, J.R. Upper Jurassic radiolarians from the Naga Ophiolite, Nagaland, northeast India. *Gondwana Res.* **2011**, *20*, 638–644. [CrossRef]
30. Yang, S.; Liang, X.; Jiang, M.; Chen, L.; He, Y.; Thet Mon, C.; Hou, G.; Thant, M.; Sein, K.; Wan, B. Slab remnants beneath the Myanmar terrane evidencing double subduction of the Neo-Tethyan Ocean. *Sci. Adv.* **2022**, *8*, eabo1027. [CrossRef]
31. Socquet, A.; Goffé, B.; Pubellier, M.; Rangin, C. Late Cretaceous to Eocene metamorphism of the internal zone of the Indo-Burma range (western Myanmar): Geodynamic implications. *Comptes Rendus Geosci.* **2002**, *334*, 573–580. [CrossRef]
32. Gramann, F. Some palaeontological data on the Triassic and Cretaceous of the western part of Burma (Arakan Islands, Arakan Yoma, western outcrops of Central Basin). *Newsl. Stratigr.* **1974**, *3*, 277–290. [CrossRef]

33. Mitchell, A.H.G. Cretaceous-Cenozoic tectonic events in the western Myanmar (Burma)-Assam region. *J. Geol. Soc. Lond.* **1993**, *150*, 1089–1102. [CrossRef]
34. Swe, W. *Outline Geology and Economic Mineral Occurrences of the Union of Myanmar*; Myanmar Geosciences Society: Yangon, Myanmar, 2012; pp. 1–215.
35. Cai, F.; Ding, L.; Zhang, Q.; Orme, D.A.; Wei, H.; Li, J.; Zhang, J.e.; Zaw, T.; Sein, K. Initiation and evolution of forearc basins in the Central Myanmar Depression. *GSA Bull.* **2020**, *132*, 1066–1082. [CrossRef]
36. Ghose, N.C.; Chatterjee, N.; Fareeduddin. *A petrographic Atlas of Ophiolite: An Example from the Eastern India-Asia Collision Zone*; Springer: New Delhi, India, 2014.
37. Xie, L.W.; Zhang, Y.B.; Zhang, H.H.; Sun, J.F.; Wu, F.Y. In situ simultaneous determination of trace elements, U-Pb and Lu-Hf isotopes in zircon and baddeleyite. *Chin. Sci. Bull.* **2008**, *53*, 1565–1573. [CrossRef]
38. Andersen, T. Correction of common lead in U-Pb analyses that do not report ^{204}Pb. *Chem. Geol.* **2002**, *192*, 59–79. [CrossRef]
39. Burma United Nations DGSE/GSE-Project BUR/72/002, Simplified Geological Map of Mindat-Saw Area 1:100000. (unpublished). 1978.
40. Liu, C.Z.; Chung, S.L.; Wu, F.Y.; Zhang, C.; Xu, Y.; Wang, J.G.; Chen, Y.; Guo, S. Tethyan suturing in Southeast Asia: Zircon U-Pb and Hf-O isotopic constraints from Myanmar ophiolites. *Geology* **2016**, *44*, 311–314. [CrossRef]
41. Sengupta, S.; Acharyya, S.K.; Van Den Hul, H.J.; Chattopadhyay, B. Geochemistry of volcanic rocks from the Naga Hills Ophiolites, northeast India and their inferred tectonic setting. *J. Geol. Soc.* **1989**, *146*, 491–498. [CrossRef]
42. Isozaki, Y.; Maruyama, S.; Furuoka, F. Accreted oceanic materials in Japan. *Tectonophysics* **1990**, *181*, 179–205. [CrossRef]
43. Chen, J.H.; Ba, D.Z. Halobia fauna from Zedang of south Xizang with a discussion on the Halobia assemblages in China. *Acta Palaeontol. Sin.* **1986**, *25*, 1–9, (In Chinese with English abstract).
44. Aikman, A.B.; Harrison, T.M.; Lin, D. Evidence for Early (>44 Ma) Himalayan Crustal Thickening, Tethyan Himalaya, southeastern Tibet. *Earth Planet. Sci. Lett.* **2008**, *274*, 14–23. [CrossRef]
45. Cai, F.; Ding, L.; Laskowski, A.K.; Kapp, P.; Wang, H.; Xu, Q.; Zhang, L. Late Triassic paleogeographic reconstruction along the Neo–Tethyan Ocean margins, southern Tibet. *Earth Planet. Sci. Lett.* **2016**, *435*, 105–114. [CrossRef]
46. Cao, H.; Huang, Y.; Li, G.; Zhang, L.; Wu, J.; Dong, L.; Dai, Z.; Lu, L. Late Triassic sedimentary records in the northern Tethyan Himalaya: Tectonic link with Greater India. *Geosci. Front.* **2018**, *9*, 273–291. [CrossRef]
47. Fang, D.-R.; Wang, G.-H.; Hisada, K.-i.; Yuan, G.-L.; Han, F.-L.; Li, D.; Tang, Y.; Pei, Q.-M.; Zhang, L.-L. Provenance of the Langjiexue Group to the south of the Yarlung-Tsangpo Suture Zone in southeastern Tibet: Insights on the evolution of the Neo-Tethys Ocean in the Late Triassic. *Int. Geol. Rev.* **2019**, *61*, 341–360. [CrossRef]
48. Li, G.W.; Liu, X.H.; Pullen, A.; Wei, L.J.; Liu, X.B.; Huang, F.X.; Zhou, X.J. In-situ detrital zircon geochronology and Hf isotopic analyses from Upper Triassic Tethys sequence strata. *Earth Planet. Sci. Lett.* **2010**, *297*, 461–470.
49. Li, X.; Mattern, F.; Zhang, C.; Zeng, Q.; Mao, G. Multiple sources of the Upper Triassic flysch in the eastern Himalaya Orogen, Tibet, China: Implications to palaeogeography and palaeotectonic evolution. *Tectonophysics* **2016**, *666*, 12–22. [CrossRef]
50. Liu, Y.M.; Dai, J.G.; Wang, C.S.; Li, H.A.; Wang, Q.; Zhang, L.L. Provenance and tectonic setting of Upper Triassic turbidites in the eastern Tethyan Himalaya: Implications for early-stage evolution of the Neo–Tethys. *Earth-Sci. Rev.* **2020**, *200*, 103030. [CrossRef]
51. Meng, J.; Gilder, S.A.; Wang, C.; Coe, R.S.; Tan, X.; Zhao, X.; He, K. Defining the Limits of Greater India. *Geophys. Res. Lett.* **2019**, *46*, 4182–4191. [CrossRef]
52. Wang, J.G.; Wu, F.Y.; Garzanti, E.; Hu, X.M.; Ji, W.Q.; Liu, Z.C.; Liu, X.C. Upper Triassic turbidites of the northern Tethyan Himalaya (Langjiexue Group): The terminal of a sediment-routing system sourced in the Gondwanide Orogen. *Gondwana Res.* **2016**, *34*, 84–98. [CrossRef]
53. Huang, Y.; Cao, H.-w.; Li, G.-m.; Brueckner, S.M.; Zhang, Z.; Dong, L.; Dai, Z.-w.; Lu, L.; Li, Y.-b. Middle–Late Triassic bimodal intrusive rocks from the Tethyan Himalaya in South Tibet: Geochronology, petrogenesis and tectonic implications. *Lithos* **2018**, *318–319*, 78–90. [CrossRef]
54. Wang, Y.Y.; Guo, L.E.; Zeng, L.S.; Chen, F.K.; Hou, K.J.; Wang, Q.; Zhao, L.H.; Guo, J.H. Multiple phases of cretaceous magic magmatism in the Gyangze-Kangma area (Tethyan Himalaya) southern Tibet. *Acta Petrol. Sin.* **2016**, *32*, 3572–3596, (In Chinese with English abstract).
55. Zhu, D.C.; Pan, G.T.; Mo, X.X.; Liao, Z.L.; Jiang, X.S.; Wang, L.Q.; Zhao, Z.D. Petrogenesis of volcanic rocks in the Sangxiu Formation, central segment of Tethyan Himalaya: A probable example of plume–lithosphere interaction. *J. Asian Earth Sci.* **2007**, *29*, 320–335. [CrossRef]
56. BYGS. *Introduction on Geological map of Longzi County (1:250000: Bureau of Yunnan Geological Survey)*; Bureau of Yunnan Geological Survey: Kunming, China, 2004; p. 270. (In Chinese)
57. Lewis, C.J.; Sircombe, K. Use of U-Pb geochronology to delineate provenance of North West Shelf Sediments, Australia. In *West Australian Basins Symposium (WABS) 2013*; Keep, M., Moss, S.J., Eds.; Perth Convention & Exhibition Centre: Perth, WA, Australia, 2013.
58. Metcalf, K.; Kapp, P. History of subduction erosion and accretion recorded in the Yarlung Suture Zone, southern Tibet. *Geol. Soc. Lond. Spec. Publ.* **2019**, *483*, 517–554. [CrossRef]
59. Webb, A.A.G.; Yin, A.; Dubey, C.S. U-Pb zircon geochronology of major lithologic units in the eastern Himalaya: Implications for the origin and assembly of Himalayan rocks. *GSA Bull.* **2013**, *125*, 499–522. [CrossRef]

60. Munasri; Maharsha, P.A. First evidence of Middle to Late Triassic radiolarians in the Garba mountains, South Sumatra, Indonesia. *Isl. Arc* **2019**, *28*, e12298.
61. Wang, Y.J.; Yang, Q.; Matsuoka, A.; Kobayashi, K.; Nagahashi, T.; Zeng, Q.G. Triassic radiolarians from the Yarlung Zangbo suture zone in the Jinlu area, Zetang County, southern Tibet. *Acta Micropalaeontol. Sin.* **2002**, *19*, 215–227, (In English with Chinese abstract).
62. Huang, W.; van Hinsbergen, D.J.J.; Dekkers, M.J.; Garzanti, E.; Dupont-Nivet, G.; Lippert, P.C.; Li, X.; Maffione, M.; Langereis, C.G.; Hu, X.; et al. Paleolatitudes of the Tibetan Himalaya from primary and secondary magnetizations of Jurassic to Lower Cretaceous sedimentary rocks. *Geochem. Geophys. Geosystems* **2015**, *16*, 77–100. [CrossRef]
63. Zhang, J.e.; Xiao, W.; Wakabayashi, J.; Windley, B.F.; Han, C. A fragment of Argoland from East Gondwana in the NE Himalaya. *J. Geophys. Res. Solid Earth* **2022**, *127*, e2021JB022631. [CrossRef]
64. Zhang, W.P.; Mo, X.X.; Zhu, D.C.; Yuan, S.H.; Wang, L.Q. Chronology and geochemistry on gabbro and basalt of the ophiolite melange in Lang County, Tibet, China. *J. Chengdu Univ. Technol.* **2011**, *38*, 538–548, (In Chinese with English abstract).
65. Zhu, D.C.; Mo, X.X.; Pan, G.T.; Zhao, Z.D.; Dong, G.C.; Shi, Y.R.; Liao, Z.L.; Wang, L.Q.; Zhou, C.Y. Petrogenesis of the earliest Early Cretaceous mafic rocks from the Cona area of the eastern Tethyan Himalaya in south Tibet: Interaction between the incubating Kerguelen plume and the eastern Greater India lithosphere? *Lithos* **2008**, *100*, 147–173. [CrossRef]
66. Shi, Y.R.; Hou, C.Y.; Anderson, J.L.; Yang, T.S.; Ma, Y.M.; Bian, W.W.; Jin, J.J. Zircon SHRIMP U–Pb age of Late Jurassic OIB-type volcanic rocks from the Tethyan Himalaya: Constraints on the initial activity time of the Kerguelen mantle plume. *Acta Geochim.* **2018**, *37*, 441–455. [CrossRef]
67. Yin, J.R.; Wan, X.Q. Jurassic ammonite morphotypes as water-depth indicator of Tethys-Himalaya sea. *Acta Palaeontol. Sin.* **1996**, *35*, 734–751, (In Chinese with English abstract).
68. Gibbons, A.D.; Barckhausen, U.; van den Bogaard, P.; Hoernle, K.; Werner, R.; Whittaker, J.M.; Müller, R.D. Constraining the Jurassic extent of Greater India: Tectonic evolution of the West Australian margin. *Geochem. Geophys. Geosystems* **2012**, *13*, Q05W13. [CrossRef]
69. Heine, C.; Muller, R.D. Late Jurassic rifting along the Australian North West Shelf: Margin geometry and spreading ridge configuration. *Aust. J. Earth Sci.* **2005**, *52*, 27–39. [CrossRef]
70. Wang, H.-Q.; Ding, L.; Cai, F.-L.; Sun, Y.-L.; Li, S.; Yue, Y.-H.; Fan, S.-Q.; Guo, X.-D.; Qasim, M. The latest Jurassic protoliths of the Sangsang mafic schists in southern Tibet: Implications for the spatial extent of Greater India. *Gondwana Res.* **2020**, *79*, 248–262. [CrossRef]
71. Westerweel, J.; Roperch, P.; Licht, A.; Dupont-Nivet, G.; Win, Z.; Poblete, F.; Ruffet, G.; Swe, H.H.; Thi, M.K.; Aung, D.W. Burma Terrane part of the Trans-Tethyan arc during collision with India according to palaeomagnetic data. *Nat. Geosci.* **2019**, *12*, 863–868. [CrossRef] [PubMed]
72. Westerweel, J.; Licht, A.; Cogné, N.; Roperch, P.; Dupont-Nivet, G.; Kay Thi, M.; Swe, H.H.; Huang, H.; Win, Z.; Wa Aung, D. Burma Terrane Collision and Northward Indentation in the Eastern Himalayas Recorded in the Eocene-Miocene Chindwin Basin (Myanmar). *Tectonics* **2020**, *39*, e2020TC006413. [CrossRef]
73. Li, R.; Ao, S.; Xiao, W.; Windley, B.F.; Zhan, M.; Huang, P.; Bhandari, S. Tectonic setting and provenance of Early Cretaceous strata in the footwall of Main Central Thrust, Eastern Nepal: Implications for the archipelago palaeogeography of the Neo-Tethys. *Geol. J.* **2021**, *56*, 1958–1973. [CrossRef]
74. Xiao, W.J.; Ao, S.J.; Yang, L.; Han, C.M.; Wan, B.; Zhang, J.E.; Zhang, Z.Y.; Li, R.; Chen, Z.Y.; Song, S.H. Anatomy of composition and nature of plate convergence: Insights for alternative thoughts for terminal India-Eurasia collision. *Sci. China Earth Sci.* **2017**, *60*, 1015–1039. [CrossRef]

minerals

MDPI

Article

Thermochronological Constraints on the Tectonic History of the Arabian–Nubian Shield's Northern Tip, Sinai, Egypt

Sherif Mansour [1,*], Khaled M. Abdelfadil [2], Noriko Hasebe [3], Akihiro Tamura [4], Kamal Abdelrahman [5], Mohamed A. Gharib [1], Mohammed S. Fnais [5] and Amer A. Shehata [1]

[1] Geology Department, Faculty of Science, Port Said University, Port Said 42522, Egypt
[2] Geology Department, Faculty of Science, Sohag University, Sohag 82524, Egypt
[3] Institute of Nature and Environmental Technology, Kanazawa University, Kanazawa 920-1192, Japan
[4] Department of Earth Sciences, Kanazawa University, Kanazawa 920-1192, Japan
[5] Department of Geology & Geophysics, College of Science, King Saud University, P.O. Box 2455, Riyadh 11451, Saudi Arabia
* Correspondence: sherif@sci.psu.edu.eg; Tel.: +20-10-9001-2146

Abstract: The effects of different regional tectonic events on the Neoproterozoic basement rocks of the Arabian–Nubian Shield in Sinai, as well as the Egyptian unstable and stable shelves, remain uncertain. Coupling fission-track thermochronometry findings with the modeling of the time–temperature history has proved to be an effective method for tackling these issues. The obtained zircon fission-track ages were differentiated into two groups from the Ediacaran–Cambrian and the Ordovician–Carboniferous periods, while the apatite fission-track data revealed two separate groups of cooling ages of the Carboniferous–Triassic and Late Cretaceous ages. The integration of these cooling ages and modeling of the time–temperature history revealed four discrete cooling pulses during the Neoproterozoic, Devonian–Carboniferous, Cretaceous, and Oligocene–Miocene eras. After integrating our findings with the regional tectonic historical and sedimentological records, these could be identified as cooling/exhumation pulses activated in response to the post-accretional event of erosion, Variscan tectonism, the disintegration of Gondwana, and the Gulf of Suez rifting, respectively. Furthermore, the southern border of the Egyptian unstable shelf was found to extend southward to South Sinai and south of the Bahariya depression.

Keywords: Sinai tectonic history; the Arabian–Nubian shield; rift flanks; Egypt thermochronology; LA-ICP-MS

Citation: Mansour, S.; Abdelfadil, K.M.; Hasebe, N.; Tamura, A.; Abdelrahman, K.; Gharib, M.A.; Fnais, M.S.; Shehata, A.A. Thermochronological Constraints on the Tectonic History of the Arabian–Nubian Shield's Northern Tip, Sinai, Egypt. *Minerals* **2024**, *14*, 1246. https://doi.org/10.3390/min14121246

Academic Editors: Jim Lee, Fan Yang, Cun Zhang and Jian Chang

Received: 28 September 2024
Revised: 24 November 2024
Accepted: 29 November 2024
Published: 6 December 2024

1. Introduction

The Arabian–Nubian Shield (ANS) was constructed between ca. 820 Ma and 530 Ma [1,2] during the East African Orogeny (EAO; ca. 900–650 Ma) through the amalgamation of island arcs and microcontinents in the Archean cratons [3]. Before the end of Cambrian time, the Egyptian ANS (Figure 1) was eroded due to delamination of the lithospheric mantle during the transition from the compressional to the extensional setting to reach a peneplained surface [4–6]. Then, a Cambrian–Carboniferous sedimentary sequence with a thickness of more than 2 km was deposited on the ANS peneplained surface [7]. Following the Carboniferous era, the Variscan tectonism influenced the entire northern part of Egypt through rock exhumations and erosions [8]. Then, an era of tectonic stability and platform sedimentation dominated until the Cretaceous [9], when the Syrian arc domal system developed as a response to the Gondwana breakup [10–12]. Classically, these Paleozoic and Mesozoic tectonic events were thought to be restricted to the Egyptian unstable shelf [13,14]. However, more recent research suggested that this unstable shelf (Figure 2) extended southward [15–18]. Afterward, the Red Sea/Gulf of Suez rift system began to form and continued through the Oligocene–Miocene, forming axial grabens and rift shoulders of the ANS Neoproterozoic basement exposures [19], which comprised the entire array of regional geological events.

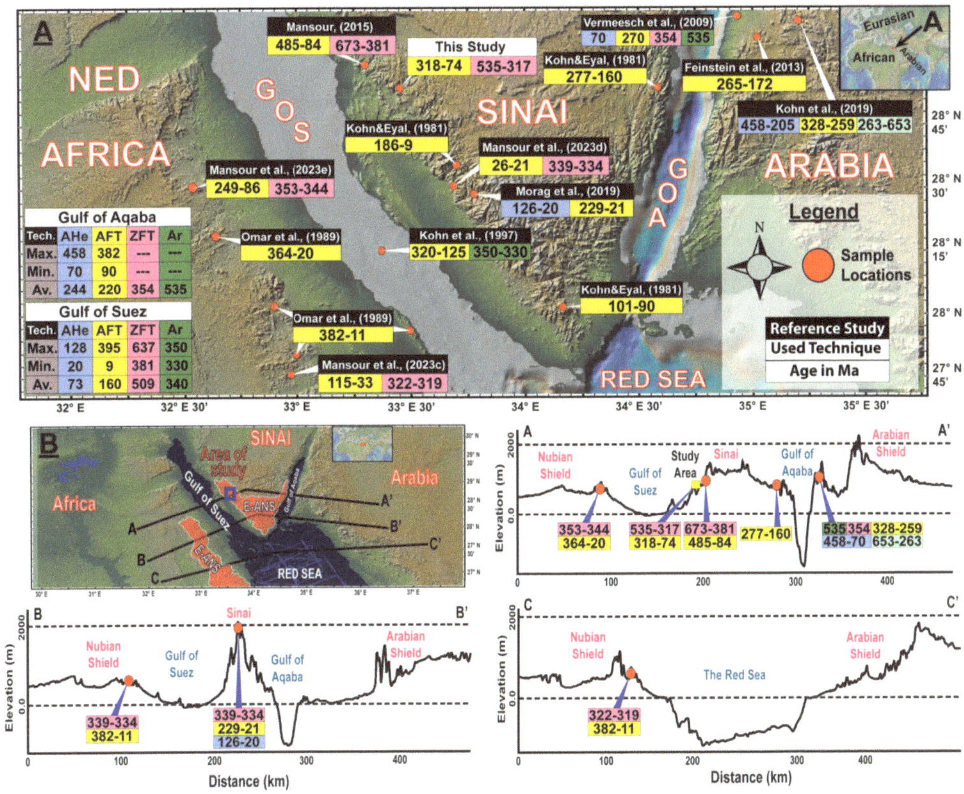

Figure 1. (**A**) Location map based on a digital elevation model (www.geomapapp.org, accessed on 23 July 2022) representing the previous thermochronological studies in Sinai and the northern ANS. At the top-right-hand corner is an inset showing the configuration of the African, Arabian, and Eurasian plates, where Ar, Ar-Ar dating; ZFT, zircon fission track; ZHe, zircon (U-Th)/He; AFT, apatite fission track; and AHe, apatite (U-Th)/He [6,16–18,20–26]. (**B**) Topographic map showing the northern Red Sea rift system representing the Egyptian ANS (E-ANS) and different cross-sections with locations of the cooling ages represented in part A of the figure. (**AA′**) Across the NED, central Suez Rift, northern South Sinai, central Aqaba Rift, and northern Arabia, where the depth to the basement reaches 3 km [27]; (**BB′**) across the NED, southern Suez Rift, southern South Sinai, southern Aqaba Rift, and northern Arabia, where the depth to the basement reaches 4 km [27]; (**CC′**) across the CED, northern Red Sea, and central Arabia (modified after [6]).

Thermochronological data from several studies were collected from different localities on the western flank of the Suez Rift, the Aqaba Rift, and the Red Sea Rift [7,16,18,19,21,22,24,25,28–31]. These studies showed a wide range of cooling ages, which extended from ca. 535 Ma to 29 ± 2 Ma. Fewer data were reported for the eastern shoulder (studied region) of the Suez Rift, and they showed cooling ages of between 674 ± 77 Ma and 11 ± 2 Ma [17,20,23,30,32].

Such a shortage of and disparities in the data for a rifted region characterized by heterogeneous topographic growth resulted in several uncertainties. These included the possible effects of the pre-rift regional tectonic activities of the Precambrian, Devonian–Carboniferous, and Cretaceous on the ANS in the southern Sinai; indefinite conclusions on whether the

South Sinai region belongs to the Egyptian stable shelf or unstable shelf (Figure 2); and the reason why the majority of the reported cooling ages were pre-dated to the rifting event. Reconstructing the thermal–tectonic history of the northern exposures of the ANS using a combination of low-temperature thermochronometers and time–temperature (t-T) modeling could effectively provide valuable insights into such geological processes in various geological settings [23,33,34].

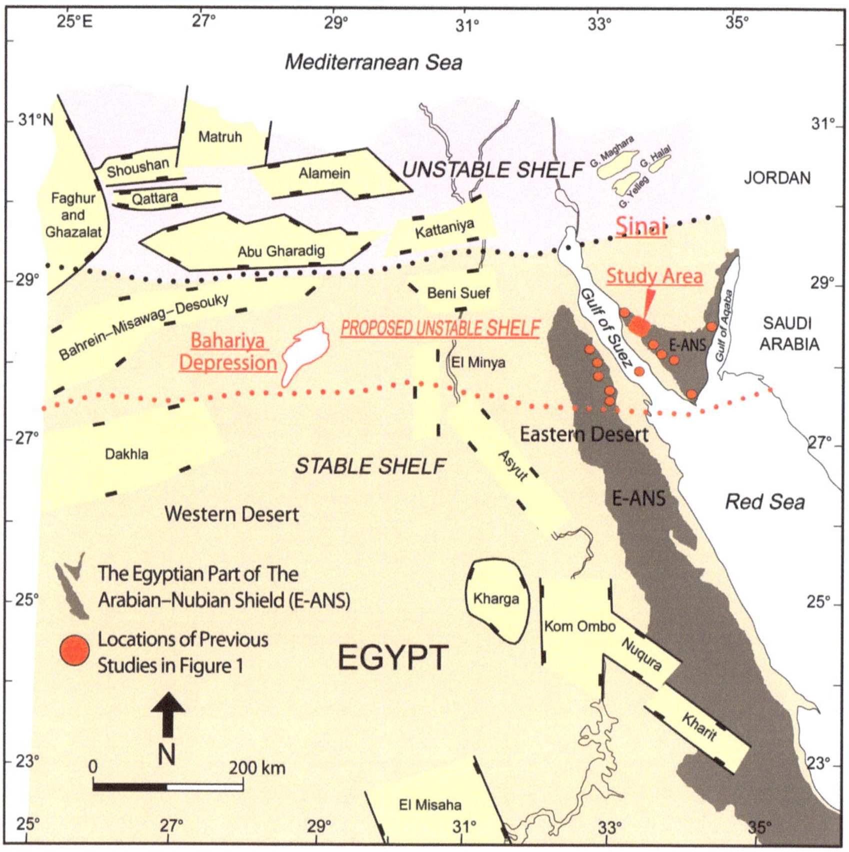

Figure 2. The unstable shelf and the stable shelf of Egypt, with the black dotted line, as proposed by Said [13,35], and the red dotted line, as proposed by this study and Moustafa et al. [15]. Representing the ANS and the main Paleozoic–Mesozoic basins (yellow) of the Western Desert and Nile Valley (modified after [15,36]).

For the purposes of this study, representative samples were collected across the Wadi Nesyrin area, which is part of the northern exposures of the Neoproterozoic ANS (Figure 3). Our samples were from different structural blocks and lithological units of the Neoproterozoic ANS suites in the region, including the metamorphic suite (four samples) with crystallization ages of ca. 870-800 Ma [1,37,38], as well as the Older granitic suite (eight samples) and the Younger granitic suite (six samples), with crystallization ages of ca. 750-550 Ma [1,37,38].

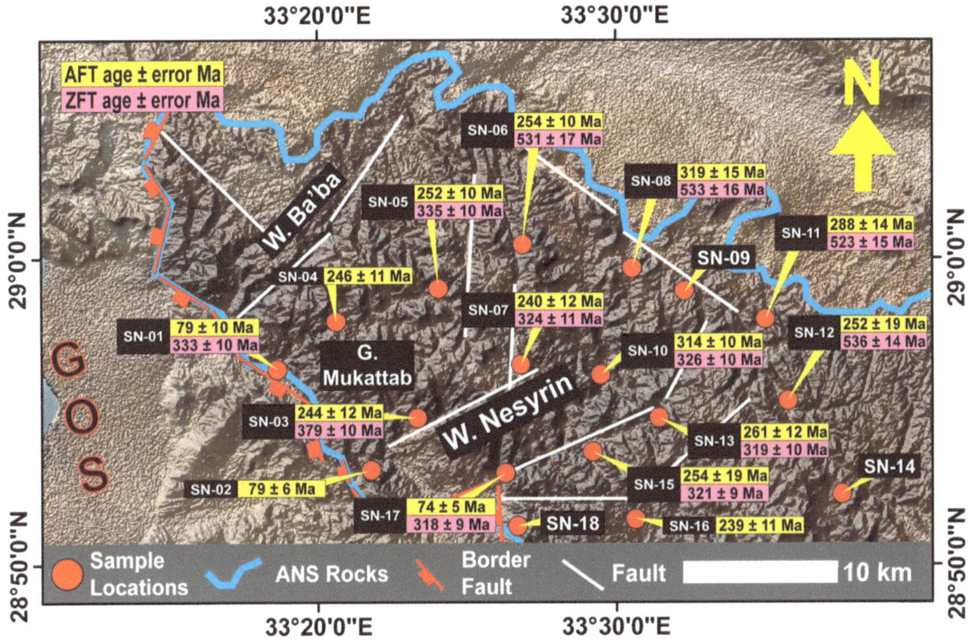

Figure 3. Simplified location and structural map for Wadi Nesyrin area with locations, ZFT ages, and AFT ages of the studied samples.

2. Geological Setting and Regional Tectonics

The ANS basement exposures terminate at southern Sinai, which is situated in a tectonically dynamic region that is affected dramatically by interactions between the African, Arabian, and Eurasian plates (Figure 1).

2.1. Rock Exposures

The ANS basement rocks in Sinai belong to the ANS, which was formed as part of the EAO during the Neoproterozoic [2,3]. Sinai basement rocks consist predominantly of the island arc metamorphic suite, which reflects the initial phase of the EAO in the region that was activated from ca. 820 Ma to 750 Ma [1,37]. The syn-orogenic granitoid suite indicates a dominant compressional regime from ca. 750 Ma to 610 Ma [1,39]. The post-orogenic granitoid and dike suite represents the final phase of the EAO, which was characterized by an extensional regime from ca. 610 Ma to 550 Ma [1,39,40].

The sedimentary sequence covers both the stable and unstable shelves (Figure 2), and the unstable shelf is characterized by exposures of asymmetric linear folds, overthrusts, and diapirisms [13,41]. The sedimentologic sequence in the vicinity of the Suez Rift started in the Cambrian and continued to the Recent, reflecting temporal changes in the depositional environments (Figure 4). Additional information regarding the sedimentary sequence, depositional settings, and relationships can be found in the Supplementary File Section S2.

2.2. Regional Tectonics

After construction during the EAO, the Neoproterozoic ANS basement complex was peneplained by the Cambrian through the post-accretional erosional event (PAEE), as shown by the Lower Cambrian fossils found in a fluviomarine Cambrian strata [42]. Then, regional intracratonic basins and long-lasting subsidence developed, which involved depositing more than 2 km of a Lower Paleozoic fluvio-marine sequence (Figure 4) [43].

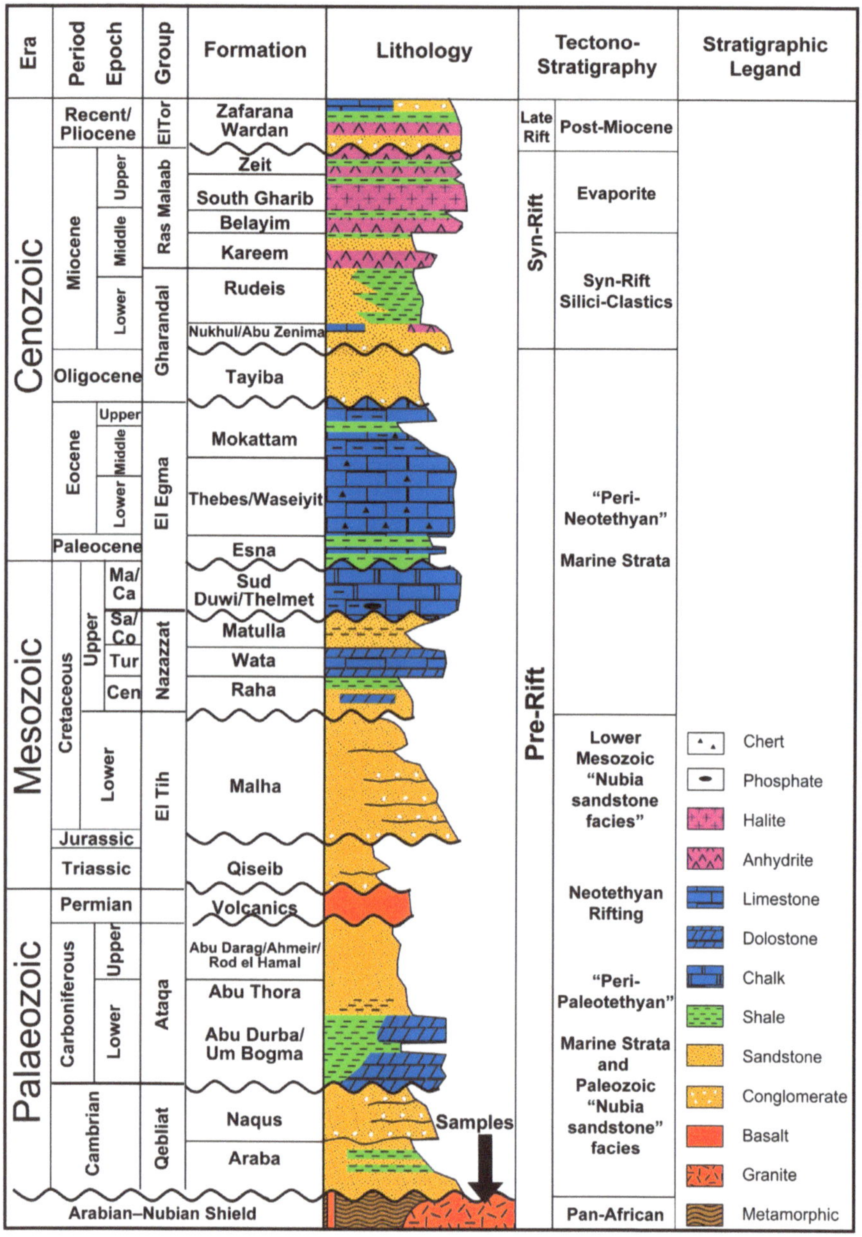

Figure 4. Generalized lithostratigraphic cross-section and tectono-stratigraphic environments for the Suez Rift region (modified after [44,45]).

During the Carboniferous, Laurasia and Gondwana collided (Africa moved northward), leading to the development of Variscan tectonism [8,46]. This tectonic activity influenced northern Africa by massive uplifts and erosions to erode most of the Lower Paleozoic sequence [16,22,28,30,47–49]. During the Cretaceous, the Gondwana disintegration caused eastward movement of the African plate with respect to the Eurasian plate producing progressive closure of the Tethys realm, and affecting northern Africa with sinistral shearings, limited volcanic activities, normal faults, domes of the Syrian Arc, and inversion of the intracratonic basins [10,11,50,51]. The Cretaceous tectonism affected the unstable shelf of northern Egypt, which classically was thought to extend from the Mediterranean margin to south of the Abu Gharading depression and North Sinai (Figure 2) [13]. However, there is accumulated evidence showing that the unstable shelf extended further to the south into what was previously interpreted as a stable shelf in the Egyptian Western Desert [15], the Eastern Desert [22], and South Sinai [16,30]. Moustafa et al. [15] found that the Bahariya region was greatly similar to the deformation of the tectonically unstable area of the northern Western Desert. Mansour et al. [16,22,30] identified tectonically induced uplifts in the Gabal Loman area (North Eastern Desert), Gabal Samra area, and Wadi Dahal area in southern Sinai (Figure 2). During the Oligocene–Miocene, the Red Sea rift system began to develop in the Gulf of Suez, which formed a subsided rift axis, uplifted flanks, normal faults, and basaltic dikes [19,52]. Finally, during the Miocene, the Gulf of Aqaba developed through the activation of the Dead Sea rift system. Since the Middle Miocene, the Gulf of Aqaba rift has become the main northern plate boundary between Africa and Arabia, with spreading rates four to eight times higher than those of the Suez Rift [24,53].

Structurally, the vicinity of the Suez Rift is primarily characterized by three fault sets (Figure 3): (1) the Suez-Rift-associated NNW–NW faults, (2) the pre-rift NE faults, and (3) the Precambrian ENE faults [49].

2.3. *The Problem and Available Information*

Previously reported thermochronological data from the northern part of the ANS revealed an extended cooling history marked by non-uniform exhumations through block-bounding faults before the rift initiation [17,18,24,29–31,54]. Collectively, the Suez Rift shoulders showed AFT cooling ages that extended from 485 ± 18 Ma to 11 ± 2 Ma [17,21,22,29,30,32], ZFT cooling ages that ranged from 674 ± 77 Ma to 334 ± 9 Ma [6,20,32], and AHe cooling ages that ranged from 126 ± 4 Ma to 19 ± 1 Ma [23]. Meanwhile, the western Red Sea shoulder yielded sphene FT cooling ages that ranged between 410 Ma and 339 Ma [28], ZFT cooling ages of that were between ca. 366 Ma and ca. 315 Ma [6,28], and AFT cooling ages that ranged between 191 ± 7 Ma and 29 ± 2 Ma [16,31].

This wide range of reported cooling ages showed that there was considerable controversy [6,17,18,23,24,28–32,54] regarding the effect and extent of the pre-rift regional tectonism and the rifting event on the ANS basement rocks in the southern Sinai. Furthermore, additional questions about the extent of the Egyptian unstable shelf's southern border were raised. This southern border was classically drawn based on the surface exposures of the Phanerozoic pre-rift structures [13,35,41].

3. Materials and Methods

A fission-track (FT) thermochronometer is based on the forming and retaining of radioactivity-induced tracks of damage within the crystal lattice. These tracks of damage are generated by the movement of the heavy fragments resulting from the spontaneous fission of ^{238}U, which creates chemically etchable latent tracks [55]. The counted number of fission tracks (radiation daughters) and the measured U concentrations (radiation parents) are used to solve the time equation, which provides a cooling age through a certain isotherm (closure temperature). This isotherm represents the higher limit of a temperature zone that is a specific-mineral-dependent called the partial annealing zone (PAZ). The PAZ for the apatite FT (AFT) technique is 110–60 °C [56,57], and for the zircon FT (ZFT) thermochronometry technique, it is 240–180 °C [58,59]. This characteristic of the temperature-related length reduction of the radiation products (fission tracks) enables a time–temperature (t-T) history reconstruction through modeling codes (e.g., HeFTy code; [60]).

For this study, representative samples were collected across the Wadi Nesyrin area on the eastern shoulder of the Suez Rift (Figure 1). Of them, 15 samples provided sufficient zircon and/or apatite grains for FT dating (Figure 3). Apatites and zircons were separated from the analyzed samples by applying the conventional methods designated by Donelick et al. [61], while the conditions of etching were chosen to be compatible with the inverse modeling code of Ketcham et al. [56]. The cooling ages included in this study were central ages with one-sigma uncertainties calculated using the IsoplotR code [62]. The software HeFTy v1.9 [60] was used for the t-T history modeling. The analytical techniques and specifications are explained in further detail in the Supporting Information File Section S3 and Table S1.

4. Results

Out of the 18 analyzed samples, 12 samples provided a proper number of zircons for ZFT thermochronometry (Table 1) and 15 samples delivered a reliable number of apatites for AFT thermochronology (Table 2). Of these, 10 specimens provided satisfactory measurable horizontal confined track lengths (HCTLs) for the subsequent modeling of the t-T histories [56,61] (Table 3). The statistical Chi-square (X^2) test for age homogeneity was passed by each of the examined samples with values that ranged between 0.92 and 1 (Tables 1 and 2).

Table 1. Zircon fission-track ages, data, and sample descriptions.

S.-No.	Elevation [m a.s.l.]	Coordinates Decimal		Lithology	^{238}U	n	ρ_s	N_s	X^2	Cen. Age	1σ
		N	E		[μg/g]		(X10^5 Track/cm^2)		[%]	[Ma]	
Group A											
SN-06	873	28.98910°	33.44455°	Syenite	276.5	15	82.5	3807	0.92	**531.1**	16.6
SN-08	654	28.97891°	33.54666°	Diorite	417.3	14	125.3	5080	0.96	**533.1**	16.2
SN-11	606	28.94663°	33.52853°	Diorite	319.2	17	93.4	4997	0.94	**522.5**	14.8
SN-12	834	28.93110°	33.56591°	Syenite	427.5	21	128.4	6985	0.95	**535.5**	13.5
Group B											
SN-01	371	28.91624°	33.33642°	Diorite	347.3	18	63.7	3911	1.00	**333.2**	10.0
SN-03	545	28.91351°	33.38155°	Migmatite	361.4	19	75.9	5685	1.00	**378.8**	10.3
SN-05	698	28.96906°	33.36418°	Syenite	283.5	18	52.6	4015	1.00	**335.3**	9.7
SN-07	819	28.94219°	33.45251°	Syenite	248.5	15	44.2	2038	0.99	**323.5**	11.3
SN-10	658	28.95223°	33.55076°	Gneiss	345.2	17	62.1	3325	1.00	**325.9**	9.8
SN-13	661	28.90790°	33.51673°	Gneiss	216.4	18	38.2	2201	0.99	**319.4**	10.4
SN-15	496	28.90193°	33.47526°	Syenite	352.8	20	62.4	3635	1.00	**320.5**	9.0
SN-17	344	28.87269°	33.40992°	Syenogranite	225.9	23	39.6	3327	1.00	**317.9**	9.1

The analyzed samples that belonged to the ANS basement rocks with formation ages between 720 and 580 Ma [2,63,64]. Groups A (older ZFT cooling ages) and B (younger ZFT ages). S.-No.: sample number, [m a.s.l.]: elevation in meters above sea-level, U: uranium concentration in μg/g, n: the number of counted zircon grains, ρ_s: density of spontaneous tracks (10^5 tr/cm^2), N_s: the number of spontaneous tracks, X^2: the Chi-square test, Cen. Age: central ages with an uncertainty of 1-sigma calculated using IsoplotR [62].

Table 2. Apatite fission-track ages, data, and sample descriptions.

S.-No.	Elevation [m a.s.l.]	Coordinates Decimal		Lithology	^{238}U	n	ρ_s	N_s	X^2	Cen. Age	1σ	Lc no.	Lc	1σ	Dpar	1σ
		N	E		[μg/g]		(X10^6 Track/cm^2)		[%]	[Ma]			(μm)		(μm)	
	Group A															
SN-01	371	28.91624°	33.33642°	Diorite	8.9	21	0.3	65	**0.99**	**79.1**	10.3	**54**	**13.4**	**1.5**	**1.4**	**0.1**
SN-02	480	28.89321°	33.36375°	Syenite	22.3	20	0.7	182	**1.00**	**78.7**	6.2	**6**	**11.9**	**1.3**	**1.5**	**0.1**
SN-17	344	28.87269°	33.40992°	Syenogranite	21.9	20	0.8	250	**1.00**	**73.7**	5.4	**80**	**12.4**	1.6	**1.5**	0.1
SN-18	432	28.83123°	33.43687°	Diorite	NA	NA	NA	NA	NA	NA	NA	NA	NA	NA	NA	NA

Table 2. *Cont.*

S.-No.	Elevation [m a.s.l.]	Coordinates Decimal N	Coordinates Decimal E	Lithology	^{238}U [μg/g]	n	ρ_s (X10^6 Track/cm^2)	N_s	X^2 [%]	Cen. Age [Ma]	1σ	Lc no.	Lc (μm)	1σ	Dpar (μm)	1σ
Group B																
SN-03	545	28.91351°	33.38155°	Migmatite	32.5	17	3.7	638	**0.99**	**244.1**	11.9	**70**	**12.5**	1.2	**1.5**	0.1
SN-04	517	28.96758°	33.41586°	Syenite	27.2	22	3.1	667	**0.99**	**246.0**	11.1	**7**	**12.1**	0.9	**1.5**	0.1
SN-05	698	28.96906°	33.36418°	Syenite	35.2	20	5.2	1120	**1.00**	**251.6**	9.9	**85**	**13.2**	1.1	**1.5**	0.1
SN-06	873	28.98910°	33.44455°	Syenite	22.0	20	2.7	681	**1.00**	**254.2**	10.4	**100**	**13.2**	1.2	**1.4**	0.1
SN-07	819	28.94219°	33.45251°	Syenite	27.8	20	3.2	519	**0.97**	**239.8**	11.9	**109**	**13.1**	1.2	**1.5**	0.1
SN-08	721	28.99807°	33.52186°	Syenite	10.4	20	2.2	464	**0.93**	**318.9**	15.3	**94**	**13.3**	1.1	**1.5**	0.1
SN-09	717	28.98177°	33.54415°	Diorite	NA	NA	NA	NA	NA	NA	NA	NA	NA	NA	NA	NA
SN-10	658	28.95223°	33.55076°	Gneiss	16.2	20	3.0	1013	**1.00**	**314.3**	10.2	**108**	**13.2**	1.2	**1.4**	0.1
SN-11	606	28.94663°	33.52853°	Diorite	23.8	20	3.6	569	**0.94**	**288.2**	14.3	**8**	**11.8**	1.2	**1.5**	0.1
SN-12	834	28.93110°	33.56591°	Syenite	40.0	20	6.1	1452	**0.92**	**251.9**	19.2	**81**	**13.2**	1.0	**1.5**	0.1
SN-13	661	28.90790°	33.51673°	Gneiss	18.2	20	2.7	698	**0.97**	**261.1**	12.4	**12**	**12.3**	1.4	**1.4**	0.1
SN-14	863	28.90091°	33.58210°	Syenite	NA	NA	NA	NA	NA	NA	NA	NA	NA	NA	NA	NA
SN-15	496	28.90193°	33.47526°	Syenite	41.2	22	6.0	1343	**0.92**	**253.7**	19.3	**103**	**13.2**	1.3	**1.5**	0.1
SN-16	664	28.88317°	33.49553°	Diorite	21.8	20	3.0	575	**0.98**	**239.2**	11.2	**5**	**12.6**	1.0	**1.4**	0.1

Table 3. Detailed length and Dpar data of the apatite horizontal confined track.

Sample	HCTLs No.	HCTLs Mean (μm)	HCTLs Std. (μm)	HCTLs Skew.	Lc Mean (μm)	Lc Std. (μm)	Lc Skew.	Dpar No.	Dpar Mean (μm)	Dpar Std. (μm)	Dpar Skew.
SN-01	54	**11.9**	1.9	−0.037	**13.4**	1.5	−0.098	315	**1.4**	0.1	0.061
SN-03	70	**10.2**	1.2	0.920	**12.5**	1.2	0.662	213	**1.5**	0.1	−0.796
SN-04 *	7	**10.2**	0.8	−0.726	**12.1**	0.9	−0.666	20	**1.5**	0.1	−0.423
SN-05	85	**11.3**	1.8	−0.141	**13.2**	1.1	0.262	255	**1.5**	0.1	−1.071
SN-06	100	**11.2**	1.8	0.267	**13.2**	1.2	0.369	300	**1.4**	0.1	0.80
SN-07	109	**11.4**	1.6	−0.008	**13.1**	1.2	−0.378	327	**1.5**	0.1	0.178
SN-08	94	**11.3**	2.8	0.231	**13.3**	1.1	0.156	282	**1.5**	0.1	−0.069
SN-10	108	**11.4**	1.7	0.386	**13.2**	1.2	0.051	324	**1.4**	0.1	1.390
SN-11 *	8	**10.5**	0.8	0.133	**11.8**	1.2	−0.021	22	**1.5**	0.1	−0.310
SN-12	81	**9.6**	1.2	−0.848	**13.2**	1.0	−0.048	243	**1.5**	0.1	−0.351
SN-13 *	12	**11.0**	1.2	0.943	**12.3**	1.4	−0.840	36	**1.4**	0.1	0.425
SN-15	103	**11.3**	2.0	0.334	**13.2**	1.3	0.333	309	**1.5**	0.1	−0.052
SN-16 *	5	**11.0**	1.3	−1.927	**12.6**	1.0	0.273	15	**1.4**	0.1	−0.144
SN-17	80	**11.8**	1.9	0.042	**12.4**	1.6	−0.178	240	**1.5**	0.1	0.123

SN-01: sample symbol, SN-10 *: sample with an insufficient HCTLs number, HCTLs no.: the number of measured horizontal confined tracks, HCTLs mean: mean horizontal confined track length, std: standard deviation, skew: skewness of distribution relative to the mean value (a measure of the asymmetry of the distribution), Lc mean: mean track length after c-axis correction, Dpar no.: the number of measured etch pit diameters, Dpar mean: mean etch pit diameter.

4.1. Zircon FT Thermochronometry

The uranium concentrations ranged across the analyzed samples from ca. 428 to 216 μg/g (Table 1). Notably, there appeared to be no discernible correlation between the ZFT ages and the uranium concentrations (Figure 5). The resultant weighted mean ages and standard errors for the ZFT thermochronometers ranged from 536 ± 14 Ma to 318; ±, 9; Ma, (Figure 3), segregated into two differentiated ZFT age groups (Table 1). Specifically, the oldest group of ages (Table 1; group A) comprised samples SN-06, SN-08, SN-11, and SN-12, which were closer to the sedimentary succession on the eastern edge of the area of study (Figure 3), with ZFT cooling ages that ranged between 536 ± 14 Ma and 523 ± 15 Ma (Table 1). These ages extended over ca. 13 Ma and overlapped within the analytical uncertainties (Table 1 and Figure 3) with an average ZFT cooling age of

530 ± 15 Ma. The younger age group contained all the other samples (Table 1 and Figure 3), which comprised the younger ZFT ages that ranged from 379 ± 10 to 318 ± 9 Ma. The ZFT ages of this age group were extended over ca. 61 Ma but overlapped within the analytical errors, indicating an average ZFT cooling age of 331 ± 15 Ma (Table 1).

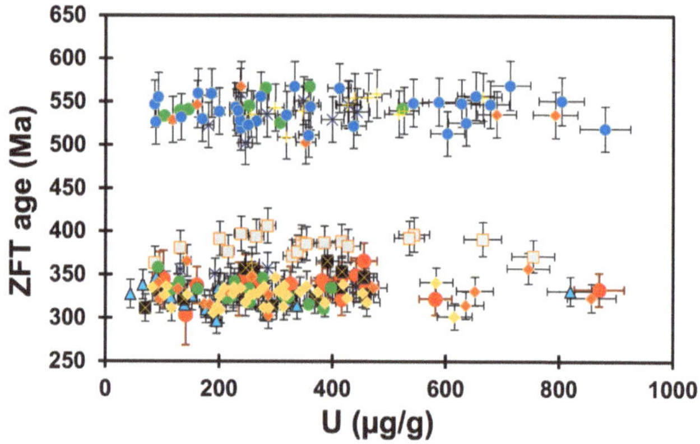

Figure 5. A plot of the ZFT age versus the uranium concentration (each color and symbol represent grains of one sample) does not show any systematic trend, thus ruling out a major effect of metamictization on the ZFT data.

4.2. Apatite FT Thermochronometry

The resulting AFT-weighted mean cooling ages ranged from 319 ± 15 Ma to 74 ± 5 Ma. This wide age range was dividable into two definite spatiotemporal groups (Table 2): Group A consisted of three specimens on the border faults of the Suez Rift that comprised Late Cretaceous cooling ages that ranged from 79 ± 10 Ma to 74 ± 5 Ma (Table 2 and Figure 3), which provided an average AFT cooling age of ca. 76 ± 7 Ma. Meanwhile, group B had the other 12 samples, which comprised Carboniferous–Triassic cooling ages that ranged from 319 ± 15 Ma to 239 ± 11 Ma (Table 2 and Figure 3), implying an average AFT age of 263 ± 13 Ma.

The HCTL means ranged from 11.9 ± 1.9 μm to 9.6 ± 1.2 μm, while the c-axis-projected HCTLs (Lc) ranged between 13.4 ± 1.5 μm and 11.8 ± 1.2 μm (Table 3). We measured a total of 2919 Dpars, which delivered values that ranged from 1.5 ± 0.1 μm to 1.4 ± 0.1 μm (Table 3), which showed an unsystematic correlation with the AFT ages (Figure 6).

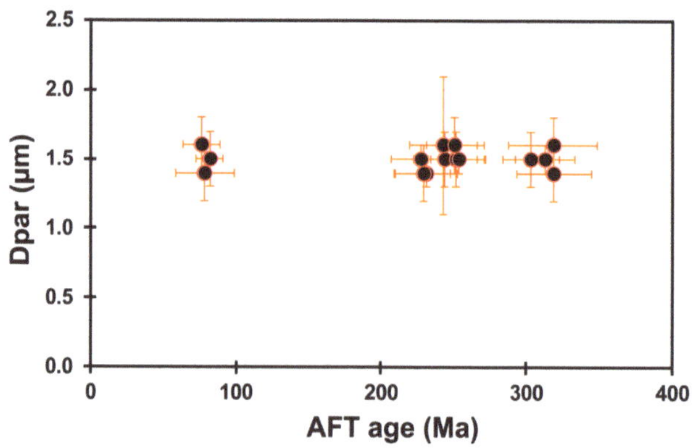

Figure 6. Dpar–AFT age plot. The AFT ages showed no significant change with a change in the Dpar values, indicating the absence of apatite chemistry differentiation in the corresponding AFT ages.

4.3. Reconstructing the Thermal History

The t-T relationship was constructed using the HeFTy code [60]. HeFTy requires t-T constraints outlined via the user to drive its Monte Carlo algorithm. During this study, the defined constraints were established following the acquired cooling ages of the ZFT and AFT, along with the acknowledged geological events. Specifically, the PAEE marks the complete erosion of the ANS before the Cambrian ended and the initiation of the rifting process in the Gulf of Suez during the Oligocene–Miocene (Figures 7 and 8). The time range of the induced constraints (width) was adjusted according to the obtained ages and the associated uncertainties. If the ZFT cooling age had not been acquired, a broad box with extended time limits was established to provide more flexibility. Detailed constraints information is shown in the Supporting Information File (Table S2). The accuracy of the obtained t-T models was highly contingent on the quantity of the measured HCTLs. Consequently, the t-T modeling was exclusively conducted on those samples that boasted an ample number of HCTLs (Table 3).

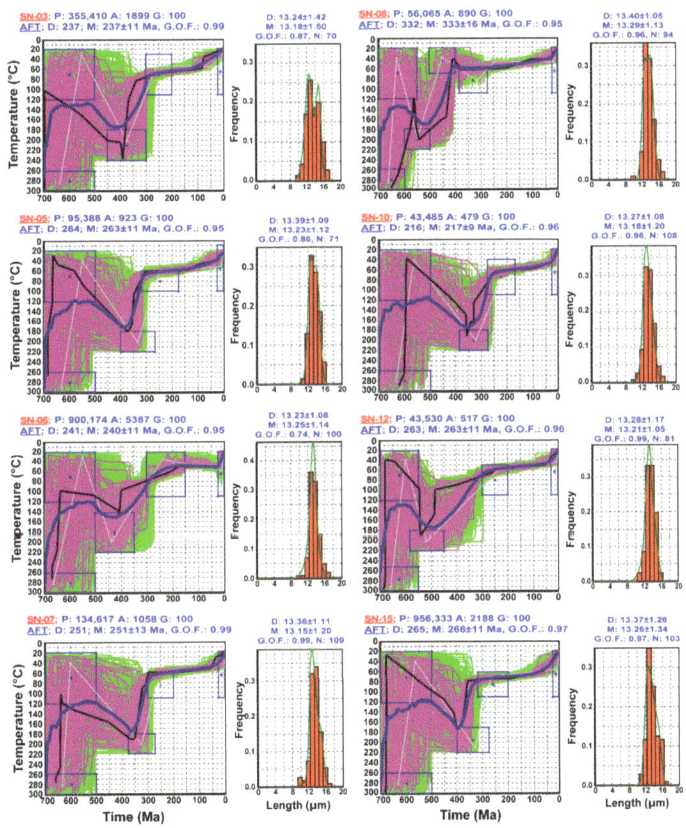

Figure 7. Thermal history modeling for older AFT age group samples showing a Neoproterozoic–Cambrian, a Devonian–Carboniferous, and Oligo–Miocene cooling events obtained using HeFTy [56]. The resulting t–T curves show four different reliability levels: green paths—acceptable fit, purple paths—good fit, black line—best fit, and blue line—the weighted mean path [56,65]. Information regarding the used constraints is provided in Table S2. P: number of inverse models' iterations, A: number of acceptable fit models, G: number of good fit models, D: determined FT age and CLs (1-σ error), M: modeled FT age and HCTLs, G.O.F.: goodness of fit, N: number of single grains and HCTLs.

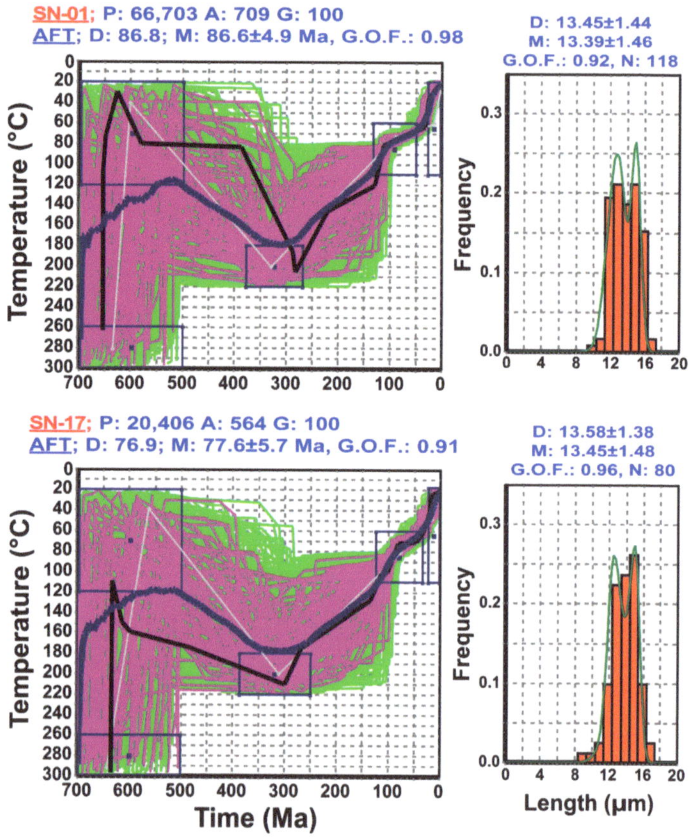

Figure 8. Thermal history modeling for the younger AFT age group samples representing a Cretaceous cooling event in addition to other events in Figure 7. For more details, see Figure 7 and Table S2.

The t-T models suggest a nearly identical cooling scenario for the studied region (Figure 7), except for samples from the younger AFT age group, which suggests an additional cooling event (Figure 8 and Table 2). During the Neoproterozoic, all modeled samples experienced their first rapid cooling event to uplift samples nearer to the Earth's surface. Hereafter, the region experienced a possible reheating (burial) event that continued through the Devonian–Carboniferous time (Figure 7) or even to the Cretaceous (Figure 8). Afterward, a second event of fast cooling influenced the region, causing sample exhumation to the PAZ of the AFT thermochronometer. Hereafter, the region experienced a slow cooling period until they were cooled/exhumed to their current elevations during the Oligocene–Miocene (Figures 7 and 8). Meanwhile, the younger AFT age group model samples show a possibility of extending the Devonian–Carboniferous cooling event to the Cretaceous (Figure 8).

The intense PAEE caused the exposed ANS basement rocks to be entirely eroded by the Cambrian, though the position of the currently exposed rocks during the Cambrian was uncertain as they could be uplifted to the Earth's surface or still buried. Unfortunately, the t-T models did not support either scenario (Figures 7 and 8). However, both the best-fit models and the weighted mean models for the samples' majority support the possibility of being exhumed during the PAEE to temperatures slightly greater than the PAZ of the AFT thermochronometer.

From all the modeled samples and the un-reset of all AFT ages, no thermal anomaly was associated with the Suez rifting. Instead, the region was exhumed from depths corresponding to ca. 60 °C (Figures 7 and 8).

5. Discussion and Interpretation

The obtained Chi-square (X^2) values ranged between 0.92 and 1 for all the analyzed samples, demonstrating a single age population for each of them (Tables 1 and 2).

The Wadi Nesyrin area reacted in response to the Devonian–Carboniferous and the Cretaceous tectonic activities by differential upliftings through the fault-bounded blocks, which is a common response style in different parts of the ANS [6,17,18,24,29–31,54]. However, identifying the fault systems that controlled these distinction responses poses a challenge owing to their elusive natures. Noticeably, samples SN-01, SN-02, and SN-17 from the Late Cretaceous AFT age group were bounded by the Suez Rift bounding faults.

The obtained thermochronological ages and information from the t-T models revealed distinct instant cooling activities, which might have represented expressions of rapid exhumation events during the Cambrian, Devonian–Carboniferous, Cretaceous, and Oligocene–Miocene if aligned with acknowledged tectonic activities [30,31,35,47] and documented shifts in stratigraphic depositional settings.

5.1. *Zircon Fission-Track Data*

The ZFT cooling ages are represented by two age groups with ages clustering around 530 ± 15 Ma and 331 ± 10 Ma (Table 1). The metamictization effect was negligible, as there was no apparent correlation between the uranium concentrations and the ZFT ages (Figure 5); consequently, there was no effect of the chemical content of the zircon grains or the crystal lattice damage on the yielded ZFT cooling ages.

The older group of ages (Table 1; group A) ranged between ca. 536 ± 14 Ma and ca. 523 ± 15 Ma. This was concordant with the PAEE, which was activated in response to the peneplaining of the ANS before the end of the Cambrian [14,42,66]. This massive erosional event was indicated by the deposition of the fluviomarine sediments of the Qebliat Group (Figure 4). The removal of this vast amount of rocks must have led to the extensive exhumations of rocks through isostatic rebounds [67,68]. Additionally, similar cooling ages were reported from different parts of the ANS [7,21,24–26,28,30].

All the other samples had younger ages (Figure 3), where the ZFT ages ranged between 379 ± 10 Ma and 318 ± 9 Ma (Table 1). These ages aligned with prior ZFT and ZHe findings reported from various localities within the ANS [6,7,21,24,28] and are attributed to tectonically driven uplifts during the Devonian–Carboniferous as a response to the Variscan tectonic event [6,7,21,24,28]. This uplifting event also aligned with the transition in the depositional regime between the Early Carboniferous marine to the Late Carboniferous erosional deposits (Figure 4) and with the regional stratigraphic evidence of the previous existence of more than 2 km of the Lower Paleozoic sequence [7].

5.2. *Apatite Fission-Track Data*

The obtained AFT cooling ages, which ranged from 319 ± 15 Ma to ca. 74 ± 5 Ma, were segregated into two temporospatially separated groups, which revealed differential tectonic–thermal histories (Table 2 and Figure 3). The histories were controlled by a localized movement through fault-bounded blocks, which is common in the ANS [6,16–18,24,29–31]. The absence of any systematic relationship between the Dpar values and the AFT ages indicates no effect of the apatites' chemical content on the yielded cooling ages.

The AFT older age group B samples (Table 2; ca. 319-239 Ma) were comparable with the various discontinuities among the different formations and the basaltic volcanics of the Permian age (Figure 4); this suggests a synchronic uplifting event. Additionally, similar cooling ages were documented in different portions of the ANS and were interpreted as tectonically driven uplifts [16,18,20–23,25]. This supports Carboniferous exhumations through the fault-bounded blocks of group B samples in response to the Variscan tectonic event.

The younger AFT group A samples (ca. 79–74 Ma) were synchronized with shifting in the depositional environment from the marine Wata Formation to the nonmarine Matulla Formation (Figure 4), which suggests a synchronic uplifting event. Additionally, similar cooling ages were reported from further parts of the ANS [6,16,18,23,25,28,31]. These AFT thermochronological ages represent older and younger events. We proposed that the older tectonic event was the Late Cretaceous closure of the Tethys. Since the t-T modeling and the track length distributions (Figures 7 and 8) indicate a long seating within the AFT PAZ upper part, the younger event had to be the Suez rifting during the Oligocene–Miocene. These findings indicate the absence of any thermal print greater than the AFT closure temperature (ca. 110°C), at least since the Cretaceous.

The measured Dpar values that ranged from 1.5 ± 0.1 μm to 1.4 ± 0.1 μm indicate the dominance of fluorapatite compositions (Table 3). The absence of a correlation between the Dpar values and AFT ages indicates no apatite compositional effects on the AFT ages (Figure 6).

5.3. Thermal History Reconstruction

The t-T modeling showed an identical thermal history represented by three rapid cooling pulses that could be introduced as rock exhumations when they were associated with our cooling ages, the acknowledged geological events, and the stratigraphic record. The recorded cooling/exhumation pulses included four events: (1) A Neoproterozoic-Cambrian event (Figures 7 and 8) supported by our ZFT older age group, the well-documented PAEE, and the sedimentological sequence transitioning during the Cambrian between fluviomarine environments. (2) A Devonian–Carboniferous event (Figure 7) supported by our ZFT younger age group and our AFT older age group. This event coincided with the Variscan tectonic event, which resulted in uplifts [8] and altered the depositional environment from the Lower Carboniferous marine sediments to the Upper Carboniferous erosional deposits (Figure 4). (3) A Cretaceous event (Figure 8) supported by our AFT younger age group and t-T modeling. This event likely resulted from the opening of the Mid-Atlantic and the consequent eastward movement of the African plate, which caused localized exhumations along the northern African margin and the Gulf of Suez region [12,16,30]. Furthermore, multiple hiatuses and the prevalence of sedimentary deposits with an erosional nature throughout the Cretaceous period (Figure 4), tectonically driven strike-slip faults, and significant cooling pulses were reported [25,28]. These differential uplifts might have differentially affected only blocks with the younger AFT ages as a response to the Cretaceous tectonic activities consistent with previous thermochronological studies [7,16–18,22,24,25,30,31]. (4) An Oligocene–Miocene event (Figures 7 and 8) supported by our t-T modeling and AFT younger age group. These ages were slightly younger than the Late Cretaceous event, indicating mixed ages between this event and a long stay in the AFT PAZ. During the opening of the northern Red Sea and the Suez, the rift flanks were exhumed from the AFT PAZ [7,16–18,22,24,25,30,31]. The t-T modeling (Figures 7 and 8) and un-reset of the AFT ages (Table 2) suggest a lack of any extraordinary thermal overprint during the Suez Rift opening; instead, flanks were exhumed from depths equivalent to ca. 60 °C (Figures 7 and 8).

5.4. Geological Implications

The obtained thermochronological ages and the t-T models revealed rapid cooling/exhumation activities during the Cambrian, Devonian–Carboniferous, Cretaceous, and Oligocene–Miocene aligned with acknowledged tectonic events and documented shifts in stratigraphic depositional settings.

The integration between the resulting thermochronological ages and the corresponding t-T models, as well as the reported tectonic processes and sedimentology record, represent four cooling/exhumation pulses: During the Neoproterozoic–Cambrian period, the PAEE was activated after an extensive continental thickening and plutonic emplacements during the EAO. The PAEE caused an isostatic rebound to result in regional exhumations [14], and then more than 2 km of the fluvio-marine succession with Lower Paleozoic ages was deposited [7]. During the Devonian–Carboniferous times, the Variscan tectonic event affected the region with rock exhumations and a consequent erosion of the Lower Paleozoic succession [7,8]. During the Permian, minor volcanic activities affected the region, followed by erosional sediments of the Triassic to Jurassic ages (Figure 4). During the Cretaceous, the Mid-Atlantic opening event affected the region with localized uplifts [11,12], which might be represented by the youngest AFT age group. During the Oligocene–Miocene, the northern Red Sea/Gulf of Suez rift system developed axial grabens and uplifts in the rift flanks of ca. 1.2 ± 0.4 km in the studied region [16,28].

The Egyptian unstable shelf has been delineated by the Phanerozoic rocks' response to the Devonian–Carboniferous and the Cretaceous tectonism through exposed structural elements (i.e., faults, folds, domes) [13,35]. However, this study and previous research revealed comparable tectonic-induced exhumations at the Neoproterozoic ANS basement rocks in the Eastern Desert [18,22,32], in southern Sinai [16,17,20,23,30], and in the subsurface structures at the Bahariya depression in the Western Desert [15] (Figure 2). This suggests that the southern border of the Egyptian unstable shelf had to move south of the Bahariya depression in the Western Desert, Gabal Loman in the Eastern Desert, and the ANS basements of South Sinai, as represented in Figure 2.

6. Conclusions

The Wadi Nesyrin area was formed during the EAO within the northern ANS. Before the end of the Cambrian, the basement complexes of the ANS were entirely eroded. Consequently, vast exhumations through the PAEE proceeded to uplift the presently exposed rocks closer to the Earth's surface. Afterward, intracratonic basins were developed and more than 2 km of the Lower Paleozoic sequence was deposited atop the basement complexes. Then, the Variscan tectonic event was activated during the Devonian–Carboniferous, which resulted in differential rock exhumations and the erosion of these Lower Paleozoic successions. Afterward, platform sedimentation predominated until the Cretaceous tectonic events activated, causing the African–Eurasian suturing, and probably inducing localized rock uplifts. Then, the Gulf of Suez initiated elevated flanks of ca. 1.2 ± 0.4 km in the studied region. Ultimately, the rift flanks were eroded and non-tectonically reshaped to form the recent landscape.

Recordings of the Paleozoic and Mesozoic tectonic events in the area of study and other regions suggest that changes occurred in the southern border of the Egyptian unstable shelf and continued to the area south of the Bahariya depression and South Sinai, as suggested in Figure 2.

Supplementary Materials: The following supporting information can be downloaded from https://www.mdpi.com/article/10.3390/min14121246/s1—Table S1: Operating conditions for the LA-ICP-MS; Table S2: The specifications of the time-temperature modelling used constraints. References [69–97] are presented in supplementary file.

Author Contributions: Conceptualization, S.M. and N.H.; methodology, S.M., A.T. and N.H.; validation, S.M., K.A., A.A.S., M.S.F. and N.H.; formal analysis, A.T., M.A.G., K.A. and S.M.; investigation, S.M., K.A., M.S.F., A.A.S. and N.H.; resources, N.H., K.M.A., M.S.F. and K.A.; data curation, S.M., A.A.S., K.M.A. and M.A.G.; writing—original draft preparation, S.M.; writing—review and editing, N.H., K.M.A., K.A., M.S.F., A.A.S. and M.A.G.; visualization, S.M., K.M.A., K.A., M.S.F. and M.A.G.; supervision, N.H.; project administration, S.M., K.M.A., A.T., K.A. and M.S.F.; funding acquisition, N.H., K.A. and M.S.F. All authors read and agreed to the published version of this manuscript.

Funding: This research was funded by Researchers Supporting Project number RSP2025R351, King Saud University, Riyadh, Saudi Arabia.

Data Availability Statement: Datasets for this research are included in this paper and the accompanying Supporting Information. In addition, all datasets are available at the Mendeley Data repository at Mansour, Sherif (2024), "Wadi Nesryin", Mendeley Data, V1, doi: 10.17632/k3sdm5c9cb.1

Acknowledgments: We acknowledge the financial support from Researchers Supporting Project-number (RSP2025R351), King Saud University, Riyadh, Saudi Arabia.

Conflicts of Interest: The authors declare no conflicts of interest.

References

1. Mansour, S.; Hasebe, N.; Meert, J.G.; Tamura, A.; Khalaf, F.I.; El-Shafei, M.K. Evolution of the Arabian-Nubian Shield in Gabal Samra Area, Sinai; Implications from Zircon U–Pb Geochronology. *J. Afr. Earth Sci.* **2022**, *192*, 104538. [CrossRef]
2. Stern, R.J. ARC Assembly and Continental Collision in the Neoproterozoic East African Orogen: Implications for the Consolidation of Gondwanaland. *Annu. Rev. Earth Planet. Sci.* **1994**, *22*, 319–351. [CrossRef]
3. Meert, J.G. A Synopsis of Events Related to the Assembly of Eastern Gondwana. *Tectonophysics* **2003**, *362*, 1–40. [CrossRef]
4. Kolodner, K.; Avigad, D.; McWILLIAMS, M.; Wooden, J.L.; Weissbrod, T.; Feinstein, S. Provenance of North Gondwana Cambrian–Ordovician Sandstone: U–Pb SHRIMP Dating of Detrital Zircons from Israel and Jordan. *Geol. Mag.* **2006**, *143*, 367–391. [CrossRef]
5. Schott, B.; Schmeling, H. Delamination and Detachment of a Lithospheric Root—ScienceDirect. *Tectonophysics* **1998**, *296*, 225–247. [CrossRef]
6. Mansour, S.; Hasebe, N.; Khedr, M.Z.; Tamura, A.; Shehata, A.A. Tectonic-Thermal Evolution of the Wadi El-Dahal Area, North Eastern Desert, Egypt: Constraints on the Suez Rift Development. *Minerals* **2023**, *13*, 1021. [CrossRef]
7. Kohn, B.P.; Eyal, M.; Feinstein, S. A Major Late Devonian-Early Carboniferous (Hercynian) Thermotectonic Event at the NW Margin of the Arabian-Nubian Shield: Evidence from Zircon Fission Track Dating. *Tectonics* **1992**, *11*, 1018–1027. [CrossRef]
8. Stampfli, G.M.; von Raumer, J.F.; Borel, G.D. Paleozoic Evolution of Pre-Variscan Terranes: From Gondwana to the Variscan Collision. In *Variscan-Appalachian Dynamics: The Building of the Late Paleozoic Basement*; Geological Society of America: Boulder, CO, USA, 2002; ISBN 978-0-8137-2364-8.
9. Alsharhan, A.S.; Nairn, A.E.M. *Sedimentary Basins and Petroleum Geology of the Middle East*; Elsevier: Amsterdam, The Netherlands, 1997; ISBN 978-0-444-82465-3.
10. Meneisy, M.Y. Mesozoic Igneous Activity in Egypt. *Qatar Univ. Sci. Bull.* **1986**, *6*, 12.
11. Sakran, S.; Shehata, A.A.; Osman, O.; El-Sherbiny, M. Superposed Tectonic Regimes in West Beni Suef Basin, Nile Valley, Egypt: Implications to Source Rock Maturation and Hydrocarbon Entrapment. *J. Afr. Earth Sci.* **2019**, *154*, 1–19. [CrossRef]
12. Shehata, A.A.; El Fawal, F.M.; Ito, M.; Aboulmagd, M.A.; Brooks, H.L. Senonian Platform-to-Slope Evolution in the Tectonically-Influenced Syrian Arc Sedimentary Belt: Beni Suef Basin, Egypt. *J. Afr. Earth Sci.* **2020**, *170*, 103934. [CrossRef]
13. Said, R. *The Geology of Egypt*, 1st ed.; Elsevier: Amsterdam, The Netherlands, 1962.
14. Said, R. *The Geology of Egypt*, 2nd ed.; A.A. Balkema: Rotterdam, The Netherlands, 1990.
15. Moustafa, A.R.; Saoudi, A.; Moubasher, A.; Ibrahim, I.M.; Molokhia, H.; Schwartz, B. Structural Setting and Tectonic Evolution of the Bahariya Depression, Western Desert, Egypt. *GeoArabia* **2003**, *8*, 91–124. [CrossRef]
16. Mansour, S.; Hasebe, N.; Abdelrahman, K.; Fnais, M.S.; Tamura, A. Reconstructing the Tectonic History of the Arabian–Nubian Shield in Sinai: Low-Temperature Thermochronology Implications on Wadi Agar Area. *Minerals* **2023**, *13*, 574. [CrossRef]
17. Kohn, B.P.; Eyal, M. History of Uplift of the Crystalline Basement of Sinai and Its Relation to Opening of the Red Sea as Revealed by Fission Track Dating of Apatites. *Earth Planet. Sci. Lett.* **1981**, *52*, 129–141. [CrossRef]
18. Omar, G.I.; Steckler, M.S.; Buck, W.R.; Kohn, B.P. Fission-Track Analysis of Basement Apatites at the Western Margin of the Gulf of Suez Rift, Egypt: Evidence for Synchroneity of Uplift and Subsidence. *Earth Planet. Sci. Lett.* **1989**, *94*, 316–328. [CrossRef]
19. Omar, G.I.; Steckler, M.S. Fission Track Evidence on the Initial Rifting of the Red Sea: Two Pulses, No Propagation. *Science* **1995**, *270*, 1341–1344. [CrossRef]
20. Rabeh, T. Geotectonic Studies of Southern Sinai and Gulf of Suez Areas, Egypt. *Ann. Geophys.* **2003**, *46*, 1325–1337.
21. Bojar, A.-V.; Fritz, H.; Kargl, S.; Unzog, W. Phanerozoic Tectonothermal History of the Arabian–Nubian Shield in the Eastern Desert of Egypt: Evidence from Fission Track and Paleostress Data. *J. Afr. Earth Sci.* **2002**, *34*, 191–202. [CrossRef]
22. Feinstein, S.; Eyal, M.; Kohn, B.P.; Steckler, M.S.; Ibrahim, K.M.; Moh'd, B.K.; Tian, Y. Uplift and Denudation History of the Eastern Dead Sea Rift Flank, SW Jordan: Evidence from Apatite Fission Track Thermochronometry: Uplift History of Dead Sea Rift Flank. *Tectonics* **2013**, *32*, 1513–1528. [CrossRef]
23. Kohn, B.P.; Feinstein, S.; Foster, D.A.; Steckler, M.S.; Eyal, M. Thermal History of the Eastern Gulf of Suez, II. Reconstruction from Apatite Fission Track and K-Feldspar Measurements. *Tectonophysics* **1997**, *283*, 219–239. [CrossRef]
24. Mansour, S. Tectonic Development of Wadi Mi'ar Area, Sinai, Egypt: Implications of Low-Temperature Thermochronology Techniques. *Alfarama J. Basic Appl. Sci.* **2023**, *4*, 727–738. [CrossRef]
25. Mansour, S.; Gharib, M.A.; Hasebe, N.; Abdelrahman, K.; Fnais, M.S.; Tamura, A. Tectonic Evolution of the Gabal Loman Area, North Eastern Desert, Egypt: Implications from Low-Temperature Multithermochronometry on the Arabian-Nubian Shield. *Front. Earth Sci.* **2023**, *11*, 15. [CrossRef]
26. Mansour, S.; Hasebe, N.; Abdelrahman, K.; Fnais, M.S.; Tamura, A. The Gulf of Suez Rifting: Implications from Low-Temperature Thermochronology. In *International Geology Review*; Taylor & Francis: Philadelphia, PA, USA, 2024; pp. 1–17. [CrossRef]
27. Omar, G.I.; Kohn, B.P.; Lutz, T.M.; Faul, H. The Cooling History of Silurian to Cretaceous Alkaline Ring Complexes, South Eastern Desert, Egypt, as Revealed by Fission-Track Analysis. *Earth Planet. Sci. Lett.* **1987**, *83*, 94–108. [CrossRef]

28. Vermeesch, P.; Avigad, D.; McWilliams, M.O. 500 m.y. of Thermal History Elucidated by Multi-Method Detrital Thermochronology of North Gondwana Cambrian Sandstone (Eilat Area, Israel). *GSA Bull.* **2009**, *121*, 1204–1216. [CrossRef]
29. Mansour, S. Long-Term Topographic Evolution of the African Plate, Causes and Consequences for Surrounding Lithospheric Plates. Ph.D. Thesis, Universität Heidelberg, Heidelberg, Germany, 2015.
30. Mansour, S.; Hasebe, N.; Azab, E.; Elnaggar, A.Y.; Tamura, A. Combined Zircon/Apatite U-Pb and Fission-Track Dating by LA-ICP-MS and Its Geological Applications: An Example from the Egyptian Younger Granites. *Minerals* **2021**, *11*, 1341. [CrossRef]
31. Morag, N.; Haviv, I.; Eyal, M.; Kohn, B.P.; Feinstein, S. Early Flank Uplift along the Suez Rift: Implications for the Role of Mantle Plumes and the Onset of the Dead Sea Transform. *Earth Planet. Sci. Lett.* **2019**, *516*, 56–65. [CrossRef]
32. Barbarand, J.; Marques, F.O.; Hildenbrand, A.; Pinna-Jamme, R.; Nogueira, C.R. Thermal Evolution of Onshore West Iberia: A Better Understanding of the Ages of Breakup and Rift-to-Drift in the Iberia-Newfoundland Rift. *Tectonophysics* **2021**, *813*, 228926. [CrossRef]
33. Mansour, S.; Glasmacher, U.A.; Krob, F.C.; Casillas, R.; Albinger, M. Timing of Rapid Cooling and Erosional Decay of Two Volcanic Islands of the Canary Archipelago: Implications from Low-Temperature Thermochronology. *Int. J. Earth Sci.* **2022**, *112*, 345–382. [CrossRef]
34. Meshref, W. Tectonic Framework of Egypt. In *Geology of Egypt*; Said, R., Ed.; A.A. Balkema: Rotterdam, The Netherlands, 1990; pp. 113–156.
35. Dolson, J.C.; Shann, M.V.; Matbouly, S.I.; Hammouda, H.; Rashed, R.M. Egypt in the Twenty-First Century: Petroleum Potential in Offshore Trends. *GeoArabia* **2001**, *6*, 211–230. [CrossRef]
36. Kröner, A.; Kröger, J.; Rashwan, A.A.A. Age and Tectonic Setting of Granitoid Gneisses in the Eastern Desert of Egypt and South-West Sinai. *Geol. Rundsch.* **1994**, *83*, 502–513. [CrossRef]
37. Abu El-Enen, M.M.; Whitehouse, M.J. The Feiran–Solaf Metamorphic Complex, Sinai, Egypt: Geochronological and Geochemical Constraints on Its Evolution. *Precambrian Res.* **2013**, *239*, 106–125. [CrossRef]
38. Stern, R.J.; Hedge, C.E. Geochronologic and Isotopic Constraints on Late Precambrian Crustal Evolution in the Eastern Desert of Egypt. *Am. J. Sci.* **1985**, *285*, 97. [CrossRef]
39. Platt, J.P.; England, P.C. Convective Removal of Lithosphere beneath Mountain Belts; Thermal and Mechanical Consequences. *Am. J. Sci.* **1994**, *294*, 307–336. [CrossRef]
40. Fergany, E.; Mekkawi, M.; Azeem, M.A.; Khalil, A. Integrated Geologic and Geophysical Studies of North Unstable Shelf Seismicity, Egypt. *Arab. J. Geosci.* **2015**, *8*, 5475–5490. [CrossRef]
41. Bosworth, W.; Huchon, P.; McClay, K. The Red Sea and Gulf of Aden Basins. *J. Afr. Earth Sci.* **2005**, *43*, 334–378. [CrossRef]
42. Perron, P.; Le Pourhiet, L.; Guiraud, M.; Vennin, E.; Moretti, I.; Portier, É.; Konaté, M. Control of Inherited Accreted Lithospheric Heterogeneity on the Architecture and the Low, Long-Term Subsidence Rate of Intracratonic Basins. *BSGF Earth Sci. Bull.* **2021**, *192*, 15. [CrossRef]
43. Frizon de Lamotte, D.; Raulin, C.; Mouchot, N.; Wrobel-Daveau, J.; Blanpied, C.; Ringenbach, J. The Southernmost Margin of the Tethys Realm during the Mesozoic and Cenozoic: Initial Geometry and Timing of the Inversion Processes. *Tectonics* **2011**, *30*, 118–134. [CrossRef]
44. Klitzsch, E. Plate Tectonics and Cratonal Geology in Northeast Africa (Egypt, Sudan). *Geol. Rundsch.* **1986**, *75*, 755–768. [CrossRef]
45. Klitzsch, E. Paleozoic. In *The Geology of Egypt*; Said, R., Ed.; Balkema: Rotterdam, The Netherlands, 1990; Volume 24, pp. 451–486.
46. Moustafa, A.R.; Yousif, M.S.M. Structural Evolution of the Eastern Shoulder of the Suez Rift: Um Bogma Area. *Neues Jahrb. Für Geol. Und Paläontologie Monatshefte* **1993**, *1993*, 655–668. [CrossRef]
47. Dewey, J.F.; Pitman, W.C.; Ryan, W.B.F.; Bonnin, J. Plate Tectonics and the Evolution of the Alpine System. *Geol. Soc. Am. Bull.* **1973**, *84*, 3137. [CrossRef]
48. Girdler, R.W. Problems Concerning the Evolution of Oceanic Lithosphere in the Northern Red Sea. *Tectonophysics* **1985**, *116*, 109–122. [CrossRef]
49. Boone, S.C.; Balestrieri, M.-L.; Kohn, B.P. Thermo-Tectonic Imaging of the Gulf of Aden-Red Sea Rift Systems and Afro-Arabian Hinterland. *Earth Sci. Rev.* **2021**, *222*, 103824. [CrossRef]
50. Steckler, M.S.; Feinstein, S.; Kohn, B.P.; Lavier, L.L.; Eyal, M. Pattern of Mantle Thinning from Subsidence and Heat Flow Measurements in the Gulf of Suez: Evidence for the Rotation of Sinai and along-Strike Flow from the Red Sea. *Tectonics* **1998**, *17*, 903–920. [CrossRef]
51. Feinstein, S.; Kohn, B.P.; Steckler, M.S.; Eyal, M. Thermal History of the Eastern Margin of the Gulf of Suez, I. Reconstruction from Borehole Temperature and Organic Maturity Measurements. *Tectonophysics* **1996**, *266*, 203–220. [CrossRef]
52. Wagner, G.A. Spaltspurenalter von Mineralen Und Natürlichen Gläsern: Eine Übersicht. *Fortschritte Der Mineral.* **1972**, *49*, 114–145.
53. Ketcham, R.A.; Donelick, R.A.; Balestrieri, M.L.; Zattin, M. Reproducibility of Apatite Fission-Track Length Data and Thermal History Reconstruction. *Earth Planet. Sci. Lett.* **2009**, *284*, 504–515. [CrossRef]
54. Gleadow, A.J.W.; Duddy, I.R. A Natural Long-Term Track Annealing Experiment for Apatite. *Nucl. Tracks* **1981**, *5*, 169–174. [CrossRef]
55. Garver, J.I. Etching Zircon Age Standards for Fission-Track Analysis. *Radiat. Meas.* **2003**, *37*, 47–53. [CrossRef]
56. Yamada, R.; Tagami, T.; Nishimura, S.; Ito, H. Annealing Kinetics of Fission Tracks in Zircon: An Experimental Study. *Chem. Geol.* **1995**, *122*, 249–258. [CrossRef]

57. Ketcham, R.A. Forward and Inverse Modeling of Low-Temperature Thermochronometry Data. *Rev. Mineral. Geochem.* **2005**, *58*, 275–314. [CrossRef]
58. Tewfik, N.; Harwood, C.; Deighton, I. The Miocene, Rudeis and Kareem Formations in the Gulf of Suez: Aspects of Sedimentology and Geochemistry. *EGPC Explor. Semin. Cairo I* **1992**, *11*, 84–113.
59. Darwish, M.; El Araby, A. Petrography and Diagenetic Aspects of Some Siliclastic Hydrocarbon Reservoirs in Relation to Rifting of the Gulf of Suez, Egypt. Geodynamics and Sedimentation of the Red Sea—Gulf of Aden Rift System. *Egypt. J. Geol.* **1994**, *3*, 155–187.
60. Donelick, R.A.; O'Sullivan, P.B.; Ketcham, R.A. Apatite Fission-Track Analysis. In *Low-Temperature Thermochronology*; Reiners, P.W., Ehlers, T.A., Eds.; De Gruyter: Berlin Germany, 2005; pp. 49–94. ISBN 978-1-5015-0957-5.
61. Vermeesch, P. IsoplotR: A Free and Open Toolbox for Geochronology. *Geosci. Front.* **2018**, *9*, 1479–1493. [CrossRef]
62. Mansour, S.; Hasebe, N.; Abdelrahman, K.; Fnais, M.S.; Gharib, M.A.; Habou, R.; Tamura, A. Development of the Arabian-Nubian Shield along the Marsa Alam-Idfu Transect, Central-Eastern Desert, Egypt: Geochemical Implementation of Zircon U-Pb Geochronology. *Geochem. Trans.* **2024**, *25*, 11. [CrossRef]
63. Mansour, S.; Hasebe, N.; Tamura, A.; Abdelrahman, K.; Fnais, M.S.; Gharib, M.A.; Khedr, M.Z. Geochronological Assessment of the Arabian-Nubian Shield Plutonic Intrusions in the Arc Assemblages along the Qift-Quseir Transect, Central Eastern Desert of Egypt. *J. Afr. Earth Sci.* **2024**, *220*, 105456. [CrossRef]
64. Ketcham, R.A.; Donelick, R.A.; Carlson, W.D. Variability of Apatite Fission-Track Annealing Kinetics; III, Extrapolation to Geological Time Scales. *Am. Mineral.* **1999**, *84*, 1235–1255. [CrossRef]
65. Seilacher, A. Paleozoic Trace Fossils. In *The Geology of Egypt*; Said, R., Ed.; Balkema: Rotterdam, The Netherlands, 1990; pp. 649–670.
66. Braun, J.; Robert, X. Constraints on the Rate of Post-Orogenic Erosional Decay from Low-Temperature Thermochronological Data: Application to the Dabie Shan, China. *Earth Surf. Process. Landf.* **2005**, *30*, 1203–1225. [CrossRef]
67. Margirier, A.; Braun, J.; Robert, X.; Audin, L. Role of Erosion and Isostasy in the Cordillera Blanca Uplift: Insights from Landscape Evolution Modeling (Northern Peru, Andes). *Tectonophysics* **2018**, *728*, 119–129. [CrossRef]
68. Kohn, B.P.; Weissbrod, T.; Chung, L.; Farley, K.; Bodorkos, S. Low-temperature Thermochronology of Francolite: Insights into Timing of Dead Sea Transform Motion. *Terra Nova* **2019**, *31*, 205–219. [CrossRef]
69. Barron, T. *The Topography and Geology of the Peninsula of Sinai (Western Portion)*; Geological Survey of Egypt Bulletin: Cairo, Egypt, 1907; Section III; pp. 123–130.
70. Hassan, A.A. A New Carboniferous Occurrencein Abu Durba, Sinai, Egypt. In Proceedings of the 6th Arab Petroleum Congress, Baghdad, Iraq, 6–13 March 1967; Volume 2, pp. 1–8.
71. Said, R. *Explanatory Notes to Accompany the Geological Map of Egypt*; Geological Survey of Egypt and Mining Authority: Cairo, Egypt, 1971.
72. Kora, M. The Paleozoic Outcrops of Um Bogma Area, Sinai. Ph.D. Thesis, Mansoura University, Mansoura, Egypt, 1984.
73. McClay, K.R.; Nichols, G.J.; Khalil, S.M.; Darwish, M.; Bosworth, W. Extensional Tectonics and Sedimentation, Eastern Gulf of Suez, Egypt. In *Sedimentation and Tectonics in Rift Basins Red Sea: Gulf of Aden*; Purser, B.H., Bosence, D.W.J., Eds.; Springer: Dordrecht, The Netherlands, 1998; pp. 223–238. ISBN 978-94-011-4930-3.
74. Weissbrod, T. The Paleozoic of Israel and Adjacent Countries. Pt 1: The Subsurfacealeozoic Stratigraphy of Southern Israel. *Isr. Geol. Surv. Bull.* **1969**, *47*, 35.
75. Klitzsch, E.; Wycisk, P. Geology of the Sedimentary Basins of Northern Sudan and Bordering Areas. *Berl. Geowiss. Abh. Reihe A Geol. Und Palaeontol.* **1987**, *75*, 97–136.
76. Abdallah, A.M.; El-Adindani, A. Stratigraphy of Upper Paleozoic Rocks of Western Side of Gulf of Suez. *Geol. Surv. Egypt Bulletin* **1963**, *25*, 1–18.
77. Abu-Zied, R. Palaeoenvironmental Significance of Early Cretaceous Foraminifera from Northern Sinai, Egypt. *Cretac. Res.* **2007**, *28*, 765–784. [CrossRef]
78. Kassem, A.; Sharaf, L.; Baghdady, A.; Abd El-Naby, A. Cenomanian/Turonian Oceanic Anoxic Event 2 in October Oil Field, Central Gulf of Suez, Egypt. *J. Afr. Earth Sci.* **2020**, *165*, 103817. [CrossRef]
79. Sultan, I.Z. On the Presence of Middle Jurassic Miospores at Gebel El Iseila, West Central Sinai, Egypt. *Bull. Fac. Sci. Alex. Univ.* **1985**, *25*, 26–40.
80. El Sharawy, M.S.; Nabawy, B.S. Geological and Petrophysical Characterization of the Lower Senonian Matulla Formation in Southern and Central Gulf of Suez, Egypt. *Arab. J. Sci. Eng.* **2016**, *41*, 281–300. [CrossRef]
81. Osman, A. Upper Cretaceous Foraminifera of Western Sinai. *Fac. Eng. Bull.* **1954**, *26*, 335–365.
82. El-Azabi, M.H. Sedimentological Characteristics, Palaeoenvironments and Cyclostratigraphy of the Middle Eocene Sequences in Gabal El-Ramliya, Maadi-Sukhna Stretch, North Eastern Desert. In Proceedings of the Egyptian 8th International Conference on the Geology of Arab World, Cairo, Egypt, 13–16 February 2006; pp. 1–31.
83. Ibrahim, A.M. Psychological Empowerment and Organizational Commitment among Employees in the Lodging Industry. *J. Hum. Resour. Hosp. Tour.* **2020**, *19*, 277–295. [CrossRef]
84. Refaat, A.A.; Imam, M.M. The Tayiba Red Beds: Transitional Marine-Continental Deposits in the Precursor Suez Rift, Sinai, Egypt. *J. Afr. Earth Sci.* **1999**, *26*, 467–506. [CrossRef]

85. El-Anwar, E. Geochemical Studies of Carbonates of the Tayiba Formation (Upper Eocene), Abu Zenima Area, West Central Sinai. *Bull. Natl. Res. Cent.* **2019**, *43*, 209. [CrossRef]
86. El-Heiny, I.; Martini, E. Miocene Foraminiferal and Calcareous Nannoplankton Assemblages from the Gulf of Suez Region and Correlations. *Géologie Méditerranéenne* **1981**, *8*, 101–108. [CrossRef]
87. Sallam, E.S.; Afife, M.M.; Fares, M.; van Loon, A.J.; Ruban, D.A. Sedimentary Facies and Diagenesis of the Lower Miocene Rudeis Formation (Southwestern Offshore Margin of the Gulf of Suez, Egypt) and Implications for Its Reservoir Quality. *Mar. Geol.* **2019**, *413*, 48–70. [CrossRef]
88. Hasebe, N.; Barbarand, J.; Jarvis, K.; Carter, A.; Hurford, A.J. Apatite Fission-Track Chronometry Using Laser Ablation ICP-MS. *Chem. Geol.* **2004**, *207*, 135–145. [CrossRef]
89. Morishita, T.; Ishida, Y.; Arai, S.; Shirasaka, M. Determination of Multiple Trace Element Compositions in Thin (> 30 μm) Layers of NIST SRM 614 and 616 Using Laser Ablation-Inductively Coupled Plasma-Mass Spectrometry (LA-ICP-MS). *Geostand. Geoanalytical Res.* **2005**, *29*, 107–122. [CrossRef]
90. Gleadow, A.J.W.; Duddy, I.R.; Green, P.F.; Lovering, J.F. Confined Fission Track Lengths in Apatite: A Diagnostic Tool for Thermal History Analysis. *Contrib. Mineral. Petrol.* **1986**, *94*, 405–415. [CrossRef]
91. Longerich, H.P.; Jackson, S.E.; Günther, D. Inter-Laboratory Note. Laser Ablation Inductively Coupled Plasma Mass Spectrometric Transient Signal Data Acquisition and Analyte Concentration Calculation. *J. Anal. At. Spectrom.* **1996**, *11*, 899–904. [CrossRef]
92. Carpenter, B.S.; Reimer, G.M. Standard Reference Materials: Calibrated Glass Standards for Fission Track Use. *Natl. Bur. Stand. Spec. Publ.* **1974**, *260*, 16.
93. Pearce, N.J.G.; Perkins, W.T.; Westgate, J.A.; Gorton, M.P.; Jackson, S.E.; Neal, C.R.; Chenery, S.P. A Compilation of New and Published Major and Trace Element Data for NIST SRM 610 and NIST SRM 612 Glass Reference Materials. *Geostand. Geoanalytical Res.* **1997**, *21*, 115–144. [CrossRef]
94. Barbarand, J.; Carter, A.; Wood, I.; Hurford, T. Compositional and Structural Control of Fission-Track Annealing in Apatite. *Chem. Geol.* **2003**, *198*, 107–137. [CrossRef]
95. Ketcham, R.A.; Carter, A.; Donelick, R.A.; Barbarand, J.; Hurford, A.J. Improved Measurement of Fission-Track Annealing in Apatite Using c-Axis Projection. *Am. Mineral.* **2007**, *92*, 789–798. [CrossRef]
96. Ketcham, R.A. *HeFTy*, Version 1.6.7. Manual. Informer Technologies, Inc.: Los Angeles, CA, USA, 2009.
97. Bosworth, W.; Stockli, D.F. Early Magmatism in the Greater Red Sea Rift: Timing and Significance. *Can. J. Earth Sci.* **2016**, *53*, 1158–1176. [CrossRef]

minerals

Article

Interpretable Machine Learning for Geochemical Anomaly Delineation in the Yuanbo Nang District, Gansu Province, China

Shuai Zhang [1], Emmanuel John M. Carranza [2,*], Changliang Fu [1,*], Wenzhi Zhang [1] and Xiang Qin [1]

1 China Aero Geophysical Survey and Remote Sensing Center for Natural Resources, Beijing 100083, China
2 Department of Geology, University of the Free State, Bloemfontein 9301, South Africa
* Correspondence: ejmcarranza@gmail.com (E.J.M.C.); fuchangliang@mail.cgs.gov.cn (C.F.)

Abstract: Machine learning (ML) has shown its effectiveness in handling multi-geoinformation. Yet, the black-box nature of ML algorithms has restricted their widespread adoption in the domain of mineral prospectivity mapping (MPM). In this paper, methods for interpreting ML model predictions are introduced to aid ML-based MPM, with the goal of extracting richer insights from the ML modeling of an exploration geochemical dataset. The partial dependence plot (PDP) and accumulated local effect (ALE) plot, along with the SHAP value analysis, were utilized to demonstrate the application of random forest (RF) modeling within both regression and classification frameworks. Initially, the random forest regression (RFR) model established the relationship between the concentrations of Au and those of elements such as As, Sb, and Hg in the study area, and from this model, the most important geochemical elements and their quantitative relationships with Au were revealed by their contributions in the modeling through PDP and ALE analyses. Secondly, the RF classification modeling established the relationships of mineralization occurrences (i.e., known mineral deposits) with geochemical elements (i.e., Au, As, Sb, Hg, Cu, Pb, Zn, and Ag), as did RFR modeling. The most important geochemical elements for indicating regional Au mineralization and the trajectories of PDP and ALE reached a consensus that As and Sb contributed the most, both in the regression and classification modeling, with regard to Au mineralization. Finally, the SHAP values illustrated the behavior of the training samples (i.e., known mineral deposits) in RF modeling, and the resulting prospectivity map was evaluated using receiver operating characteristics.

Keywords: geochemical anomaly recognition; Shapley values; interpretability technique; gold deposits; west Qinling

check for updates

Citation: Zhang, S.; Carranza, E.J.M.; Fu, C.; Zhang, W.; Qin, X. Interpretable Machine Learning for Geochemical Anomaly Delineation in the Yuanbo Nang District, Gansu Province, China. *Minerals* **2024**, *14*, 500. https://doi.org/10.3390/min14050500

Academic Editors: Behnam Sadeghi and Michael S. Zhdanov

Received: 1 March 2024
Revised: 27 April 2024
Accepted: 3 May 2024
Published: 10 May 2024

1. Introduction

The verification of anomalous metal concentrations is the foundation of geochemical exploration for mineral deposits [1], and geochemical anomalies generally result from the formation of mineral deposits or the decomposition of mineral deposits in the form of primary or secondary dispersion haloes. For the assessment of metal concentrations in a given study area through geochemical exploration, there are four general criteria proposed by [2]: (1) the magnitude of geochemical values and background; (2) the size and shape of anomalous areas; (3) geological setting; and (4) the extent to which the local environment may have influenced the metal content and the pattern of anomalies. Another criterion according to [3] is the zonality of elements in anomalies. To explore these criteria, various methods have been developed and applied, ranging from the traditional standard statistical methods to the ongoing popular machine learning (ML) methods. For example, mean $\pm$ 2 $\times$ standard deviation, probability plots [4], and histograms of frequency distributions are routine for discriminating anomalously high as well as anomalously low levels of metal concentration at early stage. Statistical analyses conducted on geochemical data often unveil linear correlations [5]. Considering the spatial structure of geochemical data, moving average [6], spatial factor analysis [7], and kriging [8] are among the popular spatial statistical methods for modeling geochemical data distribution. With regard to geochemical background, local spatial variations are common in the data, which may lead to the misclassification of weak geochemical anomalies as background. To deal with the problem of weak geochemical anomalies that are of only slightly higher concentration values compared to that of background but might be associated with mineral deposits [9], the family of fractal/multifractal models has been used widely for modeling geochemical background variation [10–12]. Other popular methods for dealing with geochemical background variation include trend surface analysis [13] and geographically weighted regression [14]. While these techniques have proven effective in identifying weak geochemical anomalies, their assumptions about data distribution (e.g., regularly and densely distributed sampling points) make them unable to depict the complexity of geochemical background variation perfectly.

Machine learning algorithms are well suited for quantifying intricate and nonlinear geochemical data, enabling the circumvention of traditional multivariate statistical methods that rely on data conforming to a predetermined multivariate probability distribution hypothesis [15]. When labeled samples are limited, unsupervised machine learning, particularly the autoencoder network and its variations, has demonstrated effectiveness in detecting geochemical anomalies [16]. The availability of sufficient labeled samples representing geochemical anomaly classes (i.e., associated with known mineral deposits) and geochemical background classes empowers exploration geochemists to establish a supervised machine learning pattern recognition framework for detecting geochemical anomalies [17]. The state-of-the-art supervised ML algorithms have the potential for the detection of geochemical anomalies. For example, the support vector machine (SVM) and its natural extension one-class support vector machine (OCSVM) have been introduced to explore geochemical anomalies in complex geological settings [17,18].

Despite machine learning, particularly its subset deep learning (e.g., convolutional neural networks), showcasing remarkable performance in specific research fields, the increasing model complexity and the utility of black-box models make it a challenge to understand data processing in certain ML algorithms [19]. This paper focuses on methods that address the interpretability of machine learning predictions regarding geochemical anomalies, aiding exploration geochemists in comprehending them for mineral exploration purposes. To this end, the partial dependence plot (PDP) and the accumulated local effect (ALE) plot were applied to assist exploration through geochemical data analysis in the framework of regression and classification by ML modeling. Moreover, the SHAP value analysis was adopted to help understand the contribution of known mineral deposits as training samples in the random forest (RF) classification modeling of geochemical data.

2. Material and Methods

2.1. Study Area

The study area of Yuanbo Nang District is in Xiahe County, Gansu Province, China. It belongs to the western sector of Qinling Orogen, which is endowed with huge potential for different mineral deposits [20]. The primary rock types in the study area are represented by the Daheba Formation and Jiangligou Formation, and the primary distinctions among the constituents of these two formations are found in the composition of the rocks, their cyclical arrangements, and variations in grain size. For instance, the 1st member of the Jiangligou Formation predominantly comprises medium-to-fine-grained plagioclase-quartz sandstone, fine-grained plagioclase sandstone, fine-grained plagioclase lithic sandstone, medium-to-fine-grained plagioclase quartz lithic sandstone, and siltstone. In contrast, the 1st member of the Daheba Formation is mainly characterized by medium-to-coarse-grained plagioclase quartz sandstone, medium-to-fine-grained plagioclase lithic sandstone, and coarse-grained quartz arenite, exhibiting rhythmic cyclic superposition. Additionally, within the same formation, there are differences in color and grain size. In the southeast sector of the study area, Neogene Formation occurs with mainly sandy conglomerate. The

Quaternary sediments distributed across the study area comprise sand, gravel, clay, and humus. The predominant structures in the study area consist of NE-trending faults, with a limited number of inferred faults oriented in the SE direction. Among the 12 mineral deposits in the study area, 10 of them exhibit Au mineralization while the remaining 2 display Sb mineralization. All of these deposits belong to the medium-to-low-temperature type of magmatic-hydrothermal deposits [21] (Figure 1).

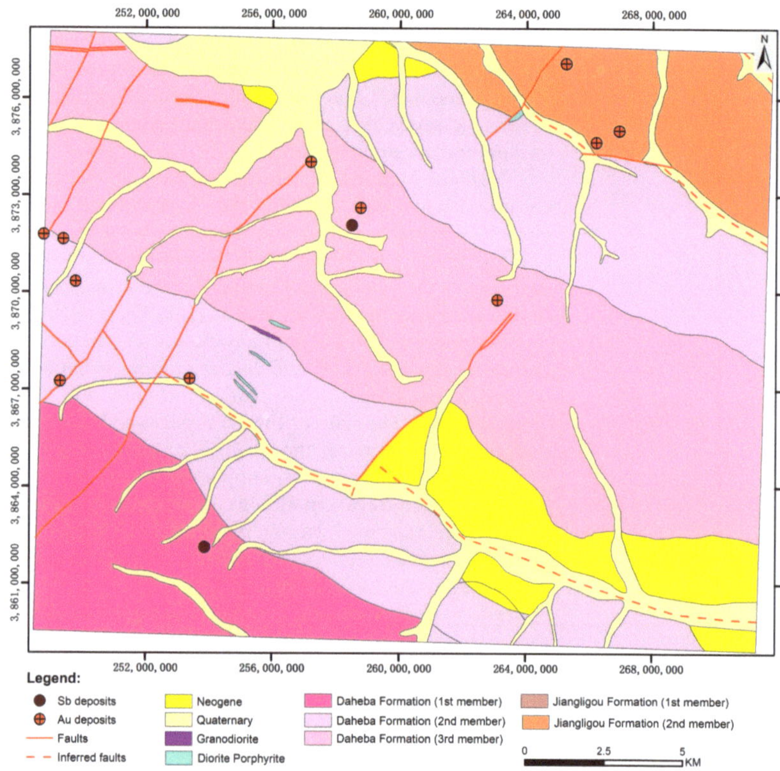

Figure 1. Simplified geological and mineral map of the study area.

The geochemical dataset utilized in this study includes 1859 stream sediment samples, which were collected at a scale of 1:50,000 by the Third Institute of Geology and Mineral Exploration, Gansu Provincial Bureau of Geology and Mineral Exploration and Development. Inductively coupled plasma–mass spectrometry was employed to analyze the samples for 16 elements (i.e., Au, As, Sb, Hg, Co, Cr, Cu, Ni, Pb, Zn, Sn, Ag, Mo, Cd, W, and Bi). Among these samples, eight elements (i.e., Au, As, Sb, Hg, Cu, Pb, Zn, and Ag) were identified as potentially linked to gold (Au) and antimony (Sb) mineralization within the region. These elements were chosen for the purpose of detecting geochemical anomalies.

2.2. Random Forest

RF has raised much interest in the field of mineral prospectivity mapping (MPM) because its results are easy to understand, it needs a few parameters to be set, and its processing speed during training is satisfactory [22]. As an ensemble learning method, RF comprises a collection of decision trees [23]. CART (classification and regression trees), proposed by [24], forms the basis of the RF method. With a training dataset, an ensemble of decision trees can be built, and prediction results can be obtained using the mode of predictions (in a classification framework) or the mean of predictions (in a regression framework) by individual trees [25]. In the process of constructing an RF model, independent trees are built by the unique bootstrap sampling of the training dataset, whereby the bootstrap sampling function randomly divides the training dataset into homogenous subsets. Then, the random subsets of samples are utilized for growing and training every tree, and the remaining subset (i.e., out-of-bag (OOB) samples) are used for validation and assessment of model accuracy [25,26]. An important feature of RF is its capacity to assess the relative importance of the input variables by building alternating decision trees including and excluding every variable in the calculation of mean-square error (MSE) and node purity. The estimation of the relative importance of a variable is performed by assessing the increment of the prediction error when OOB data for that variable are permutated, while all the other variables are left unchanged [23]. In this study, RF modeling for the delineation of geochemical anomalies was performed in both the regression and classification frameworks.

2.3. Partial Dependence Plot (PDP)

The PDP depicts changes in average predicted values as specified feature(s) vary over their marginal distribution [27]. It has made a great contribution to the understanding of various supervised ML applications across a variety of disciplines. A PDP can demonstrate the relationship (e.g., linear, monotonic, or complex) between a target variable and predictor variables. Let $S \in \{1, \ldots, p\}$ and let C be the complement set of S. Then, the PDP for regression is defined as follows:

$$\hat{f}_{x_S}(x_S) = E_{x_C}\left[\hat{f}(x_S, x_C)\right] = \int \hat{f}(x_S x_C) d\mathrm{P}(x_C) \tag{1}$$

where S and C are subsets of predictors; x_S denotes the features for which PDPs are made; and x_C represents the complement features used in the ML model $\hat{f}$; x_S and x_C comprise the total feature space. The PDP marginalizes the outcome of ML modeling over the distribution of a feature in a set of C. Thus, the function depicts the relationship between the features in S and the predictions.

In practice, the estimation of $\hat{f}_{x_S}$ is performed by calculating averages in the training data via the Monte Carlo method; thus,

$$\hat{f}_{x_S}(x_S) = \frac{1}{n}\sum_{i=0}^{n} \hat{f}\left(x_S, x_C^{(i)}\right) \tag{2}$$

where $\left\{x_C^{(1)}, \ldots, x_C^{(n)}\right\}$ represents the different values of x_C observed in the training data, and n denotes the number of instances in the dataset. This equation yields the average marginal effect on the prediction for a given value of x_S. However, the PDP assumes that the features in C are not correlated with the features in S. The violation of this assumption may lead to very unlikely or impossible data points in the PDP. In the context of classification, where the ML model outputs probabilities, the PDP shows the probability for a certain class given different values for a feature in S [27].

2.4. Accumulated Local Effect (ALE) Plot

An ALE plot, like PDP, describes the average influence of features on the prediction by an ML model. However, it is deemed to be a faster and more unbiased approach than the PDP. Using either a PDP or an ALE plot, a complex prediction function can be reduced to a function that depends only on one (or two) features by averaging the influence of the other features. However, the PDP averages the prediction over the marginal distribution. In contrast, rather than averaging the prediction directly, the ALE plot calculates the differences in prediction conditioned on a feature in S, and it integrates the derivative over the feature in S for estimating the effect; thus,

$$\hat{f}_{x_S ALE}(x_S) = \int_{z_{0,1}}^{x_S} E_{X_c|X_S}\left[\hat{f}^s(X_S, X_c)|X_S = z_S\right] dz_S - constant = \int_{z_{0,1}}^{x_S} \int_{x_C} \hat{f}^S(z_s, x_c)\mathbb{P}(x_C \mid z_S) dx_C dz_S - constant \tag{3}$$

The formula for ALE indicates that it averages the differences in predictions, not the prediction itself, so it is a gradient. In practice, the differences in predictions are replaced by the differences in prediction over an interval (Figure 2).

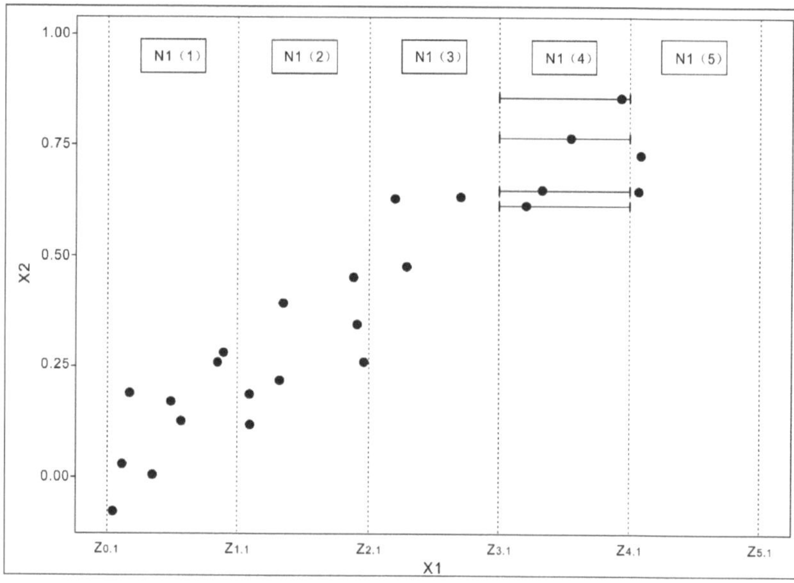

Figure 2. Schematic example of ALE calculation for feature X1 (correlated with X2). First, X1 is divided into intervals (dotted vertical lines). For the data points within an interval, the differences in the prediction when X1 is replaced with the upper and lower limits of the interval (solid horizontal lines) are calculated, accumulated, and centered, yielding the ALE curve (modified from [23]).

2.5. SHAP Analysis

The SHAP analysis, proposed by [28] is a popular method for characterizing the performance of an ML model based on the game theory [29] and local explanations [30]. It was originally proposed as an axiomatic description of a fair distribution of a total surplus for all players, which can be used in predictive models in which every variable resembles a player.

Let us take a coalition game as a tuple $\langle \mathrm{N}, \mathrm{v} \rangle$ where $\mathrm{N} = \{1, 2, \ldots, n\}$ represents a finite set of n players and $\mathrm{v} : 2^N \rightarrow \mathrm{IR}$ serves as a characteristic function with $v(\Phi) = 0$, defined over S (i.e., subsets of N) that characterizes the payoff of each subset of N. The purpose of these definitions in such a coalition game is to obtain a solution for allocating the total amount of payoff to players, i.e., $v(\mathrm{N})$ to every player. Here, $\Phi_i(v)$ denotes the allotted payoff to player i in terms of characteristic function v. Thus, a Shapley value represents a distribution solution Φ with unique properties [31].

For a game $\langle \mathrm{N}, v \rangle$, the Shapley value of a player i is defined as follows:

$$\Phi_i(v) = \sum_{\mathrm{S} \subseteq \mathrm{N} - \{i\}} \frac{(|\mathrm{N}| - |\mathrm{S}| - 1)!|\mathrm{S}|!}{|\mathrm{N}|!} \left[v\left(\mathrm{S} \bigcup \{i\} \right) - v(\mathrm{S}) \right] \tag{4}$$

Thus, the Shapley value of player i is the weighted sum of the contribution of i in each subset of N. The contribution of i to a subset S is defined by $v(\mathrm{S} \bigcup \{i\}) - v(\mathrm{S})$.

Although the Shapley value has been originally formulated in the context of the game theory with straightforward mathematical definition, it has been recently utilized for explaining complex models generated by ML algorithms, but the connotation of taking it as a tool for model interpretation has not been illustrated yet [31].

3. Results and Discussion

Since eight elements (Au, As, Sb, Hg, Cu, Pb, Zn, and Ag) were analyzed in this study (Figure 3), the parameters for the number of trees (k) and the number of features (m) randomly sampled at each split were set as k = 1000 and m = 3 after a trial-and-error process. Both k and m settings were utilized for regression and classification in RF modeling.

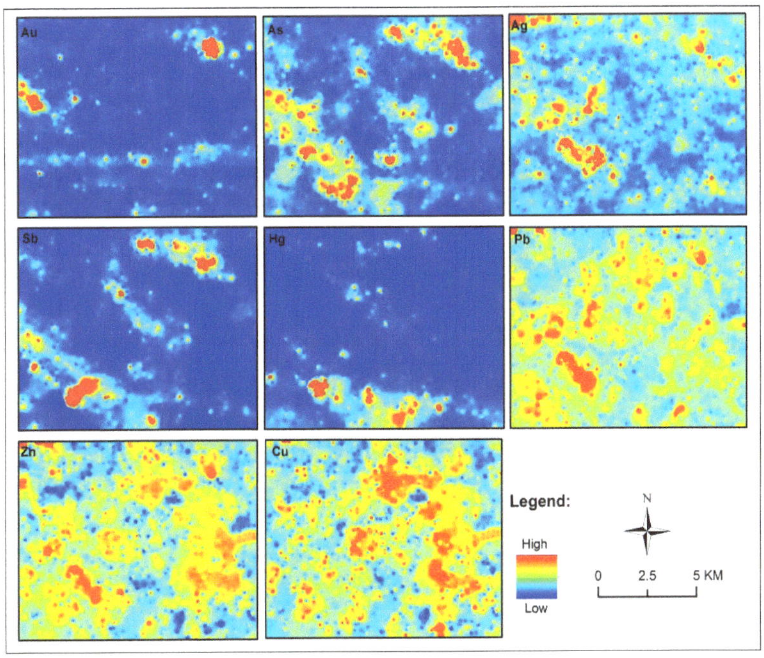

Figure 3. Geochemical elements used in RF modeling.

3.1. Regression by RF Modeling

The regression analysis in this study aims to find associations between known mineralization and potentially relevant geochemical elements. As the study area is dominated by Au mineralization, Au served as the dependent variable, and the other seven elements served as independent variables, forming the inputs to random forest regression (RFR) modeling. Based on the ranks of variable importance, As and Sb were found to be the most significant contributors to the RFR in terms of MES increments and node purity (measured by the Gini index). This suggests that As and Sb are closely linked to the distribution of Au (Table 1). Further information can be provided by the ALE plot and PDP, as these plots can assist in describing the effects of features on the prediction by an ML model. The most important geochemical elements, As and Sb, had positive relationships with Au, and they exhibited similar trajectories (Figures 4 and 5). For the PDP, the greatest issue is the assumption of independence, namely, that the features for calculating the PDP should not be correlated with other features. However, the ALE plot calculates the differences in predictions, and it isolates the effects of the features of interest, thereby blocking the effect of other features. Thus, overall, the trajectories of all geochemical elements in the PDP were consistent with those in the ALE plot, proving the robustness of the PDP and ALE plot in this case study.

Table 1. Variable importance for RFR regression.

Variables	As	Sb	Hg	Cu	Pb	Zn	Ag
%IncMSE	5.13	3.61	0.94	1.54	0.37	1.91	0.02
IncNodePurity	2914	1590	1381	1269	1206	1124	1134

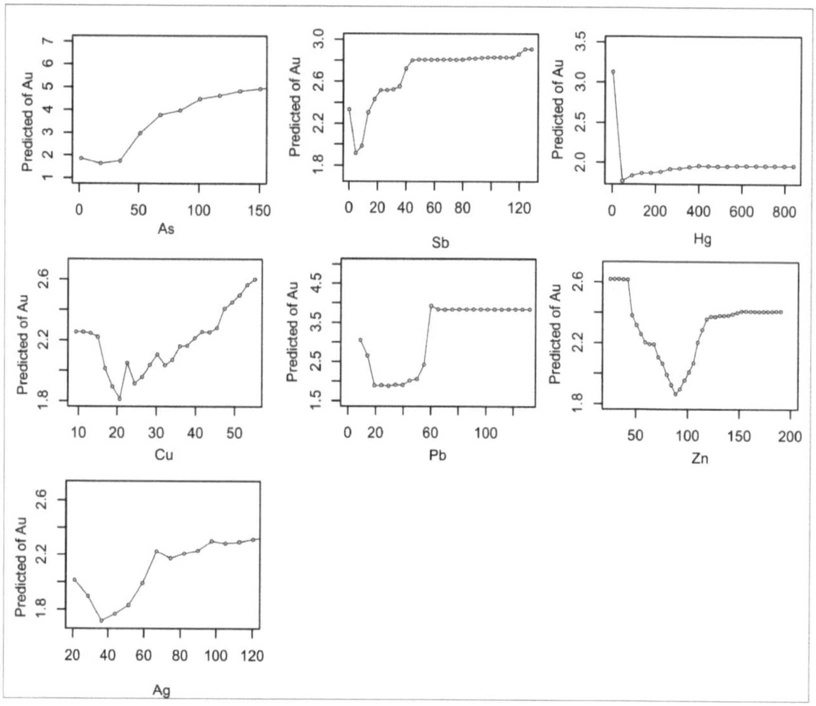

Figure 4. Partial dependence plot of geochemical element for RFR modeling.

Moreover, the ALE plots were centered with the average effect over the prediction equal to 0 by subtracting a constant. The value of the ALE can be interpreted as the main effect of the feature at a certain value compared to the average prediction of data. An ALE estimate of positive value (e.g., 0.3) means the prediction of Au is higher by 0.3 than the average prediction. Therefore, they provide potential thresholds for delineating geochemical anomalies (e.g., variable values in an ALE plot when Au = 0). As the geochemical variables As and Sb were found to be the most important variables in the RFR modeling, the thresholds of 42 for As and 9 for Sb (i.e., thresholds of ALE of Au = 0) were selected from the respective ALE plots (Figure 5) and used for geochemical anomaly delineation (Figure 6). The geochemical anomalies of As delineated 12 of the 13 known mineralized locations in the study area, and they showed a northwest trend. In contrast, the geochemical anomalies of Sb delineated fewer of the 13 known mineralized locations, but their consistent northwest trend implies 3 similar prospective areas for further exploration (Figure 6). Moreover, in the scenario of MPM using supervised learning, the selection of negative samples (e.g., non-deposits) for supervised MPM remains a challenge [32,33]. Here, areas excluding As and Sb anomalies provide preferred locations for creating negative samples. Thus, in this study, the 13 known mineralized locations were utilized as positive samples, and areas excluding As and Sb anomalies (i.e., areas with Sb values of >9 and As values of >42) were outlined and used as areas for the selection of randomly negative samples (13 non-deposits) for the subsequent RF classification modeling (Figure 6).

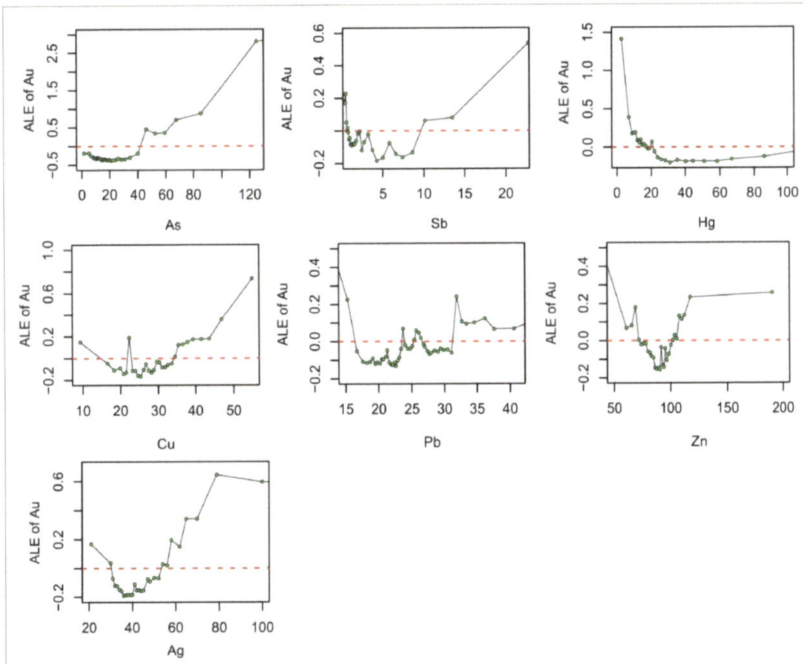

Figure 5. ALE of geochemical element for RFR modeling.

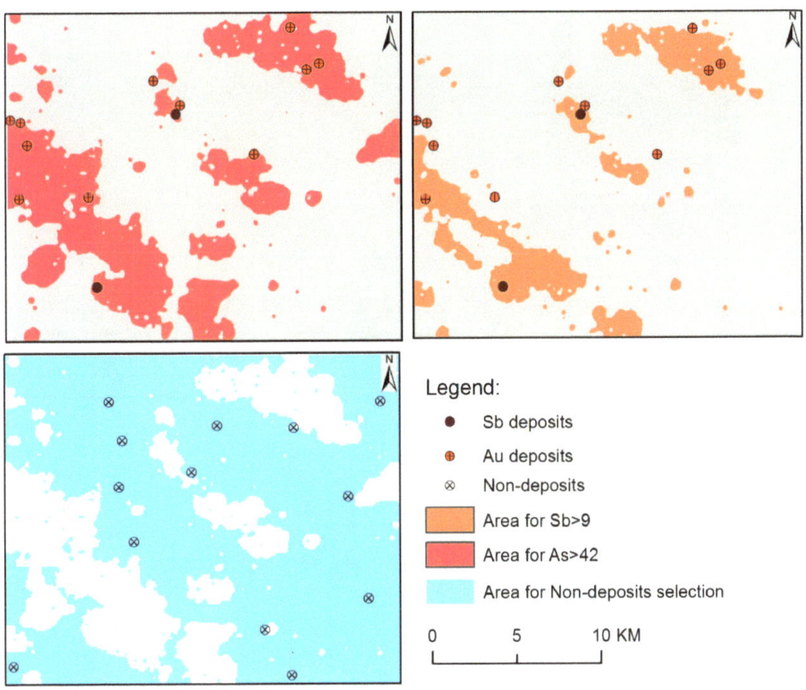

Figure 6. Non-deposit area selection based on ALE plots of As and Sb.

3.2. *Classification by RF Modeling*

The RFR modeling provided insights into the relationships of As, Sb, Hg, Cu, Pb, Zn, and Ag with the target Au mineralization in the study area. However, there are scenarios where mineralization is absent in the case of false anomalies with high magnitude [34]. For example, areas with high values of Au may not mean favorability of an area for Au mineralization. It is necessary to discover the correlation of geochemical elements with known mineralized locations. RF classification was applied to correlate the known mineralization areas and the concentrations of the studied geochemical elements (i.e., Au, As, Sb, Hg, Cu, Pb, Zn, and Ag).

In the RF modeling, the importance of evidential maps was evaluated by the mean decrease in accuracy and the mean decrease in Gini indices. The former is calculated from OOB data and indicates the reduction in accuracy across the entire RF modeling, while the latter measures the average increase in purity achieved by utilizing splits of a particular variable.

The results of RF modeling indicate that As and Sb are the most important geochemical elements, which is consistent with the results of RFR modeling (Figure 7). Au ranked fourth behind As, Sb, and Ag. In contrast to the PDP and the ALE plot, in RFR modeling, Au, As, Ag, and Sb not only demonstrated similar trajectories in the PDP and the ALE plot (Figures 8 and 9) but also showed significant differences to the other geochemical variables (i.e., Hg, Cu, Pb, and Zn). Geological/geochemical processes can be recognized by a continuous range of variable responses and the corresponding relative increase or decrease in element concentration, and high-density surveys are capable of detecting local scale processes [5]. The behaviors of the geochemical variables in RF modeling and the spatial distributions of their concentrations (Figure 3) imply that Au, As, Ag, and Sb had a positive influence on the formation of known mineralization. The coincidence of the locations of known mineralization with smaller and less intense anomalies may be more indicative of the occurrence of mineral deposits [1]. This prompts the introduction of interpretable ML for geochemical anomaly detection.

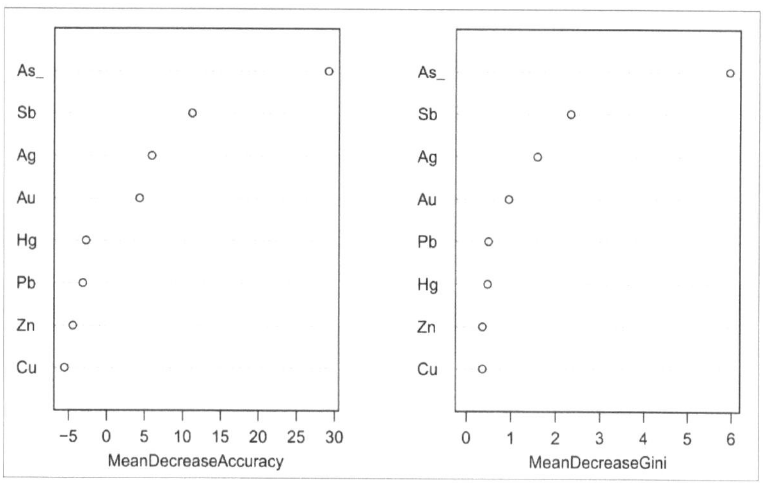

Figure 7. The relative importance of geochemical variables used in RF modeling.

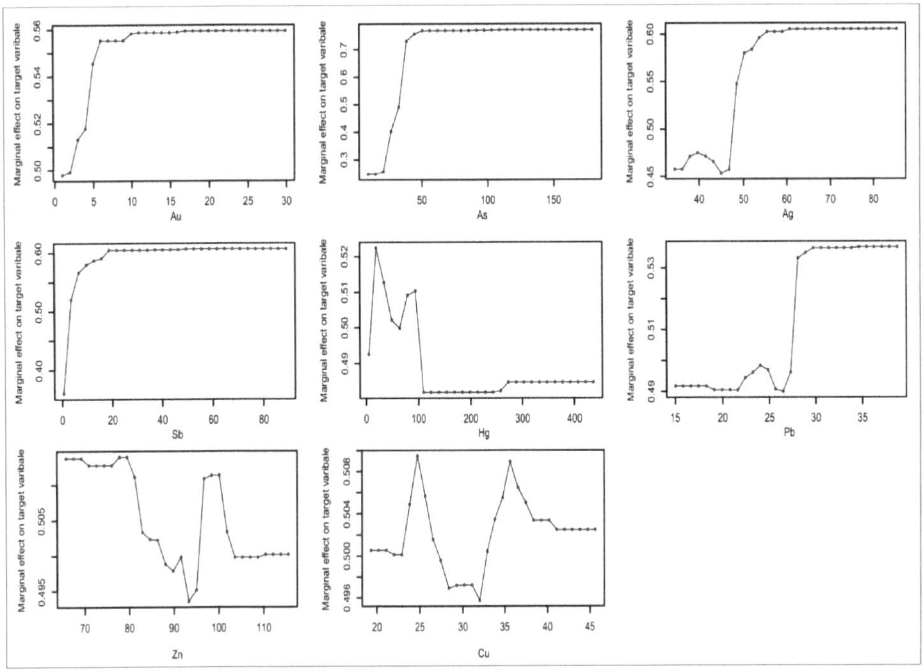

Figure 8. Partial dependence plot of geochemical elements used in RF modeling.

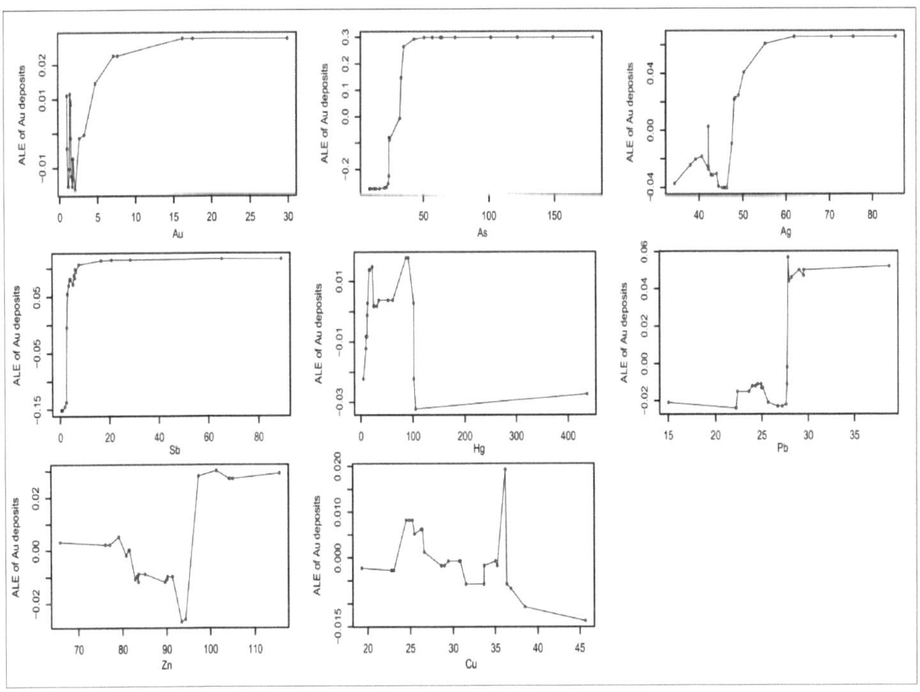

Figure 9. ALE plots of geochemical elements used in RF modeling.

Besides the PDP and the ALE plot, which shed light on the relationships between the geochemical elements of interest and known mineralization (i.e., Au element concentration in RFR modeling and Au deposits in RF modeling), the Shapley values were adopted to illustrate the behavior of the training samples in ML-based prediction. In the context of RF modeling for MPM and given the current set of geochemical variables, the Shapley value serves as an approximation of how much a particular value of a geochemical variable contributes to the disparity between the actual prediction and the average prediction.

Figure 10 shows the contributions of each feature of an instance (i.e., training sample) in RF modeling toward known mineralization. Figure 10a,b depict two of the negative training samples (i.e., non-deposits), whereas Figure 10c,d depict the training samples taken from locations of known mineralization. Both the negative and positive training samples imply that As, Sb, and Ag contributed the most. This aligns with the importance of the variable in terms of accuracy and the Gini index (Figure 7). Moreover, the geochemical variables of negative training samples showed a negative contribution, in contrast to that of the positive training samples, and the Shapley values of the positive samples varied, with As (labeled 5), Sb (labeled 12) and Ag (labeled 4, 5, 6, and 8) having negative contributions (Figures 11 and 12). These findings suggest that the Shapley value holds promise as a tool for evaluating the efficacy of a training dataset in machine learning-based mineral prospectivity mapping (MPM), particularly in the context of identifying non-deposit locations for use as negative training samples.

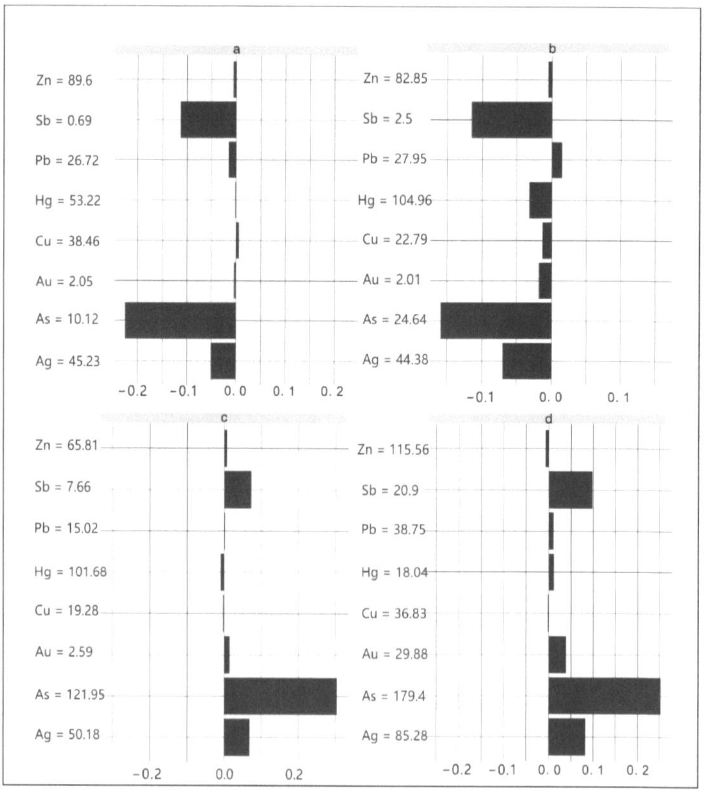

Figure 10. Examples of Shapley value of training samples ((**a**,**b**): negative samples; (**c**,**d**): positive sample) in RF modeling toward known mineralization.

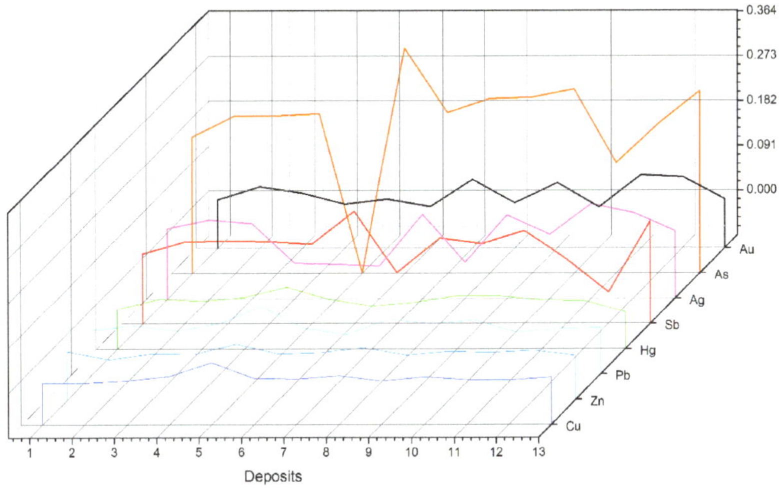

Figure 11. Shapley values of positive training samples for known mineral deposits in RF modeling.

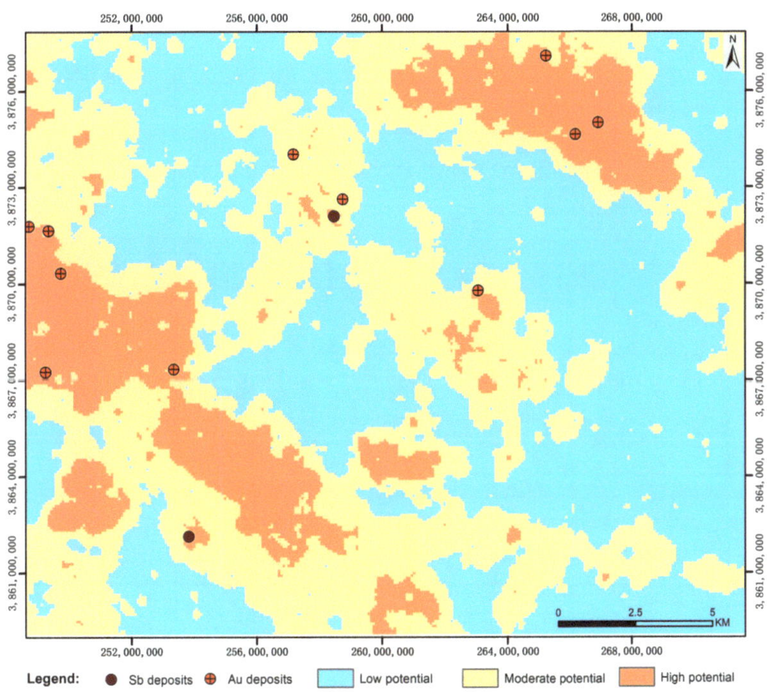

Figure 12. Ternary class gold prospectivity map obtained from RF modeling.

The final prospectivity map shows that areas of high potential delineated most of the known mineral deposits (Figure 12). The performance of random forest (RF) modeling was assessed using the receiver operating characteristic (ROC) curve, which illustrates the true-positive rate versus the false-positive rate, with the area under the curve (AUC) indicating the overall performance of the RF modeling [35]. By utilizing the probability of mineralization occurrence to compute the AUC, we obtained a value of 0.917 for deposits (i.e., positive training samples) and 0.182 for non-deposits. These results suggest that RF modeling effectively distinguished between known mineral deposit locations and the selected non-deposit locations (Figure 13). The relationship between the distribution of prospectivity values obtained and the known mineral deposits was evaluated using prediction–area (P-A) plots and concentration–area (C-A) fractal analysis. The P-A plot illustrates the ratio of the area covered by prospectivity classes (relative to the total study area) against the ratio of known mineral occurrences (relative to the total number of known occurrences) predicted by those respective prospectivity classes [36]. The results indicate that RF modeling identified 9.7% of the study area, accurately capturing 84.6% of the known mineral deposits. In contrast, the C-A fractal analysis identified three probability categories: high potential (probability > 0.80), moderate potential (0.22 < probability < 0.80), and low potential (probability < 0.22) (Figure 14).

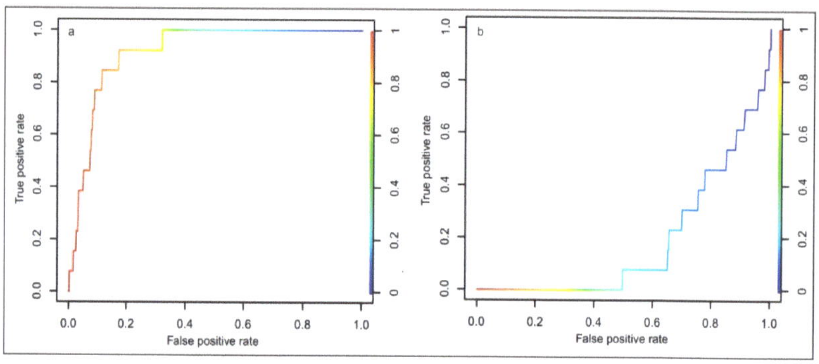

Figure 13. ROC curves for (**a**) deposits (AUC = 0.917) and (**b**) non-deposits (AUC = 0.182).

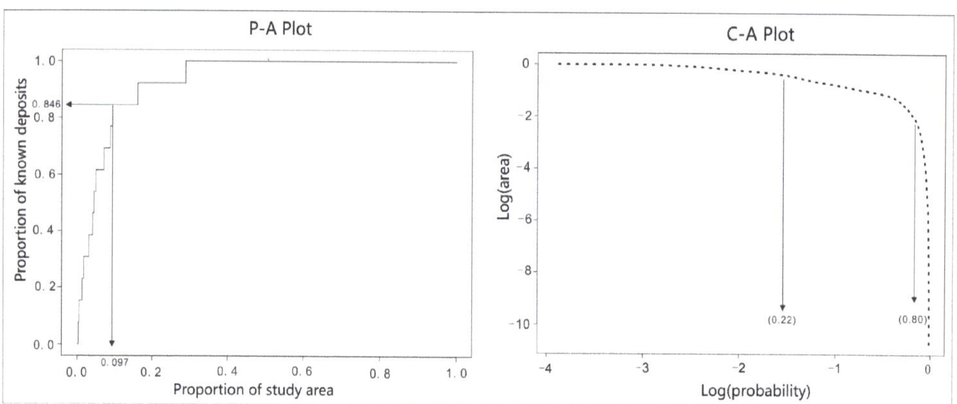

Figure 14. P-A and C-A plots for gold prospectivity map.

4. Conclusions

Machine learning algorithms have demonstrated their effectiveness in mineral prospectivity mapping (MPM). Nonetheless, the black-box nature of the majority of ML algorithms results in limited insights into the relationship between input variables (i.e., evidence maps) and the target variable (i.e., mineral deposits). In this study, PDP and ALE plot analyses were utilized to visualize the impacts of geochemical variables on documented mineralization within the context of regression and classification models. Additionally, SHAP analysis was employed to elucidate the performance of individual training samples in machine learning modeling. The utilization of these interpretive tools effectively yielded quantitative insights from the concentrations of geochemical elements for delineating geochemical anomalies. These tools enhanced the credibility of derived geochemical anomaly information, suggesting that research on interpretable machine learning-based predictions can further advance the practice of mineral prospectivity mapping (MPM).

The work discussed here involved a combination of mathematical concepts with the identification of geochemical anomalies, that is, the transformation of mathematical parameters into geochemical parameters to improve the recognition and assessment of geochemical anomalies. Nevertheless, in the study area, the interpretation of geochemical anomalies using the proposed methodology can probably be enriched by additional analyses, e.g., combined petrographic and ore microscopic analyses of known mineralization, as these would provide a more holistic understanding of the geological processes at play and reveal further insights into the mineralogical composition and microstructural features of the studied area. However, this is a recommendation for further study.

Author Contributions: Conceptualization, S.Z.; writing—original draft preparation, S.Z. and E.J.M.C.; writing—review and editing, E.J.M.C.; supervision, C.F.; visualization, X.Q.; project administration, W.Z.; funding acquisition, C.F. All authors have read and agreed to the published version of the manuscript.

Funding: This research was funded by the China Geological Survey Project (Grant no. DD20191011).

Data Availability Statement: Data available on request due to restrictions.

Acknowledgments: We thank the anonymous reviewers for their constructive comments.

Conflicts of Interest: Shuai Zhang is an employee of China Aero Geophysical Survey and Remote Sensing Center for Natural Resources. The paper reflects the views of the scientists and not the institution.

References

1. Goldberg, I.S.; Abramson, G.Y.; Los, V.L. Exploration Criteria for Appraising Geochemical Anomalies through Mapping Geochemical Systems. *Geochem. Case Hist. Geochmical Explor. Methods* **2007**, 963–968.
2. Rose, A.W.; Hawkes, H.E.; Webb, J.S. *Geochemistry in Mineral Exploration*, 2nd ed.; Academic Press: London, UK, 1979; p. 657.
3. Grigoryan, S. Primary geochemical halos in prospecting and exploration of hydrothermal deposits. *Int. Geol. Rev.* **1974**, *16*, 12–25. [CrossRef]
4. Reimann, C.; Filzmoser, P.; Garrett, R.G. Background and threshold: Critical comparison of methods of determination. *Sci. Total Environ.* **2005**, *346*, 1–16. [CrossRef]
5. Grunsky, E.C.; Grunsky, E.C.; Caritat, P.D.; Caritat, P.D. State-of-the-art analysis of geochemical data for mineral exploration. *Geochem. Explor. Environ. Anal.* **2019**, *20*, 217–232. [CrossRef]
6. Cheng, Q.; Agterberg, F.P.; Bonham-Carter, G.F. A spatial analysis method for geochemical anomaly separation. *J. Geochem. Explor.* **1996**, *56*, 183–195. [CrossRef]
7. Grunsky, E.C.; Agterberg, F.P. Spatial and multivariate analysis of geochemical data from metavolcanic rocks in the Ben Nevis area, Ontario. *Math. Geol.* **1988**, *20*, 825–861. [CrossRef]
8. Jimenez-Espinosa, R.; Sousa, A.J.; Chica-Olmo, M. Identification of geochemical anomalies using principal component analysis and factorial kriging analysis. *J. Geochem. Explor.* **1993**, *46*, 245–256. [CrossRef]
9. Zuo, R.; Carranza, E.J.M.; Wang, J. Spatial analysis and visualization of exploration geochemical data. *Earth-Sci. Rev.* **2016**, *158*, 9–18. [CrossRef]
10. Zuo, R.; Wang, J. Fractal/multifractal modeling of geochemical data: A review. *J. Geochem. Explor.* **2016**, *164*, 33–41. [CrossRef]
11. Cheng, Q.; Agterberg, F.; Ballantyne, S. The separation of geochemical anomalies from background by fractal methods. *J. Geochem. Explor.* **1994**, *51*, 109–130. [CrossRef]
12. Ford, A.; Blenkinsop, T.G. Combining fractal analysis of mineral deposit clustering with weights of evidence to evaluate patterns of mineralization: Application to copper deposits of the Mount Isa Inlier, NW Queensland, Australia. *Ore Geol. Rev.* **2008**, *33*, 435–450. [CrossRef]
13. Wang, H.; Zuo, R. A comparative study of trend surface analysis and spectrum–area multifractal model to identify geochemical anomalies. *J. Geochem. Explor.* **2015**, *155*, 84–90. [CrossRef]
14. Tian, M.; Wang, X.; Nie, L.; Zhang, C. Recognition of geochemical anomalies based on geographically weighted regression: A case study across the boundary areas of China and Mongolia. *J. Geochem. Explor.* **2018**, *190*, 381–389. [CrossRef]
15. Yin, B.; Zuo, R.; Xiong, Y.; Li, Y.-S.; Yang, W. Knowledge discovery of geochemical patterns from a data-driven perspective. *J. Geochem. Explor.* **2021**, *231*, 106872. [CrossRef]
16. Zhang, S.; Xiao, K.; Carranza, E.J.M.; Yang, F.; Zhao, Z. Integration of auto-encoder network with density-based spatial clustering for geochemical anomaly detection for mineral exploration. *Comput. Geosci.* **2019**, *130*, 43–56. [CrossRef]

17. Gonbadi, A.M.; Tabatabaei, S.H.; Carranza, E.J.M. Supervised geochemical anomaly detection by pattern recognition. *J. Geochem. Explor.* **2015**, *157*, 81–91. [CrossRef]
18. Chen, Y.; Wu, W. Application of one-class support vector machine to quickly identify multivariate anomalies from geochemical exploration data. *Geochem. Explor. Environ. Anal.* **2017**, *17*, 231–238. [CrossRef]
19. Saeed, W.; Omlin, C.W.P. Explainable AI (XAI): A Systematic Meta-Survey of Current Challenges and Future Opportunities. *arXiv* **2021**, arXiv:abs/2111.06420. [CrossRef]
20. Mao, J.; Qiu, Y.; Goldfarb, R.J.; Zhang, Z.; Garwin, S.L.; Fengshou, R. Geology, distribution, and classification of gold deposits in the western Qinling belt, central China. *Miner. Depos.* **2002**, *37*, 352–377. [CrossRef]
21. Shi, Y.X.; Li, K.N.; Shi, H.L.; Jin, D.G.; Liang, Z.L. *Report on the prospect survey of gold mines in Xiahe-Hezuo district*; Third Institute Geological and Mineral Exploration of Gansu Provincial Bureau of Geology and Mineral Resources: Gansu, China, 2017; p. 155. (In Chinese)
22. Wang, J.; Zuo, R.; Xiong, Y. Mapping mineral prospectivity via semi-supervised random forest. *Nat. Resour. Res.* **2019**, *29*, 189–202. [CrossRef]
23. Breiman, L. Random forests. *Mach. Learn.* **2001**, *45*, 5–32. [CrossRef]
24. Breiman, L.; Friedman, J.; Olshen, R.A.; Stone, C.J. *Classification and Regression Trees*; Chapman & Hall: New York, NY, USA, 1984.
25. Rasaei, Z.; Bogaert, P. Spatial filtering and Bayesian data fusion for mapping soil properties: A case study combining legacy and remotely sensed data in Iran. *Geoderma* **2019**, *344*, 50–62. [CrossRef]
26. Khanal, S.K.; Fulton, J.P.; Klopfenstein, A.; Douridas, N.; Shearer, S.A. Integration of high resolution remotely sensed data and machine learning techniques for spatial prediction of soil properties and corn yield. *Comput. Electron. Agric.* **2018**, *153*, 213–225. [CrossRef]
27. Molnar, C. *Interpretable Machine Learning: A Guide for Making Black Box Models Explainable*; Lulu.com.: Morrisville, NC, USA, 2019. Available online: https://www.bookstack.cn/read/interpretable-ml-book/8935d4eb447a2642.md (accessed on 2 February 2024).
28. Lundberg, S.M.; Lee, S.-I. A Unified Approach to Interpreting Model Predictions. *arXiv* **2017**, arXiv:abs/1705.07874.
29. Shapley, L. A value for n-person games. In *The Shapley Value: Essays in Honor of Lloyd, S. Shapley*; Roth, A., Ed.; Cambridge University Press: Cambridge, UK, 1988; pp. 31–40. [CrossRef]
30. Ribeiro, M.T.; Singh, S.; Guestrin, C. Why Should I Trust You? Explaining the Predictions of Any Classifier. In Proceedings of the 22nd ACM SIGKDD International Conference on Knowledge Discovery and Data Mining, San Francisco, CA, USA, 13–17 August 2016; Association for Computing Machinery: New York, NY, USA, 2016; pp. 1135–1144.
31. Ma, S.; Tourani, R. Predictive and Causal Implications of using Shapley Value for Model Interpretation. *J. Mach. Learn. Res.* **2020**, arXiv:2008.05052.
32. Carranza, E.J.M.; Laborte, A.G. Data-Driven Predictive Modeling of Mineral Prospectivity Using Random Forests: A Case Study in Catanduanes Island (Philippines). *Nat. Resour. Res.* **2015**, *25*, 35–50. [CrossRef]
33. Zuo, R.; Wang, Z. Effects of Random Negative Training Samples on Mineral Prospectivity Mapping. *Nat. Resour. Res.* **2020**, *29*, 3443–3455. [CrossRef]
34. Levinson, A.A. *Introduction to Exploration Geochemistry*; Applied Publishing Ltd.: Calgary, AB, Canada, 1974; p. 611.
35. Zhang, S.; Xiao, K.; Carranza, E.J.M.; Yang, F. Maximum Entropy and Random Forest Modeling of Mineral Potential: Analysis of Gold Prospectivity in the Hezuo–Meiwu District, West Qinling Orogen, China. *Nat. Resour. Res.* **2018**, *28*, 645–664. [CrossRef]
36. Yousefi, M.; Carranza, E.J.M. Prediction–area (P–A) plot and C–A fractal analysis to classify and evaluate evidential maps for mineral prospectivity modeling. *Comput. Geosci.* **2015**, *79*, 69–81. [CrossRef]

minerals

MDPI

Article

In-Situ Geochemical and Rb–Sr Dating Analysis of Sulfides from a Gold Deposit Offshore of Northern Sanshandao, Jiaodong Peninsula, North China: Implications for Gold Mineralization

Jiepeng Tian [1,2], Jinhui Wang [3,*], Tongliang Tian [4], Ligong Wang [3], Yingpeng Wang [3], Xiaowei Yu [3], Wen Zhang [3], Tianlong Ren [3] and Bin Sun [3]

1. Key Laboratory of Building Structural Retrofitting and Underground Space Engineering, Ministry of Education, Jinan 250101, China
2. School of Civil Engineering, Shandong Jianzhu University, Jinan 250101, China
3. Shandong Institute of Geological Survey, Jinan 250014, China
4. Shandong Institute of Geophysical and Geochemical Exploration, Jinan 250013, China

* Correspondence: jinhuiwang1979@163.com

Abstract: The gold deposit offshore of Northern Sanshandao is an ultra-large-scale gold deposit discovered in the Jiaodong ore area in recent years. This deposit is a fractured-zone altered-rock-type gold deposit; however, its ore genesis and precise mineralization processes are still highly controversial. Based on petrographical observation, the trace elements, sulfur isotopes, and rubidium–strontium isotopes of the gold-bearing pyrite were analyzed using LA-MC-ICP-MS to obtain the source of the ore-forming fluids and ore genesis. The results show that Au has a good positive correlation with Ag, As, and Cu. It is speculated that the As in the pyrite of the gold deposit offshore of Northern Sanshandao is in the form of As^-, replacing S^- and entering the pyrite, causing its lattice defects, and thus promoting the entry of Au^+ into the gold-bearing pyrite. The Co/Ni ratios mainly range between 0.1 and 10, indicating that the mineralization process has experienced different forms of hydrothermal evolution and the mixing of different fluids. The results of the in-situ sulfur isotope analysis show that pyrite $\delta^{34}S$ in the mineralization period is characterized by a high sulfur value. The authors of this study believe that the initial sulfur isotope composition has mantle-derived components. The large-scale, deep cutting, and high degree of fragmentation in the Sanshandao fault zone are conducive to the interaction between fluids and rocks, as well as the mixing and addition of seawater, resulting in the characteristic high $\delta^{34}S$ value. The Sr isotopic compositions indicate a crust–mantle mixing attribute of the mineralized material source. The Rb–Sr isochron age of the pyrite is 118.5 ± 0.65 Ma, which represents the age of gold mineralization. According to the characteristics of the trace elements and sulfur isotopes, it is inferred that the gold deposit minerals offshore of Northern Sanshandao originated from deep magmatic-hydrothermal reservoirs, and the mixing of seawater and Au–As-rich hydrothermal fluids was the formation mechanism of huge amounts of gold precipitation.

Keywords: gold deposit; offshore of Northern Sanshandao; in-situ sulfur isotopes; pyrite trace elements; Rb–Sr isochron age

Citation: Tian, J.; Wang, J.; Tian, T.; Wang, L.; Wang, Y.; Yu, X.; Zhang, W.; Ren, T.; Sun, B. In-Situ Geochemical and Rb–Sr Dating Analysis of Sulfides from a Gold Deposit Offshore of Northern Sanshandao, Jiaodong Peninsula, North China: Implications for Gold Mineralization. *Minerals* **2024**, *14*, 456. https://doi.org/10.3390/min14050456

Academic Editor: Maria Economou-Eliopoulos

Received: 28 March 2024
Revised: 22 April 2024
Accepted: 23 April 2024
Published: 26 April 2024

1. Introduction

The Jiaodong gold province is located on the southeastern edge of the North China Craton (NCC), which is China's most productive gold province and the third-largest gold mining area in the world [1–7]. The gold deposit offshore of Northern Sanshandao is an ultra-large-scale gold deposit that was discovered in the Jiaodong ore area in recent years [8]. It is located under the sea north of the Sanshandao gold deposit. This gold deposit is a deep extension of the ore body of the Sanshandao deposit and a typical case for the study of differences in mineralization at different depths in the Sanshandao ore-bearing fault zone.

Researchers have carried out some work on the geological characteristics of ore deposits, ore-controlling structures, and their structural-magmatic backgrounds [8–15]. Previous petrographic and mineralogy studies indicate that the main alteration types are pyritic sericization, sericization, and potassic alteration [8]. Other studies have suggested the unit cell parameters and thermoelectric characteristics of pyrite, and that their alteration is strong in the range of −1200 to −1400 m. In the range of −1200 to −1600 m, sericitolite and pyrite–sericite–quartz-altered rock are rich in As, Ag, Sb, and other low-temperature elements, which have great ore-prospecting potential for ore prospecting in deep areas [10,11]. Regarding ore-forming fluids, it is believed that they have the characteristics of medium temperature, medium–low salinity, medium acidity, and richness in carbonaceous materials [12]. Fluid immiscibility is generally developed in the early and main stages of mineralization. On the basis of fluid immiscibility, the interaction between fluid and surrounding rocks is superimposed, thus enriching and forming gold deposits [12]. During the hydrothermal alteration process, Si, Mg, Ca, and other elements were added, Na was precipitated, and the K content did not change significantly. This is consistent with the hydrothermal alteration characteristics of the Sanshandao gold mineralization belt proposed by researchers [16,17]. However, there has not yet been a systematic study of the gold occurrence status, gold precipitation mechanism, or mineralization age. It is unknown whether there are differences in the mineralization mechanisms between the marine gold deposits offshore and the gold deposits onshore. Therefore, the authors of this study selected the gold deposit offshore of Northern Sanshandao, Jiaodong Peninsula, as the main research object, and researched the geological characteristics of the gold deposit, pyrite trace element characteristics, sulfur isotope composition, and mineralization chronology to understand the alteration rock type of the fracture zone. Theoretical research on gold deposits has become more abundant, which has improved the research level of gold deposits in the Sanshandao gold mineralization belt, providing a theoretical basis for searching for similar types of gold deposits in the area.

2. Regional Geological Background

The gold deposit offshore of Northern Sanshandao is located in the northwest of the Jiaodong Peninsula. In the shallow sea area in the north of Sanshandao Village, the mining area is covered by seawater, with a water depth of 10–15 m. The thickness of the Quaternary strata under the sea is generally 35–40 m, and the thickest position can reach 60 m. The lithology is mainly coarse, medium, and fine sand, and marine deposit silt. The gold deposit belongs to the Jiaoxibei (Laizhou–Zhaoyuan) ore area, and the Precambrian metamorphic rock series strata and Mesozoic intrusive rocks are mainly exposed in this area (Figure 1). The Precambrian metamorphic rock series mainly includes Mesoarchean granulites, metamorphic supracrustal rocks of the Neoarchean Jiaodong Group, and Neoarchean granitic gneiss that intrudes into them. The intrusive rock mass is mainly composed of Late Jurassic granites (165–150 Ma) and Early Cretaceous (132–123 Ma) porphyritic granodiorite. The late Jurassic granites are mainly monzonitic granite, biotite granite, and biotite granite. Early Cretaceous porphyritic granodiorite, as well as lamprophyre, diabase porphyry, quartz diorite porphyry, diorite porphyry, and other dikes, intrude into the Late Jurassic granites. The fault structure is divided into three groups: north–northeast-trending, northeast-trending, and north–northwest-trending. The north–northeast trending fault zone is the main ore-controlling structure in this area. There are three NNE- and NE-trending ore-controlling faults in the area: the Sanshandao fault zones, the Jiaojia fault zones, and the Zhaoyuan–Pingdu fault zones (Figure 1). The gold deposit offshore of Northern Sanshandao is located in the direction in which the Sanshandao fault zones extend to the sea.

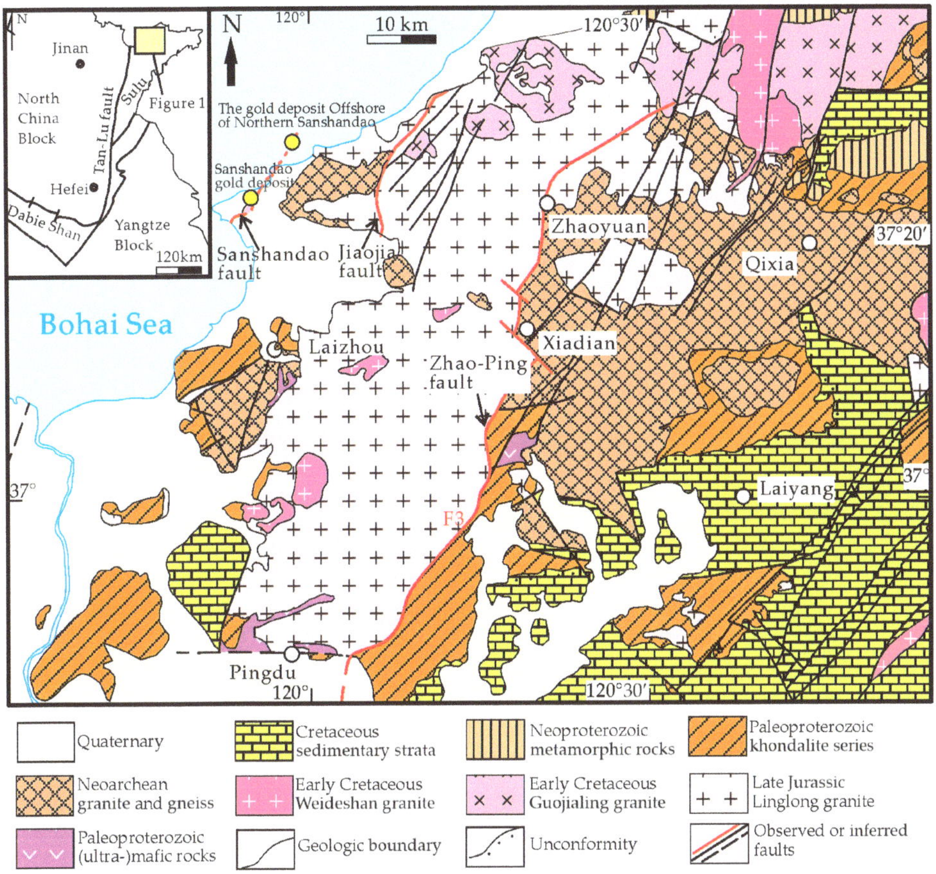

Figure 1. Map showing the regional geology in the Jiaoxibei ore area.

3. Ore Deposit Geology

The magmatic rocks are mainly composed of Early Cretaceous porphyritic granodiorite, Late Jurassic granite, and Neoarchean metagabbro (Figure 2). The Late Jurassic biotite granodiorite is intruded by Early Cretaceous porphyritic granodiorite, and the veinstone mainly consists of lamprophyre, diabase porphyry, quartz diorite porphyry, diorite porphyry, etc. Metagabbro is mainly distributed in the hanging wall of the Sanshandao fault.

The Sanshandao fault is the ore-controlling fault, which strikes about 35° and dips to the southeast. Faults cut the Late Jurassic granite. The alteration and petrogenesis zones are divided along the vertical direction with the main fracture surface as the boundary. From top to bottom, the layers of the rock mass are fine–medium grained metagabbro, monzonitic granite, pyrite sericite cataclastic rocks, pyrite sericite granodiorite clastic rocks, fault gouge, pyrite sericite quartzite cataclastic rocks (pyrite–sericite belt), pyrite sericite granodiorite clastic rocks, pyrite sericite cataclastic rocks, and porphyritic granodiorite, with a gradual transitional contact between the altered rock (Figure 3). Local NW-trending small faults offset the Sanshandao fault.

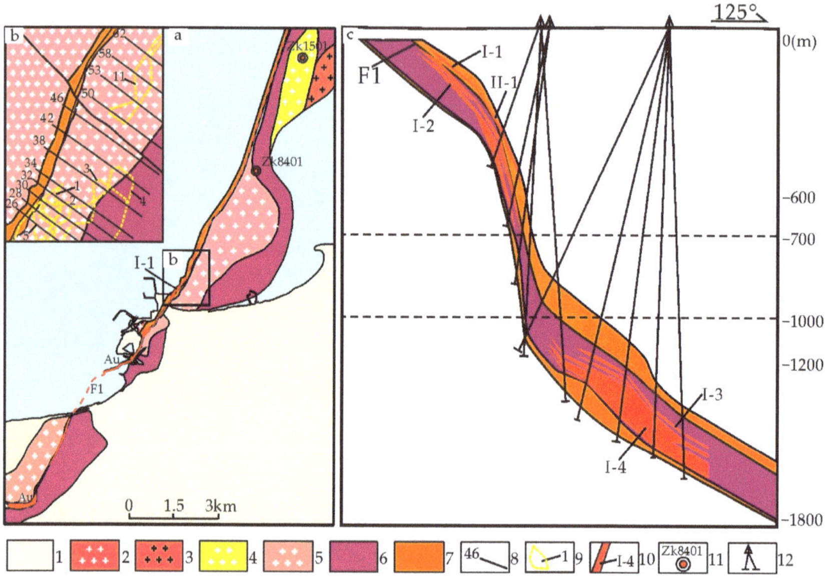

Figure 2. Geological map of the Sanshandao metallogenic belt (**a**,**b**) and the exploration line profile of the deposit (**c**). (1) Quaternary sediments; (2) Early Cretaceous porphyritic granodiorite; (3) Late Jurassic monzonitic granite; (4) Late Jurassic biotite granodiorite; (5) Late Jurassic biotite granite; (6) Neoarchean metagabbro(amphibolite); (7) structural metamorphic belt; (8) exploration line and number; (9) horizontal projection range of gold ore body; (10) location and number of gold ore bodies; (11) drilling location and number; (12) drilling starting and ending position and path.

The metallic minerals are pyrite, galena, and sphalerite, followed by chalcopyrite, arsenopyrite, pyrrhotite, etc. The non-metallic minerals include quartz, sericite, feldspar, and carbonate minerals (calcite, dolomite, siderite, etc.). The prevailing electrum with an Ag content of 6.40 wt% is accompanied by native gold with only 4.30 wt%. The main ore textures are granular metamorphic, fragmented, and metasomatic structures, and a few emulsion droplet structures (Figure 4). The ore structures are mainly disseminated, followed by network vein, breccia, and fine vein structures.

Based on observation of the core samples under a microscope and their identification results, and combined with the mineral texture, structural characteristics, and generation sequence, the mineralization period can be divided into the following four stages: The pyrite–sericite stage (I) is the early stage of mineralization mainly characterized by the irregular granular or euhedral coarse-grained distribution of pyrite, which is disseminated or distributed in a star-like pattern in the sericite. In the quartz–pyrite stage (II), the pyrite is mostly fine-to-medium-grained, isomorphic granular, and significantly more abundant. The coexisting quartz and pyrite form small granular or scattered shapes, fine veins, network veins, etc. This stage is the main stage of mineralization, and gold is distributed throughout the crystals and fissures of pyrite. The gold–quartz–base-metal-sulfide stage (III) is the late stage of mineralization, where base metal sulfide is fine-grained, vein-like, and has a dip distribution. Mineral filling is significant in the middle and late stages of mineralization. The quartz–carbonate stage (IV) is the final stage of mineralization. Carbonate minerals such as quartz and calcite are mostly interspersed in the ore in the form of fine veins, and no gold mineralization occurs.

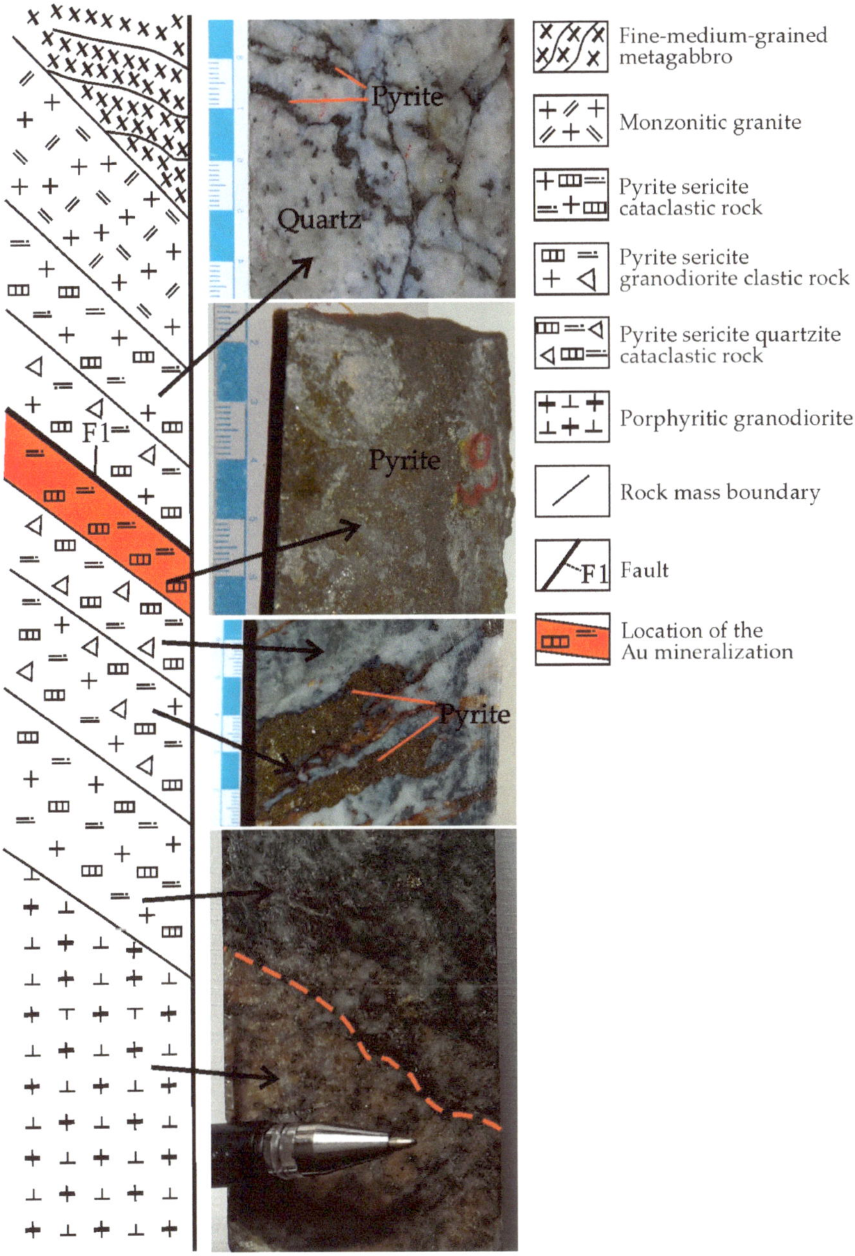

Figure 3. Lithologic zoning of fault alteration zone (modified from [18]).

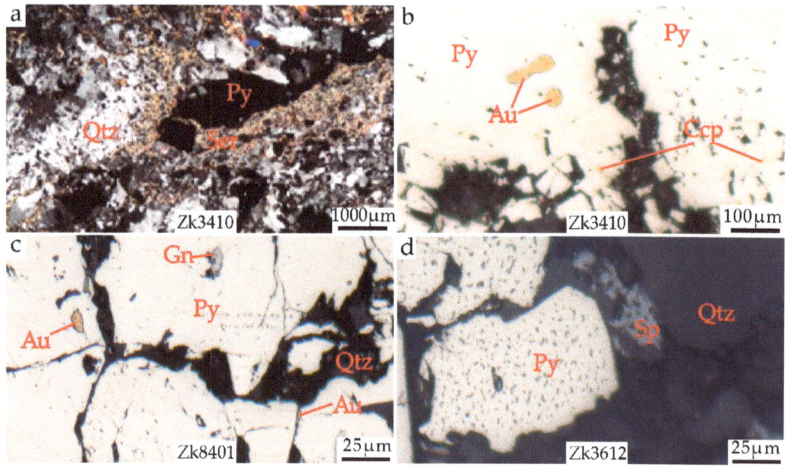

Figure 4. Photomicrographs showing the mineralization and structure of the ore. (**a**) Pyrite–sericite–quartzite in borehole ZK3410; (**b**) encapsulated gold of pyrite in borehole ZK3410; (**c**) encapsulated and fissure gold of pyrite in borehole ZK8401; (**d**) automorphic pyrite in borehole ZK3612. Abbreviations: Py, pyrite; Qtz, quartz; Ser, Sericite; Au, gold; Gn, galena; Sp, sphalerite; Ccp, chalcopyrite.

4. Sampling and Analytical Methods

Typical core samples from the gold deposit offshore of Northern Sanshandao were selected for this study, and the selected samples represent different spatial locations. The samples were polished and sliced for observation and analysis under a microscope. Then, laser ablation inductively coupled plasma mass spectrometry (LA-ICP-MS) was carried out.

4.1. Backscattered Electron (BSE) Imaging and Wavelength Dispersive Elemental Maps

Backscattered electron (BSE) imaging and wavelength dispersive elemental mapping were carried out at the Key Laboratory of Gold Metallogenic Processes and Deep Prospecting of the Ministry of Natural Resources, Shandong Institute of Geological Sciences (SIGS). The samples were analyzed using electron probe microanalysis and wavelength dispersion spectrometry (EPMA-WDS), and the instrument model was JEOLJXA-8230. The acceleration voltage was 15 kV, the current was 2×10^{-8} A, and the beam spot diameter was 0.5 μm. The data collection time was 20–60 s. The ZAF method was used to correct the data, and the analysis accuracy was less than 1%. All standard samples used are standard samples from Canada's Astimex Company. Different standards are used for different elements. Au and Ag use pure metal standards; S and Fe use pyrite; Co uses chloanthite; and Ni uses nickel pyrite; the standard sample used for Cu is cuprite; the standard sample used for Pb is galena.

4.2. In-Situ Trace Element Analysis of Pyrite Using LA-ICP-MS

Trace element analysis of pyrite was conducted using LA-ICP-MS at the Wuhan Sample Solution Analytical Technology Co., Ltd., Wuhan, China. Laser sampling was performed using a GeolasPro laser ablation system, which consisted of a COMPexPro 102 ArF excimer laser (wavelength of 193 nm and maximum energy of 200 mJ) and a MicroLas optical system. An Agilent 7900 ICP-MS instrument was used to obtain the ion signal intensity. Helium was used as the carrier gas. Argon was applied as the supplementary gas and was mixed with the carrier gas through a T-connector before entering the ICP. The laser ablation system contained a "wire" signal-smoothing device [19]. In this study, the laser spot size was set to 5 μm and the frequency to 44 Hz. The trace element composition of pyrite was calibrated against various standards (NIST 610 and NIST 612) without the use of internal standards [20]. The accuracy of the calibration method was verified using MASS-1 (USGS) sulfide standards as unknown samples. Each analysis consisted of approximately 20–30 s of background acquisition, followed by 50 s of data collection from the sample. Offline selection and integration of the background and analytical signals, temporal drift correction, and quantitative calibration for trace element analysis were performed using Excel-based ICPMS-DataCal software [20].

4.3. In-Situ S Isotope Analysis of Pyrite Using LA-MC-ICP-MS

In-situ sulfur isotope analyses of pyrite were performed on a Neptune Plus MC-ICP-MS (Thermo Fisher Scientific, Bremen, Germany) equipped with a Geolas HD excimer ArF laser ablation system (Coherent, Göttingen, Germany) at the Wuhan Sample Solution Analytical Technology Co., Ltd, Hubei, China. In the laser ablation system, helium was used as the carrier gas for the ablation bath, and mixed with argon (supplementary gas) at the end of the ablation bath. The single-spot ablation mode was used. A large spot size (44 μm) and a slow pulse frequency (2 Hz) were used to avoid the downhole chunking effect reported by [21]. One hundred laser pulses were completed in a single analysis. A new signal-smoothing device was used downstream of the sampling unit to effectively eliminate short-term variations in the signal, especially for the slow pulse frequency condition [19]. The laser fluence was constant at 3–5 J/cm^2.The Neptune Plus was equipped with nine Faraday cups with 10^{11} Ω resistors. ^{32}S, ^{33}S, and ^{34}S isotopes were collected in Faraday cups using the static mode. The newly designed X-swept cone and Jet sample cone in the Neptune Plus were used to increase the signal intensity. Nitrogen (4 mL/min) was added to the central airflow to reduce multimer interference. All measurements were performed using medium-resolution rotational power (defined by the peak edge widths ranging from 5% to 95% of the full peak height), which was always greater than 5000.

A standard-sample-bracketing method (SSB) was employed to correct for instrumental mass fractionation. To avoid the matrix effect, a pyrite standard PPP-1 was chosen as the reference material for correcting the natural pyrite. In addition, the in-house references of a pyrrhotite SP-Po-01 ($\delta^{34}S_{V\text{-}CDT} = 1.4 \pm 0.4$), a chalcopyrite SP-CP-01 ($\delta^{34}S_{V\text{-}CDT} = 5.45 \pm 0.3$), and two synthetic Ag_2S standards, IAEA-S-2 ($\delta^{34}S_{V\text{-}CDT} = 22.58 \pm 0.39$) and IAEA-S-3 ($\delta^{34}S_{V\text{-}CDT} = -32.18 \pm 0.45$), were analyzed repeatedly as unknown samples to verify the accuracy of the calibration method. All data reduction for the MC-ICP-MS analysis of S isotope ratios was conducted using "Iso-Compass" software [22].

4.4. Rb–Sr Dating

Isotope measurements were accomplished on an IsoProbe-T mass spectrometer at the Beijing Institute of Geological Research of the Nuclear Industry, equipped with nine Faraday cups and four ion-counting detectors. Rubidium and strontium isotopes were measured using static multi-reception. The fractionation effects during the strontium isotope measurements were regularized and corrected using $^{88}Sr/^{86}Sr = 8.375219$. The strontium isotope composition and strontium concentration were also determined using a computational method established in our laboratory.

We accurately weighed 0.1–0.2 g of the sample powder in a low-pressure closed dissolution jar, added rubidium–strontium diluent, and dissolved it with mixed acid (HF + HNO_3) for 24 h. When the sample was completely dissolved and evaporated, we added 6.0 mol/L of hydrochloric acid to convert into chloride for evaporation. Then, it was dissolved with 0.5 mol/L of hydrochloric acid solution and separated using centrifugation. The supernatant was added to a cation exchange column (φ 0.5 × 15.0 cm, AG50W × 8 (H^+) 100–200 mesh). The rubidium was drenched with a 2.0 mol/L hydrochloric acid solution, and the strontium was drenched with a 3.0 mol/L hydrochloric acid solution and then evaporated to dryness for mass spectrometry analysis.

Rubidium and strontium isotope analyses were performed using a PHOENIX thermal ionization mass spectrometer with a single-band, M+, adjustable multi-Faraday receiver. Mass fractionation was corrected with $^{86}Sr/^{88}Sr = 0.1194$, with the following standard measurements: $^{87}Sr/^{86}Sr$ = 0.710248 ± 8 for the NBS987 specimen, $^{87}Sr/^{86}Sr$ = 1.20022 ± 8 for the NBS607 specimen with Rb 522.550 ug/g and Sr 65.511 ug/g; and $^{87}Sr/^{86}Sr$ = 1.20022 ± 8 with a laboratory process background of Rb 2.0 × 10^{-10} g and Sr 2.0 × 10^{-10} g. The detailed experimental separation process and instrument test parameters followed those of previous researchers [23,24].

5. Results

5.1. Wavelength Dispersive Elemental Maps

The analysis results of the wavelength dispersive elemental maps show that if the content of the element in a certain area of the sample is high, the bright spots in the corresponding area of the image are dense. The bright spots of elements such as Au, Ag, and Ni are denser, and the content is higher (Figure 5). Only the As element had an obvious ring structure, and Au, Ag, Cu, and other elements did not show an obvious ring structure (Figure 5d–i). The results indicate that Au–Ag–Cu–Co–Ni in pyrite changes uniformly and has a good positive correlation.

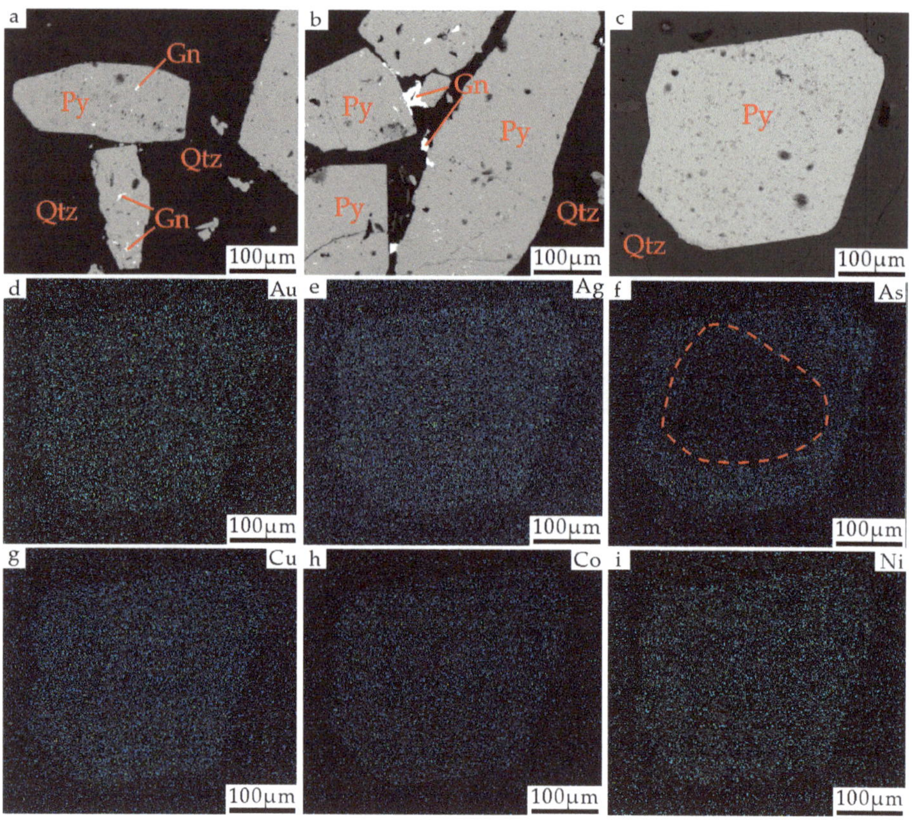

Figure 5. Photomicrographs of BSE images (**a**–**c**), and wavelength dispersive elemental maps (**d**–**i**). Pyrite is in dark gray. Galena is present as inclusions within pyrite or along pyrite margins. In the BSE images, the galena in pyrite is light gray and white (**a**,**b**). Wavelength dispersive elemental maps (**d**–**i**) correspond to the area of the BSE image in (**c**). Abbreviations: Qz = quartz; Py = pyrite; Gn = galena.

5.2. In-Situ Trace Element Compositions

Pyrite samples from three boreholes were analyzed for trace elements, and the results are shown in Table 1 for a total of 30 analytical points.

Table 1. Trace element contents of pyrite from the gold deposit offshore of Northern Sanshandao (ppm).

Sample Serial Number	Co	Ni	Cu	Zn	As	Ag	Sb	Au	Pb	Bi
ZK8401-1	3.37	-	0.10	1.32	125	0.19	-	0.01	1.87	9.35
ZK8401-2	5.22	0.06	2.24	1.04	68	0.42	0.02	0.04	8.40	10
ZK8401-3	3.08	1.29	1.21	1.09	30	0.67	0.17	0.04	15	6.94
ZK8401-4	134	42	2.14	0.66	25	0.41	0.03	0.03	4.89	2.72
ZK8401-5	130	38	5.16	1.26	22	2.58	0.14	0.14	77	43
ZK8401-6	457	19	1.05	1.33	9	1.70	0.04	0.03	107	10
ZK8401-7	7	0.37	0.24	1.36	35	0.07	-	-	1.45	4.93
ZK8401-8	180	30	0.49	0.79	25	0.65	0.06	0.02	5.40	19
ZK8401-9	89	58	0.24	0.70	12	-	-	-	0.65	2.02
ZK8401-10	274	27	1.26	1.30	24	0.79	0.08	0.02	35	13.20
ZK3612-1	20	1.50	31	4216	109	40	12	0.23	18,400	0.20
ZK3612-2	47	24	23	2.90	8019	51	13	0.99	21,290	0.15
ZK3612-3	24	25	11	1.50	7915	18	8.04	0.64	3750	0.03
ZK3612-4	0.29	0.38	6.95	1.02	52	14	3.58	0.14	4386	0.02
ZK3612-5	10	0.62	5.29	0.94	49	7.66	1.56	0.12	247	-
ZK3612-6	0.23	0.55	0.19	1.49	0.70	0.08	0.47	0.03	7.86	-
ZK3612-7	1.06	10	277	37	920	551	127	3.84	57,880	0.62
ZK3612-8	0.02	0.13	13	0.94	166	33	7.13	0.15	9332	0.17
ZK3612-9	0.06	3.02	59	1.83	4103	197	22	1.42	45,896	0.58
ZK3612-10	0.25	0.87	88	15	468	6.12	1.97	0.25	492	0.05
ZK3410-1	1.45	6.50	1.01	1.66	230	0.44	0.28	0.08	10	7.02
ZK3410-2	3.76	4.51	0.55	1.24	136	0.40	0.12	-	3.87	6.52
ZK3410-3	34	17	1.08	1.39	190	2.09	0.11	0.01	15	43
ZK3410-4	4.65	5.60	0.56	1.14	87	0.08	-	-	2.04	1.81
ZK3410-5	6.45	7.46	0.41	1.03	160	0.15	0.05	0.01	3.29	4.88
ZK3410-6	23	13	3.53	0.55	438	3.94	2.31	0.37	18	10
ZK3410-7	1.41	2.78	0.51	1.29	143	0.30	0.24	0.03	5.43	1.43
ZK3410-8	27	9.20	20	1.32	270	5.55	4.07	0.17	100	30
ZK3410-9	61	13	41	2.73	206	4.88	3.27	0.06	80	17
ZK3410-10	2.43	2.03	3.37	1.17	574	2.79	2.36	0.07	23	4.17

Note: - represents contents less than 0.1 ppm.

Pyrite analytical results from borehole ZK8401 are as follows: The Co content was 3.08−457.00 ppm, with an average of 129.00 ppm. The Ni content was 0.06−58.00 ppm, with an average of 23.97 ppm, except for one point that was lower than the detection limit. The Co/Ni ratio ranged from 1.54 to 87.00, with a large variation, and an average value of 17.35. The Cu content was 0.10−5.16 ppm, with an average of 1.41 ppm. The content of Zn was 0.66–1.36 ppm, with an average of 1.09 ppm. The content of As was 8.74–125.00 ppm, with an average of 37.50 ppm. The content of Ag was lower than the detection limit at all but one point, at which it was 0.07–2.58 ppm, with an average of 0.83 ppm. The content of Sb was below the detection limit at all but three points, at which it was 0.02–0.17 ppm, with an average of 0.08 ppm. The content of Au was below the detection limit at all but two points, at which it was 0.01–0.14 ppm, with an average of

0.04 ppm. The content of Pb was 0.65–107.00 ppm with an average of 25.66 ppm, and the content of Bi was 2.02–43.00 ppm, with an average of 12.11 ppm.

Pyrite analytical results from borehole ZK3612 are as follows: The content of Co was 0.02–47.00 ppm, with an average of 10.27 ppm. The content of Ni was 0.13–25.00 ppm, with an average of 6.58 ppm. The Co/Ni ratio ranged from 0.02 to 16.21, with a large variation, and an average value of 3.42. The content of Cu was 0.19–277.00 ppm, with an average of 51.48 ppm. The content of Zn was 0.94–4216.00 ppm, with an average of 418.87 ppm. The content of As was 0.70–8019.00 ppm, with an average of 2,179.97 ppm. The content of Ag was 0.08–551.00 ppm, with an average of 91.79 ppm. The content of Sb was 0.47–127.00 ppm, with an average of 19.77 ppm, except for four points that were below the detection limit. The content of Au was 0.03–3.84 ppm, with an average of 0.78 ppm. The content of Pb was 7.86–57,880.00 ppm, with an average of 16,167.72 ppm. The content of Bi was 0.02–0.62 ppm, with an average of 0.23 ppm, except for two points below the detection limit.

Pyrite analytical results from borehole ZK3410 are as follows: The content of Co was 1.41–61.00 ppm, with an average of 16.45 ppm. The content of Ni was 2.03−17.00 ppm, with an average of 8.17 ppm. The Co/Ni ratio ranged from 0.22 to 4.61, with a large variation, and an average value of 1.57. The content of Cu was 0.41−41.00 ppm, with an average of 7.17 ppm. The content of Zn was 0.55−2.73 ppm, with an average of 1.35 ppm. The content of As was 87.00−574.00 ppm, with an average of 243.00 ppm. The content of Ag was 0.08−5.55 ppm, with an average of 2.06 ppm. The content of Sb was 0.05−4.07 ppm, with an average of 1.42 ppm, except for one point below the detection limit. The content of Au was 0.01−0.37 ppm, with an average of 0.10 ppm, except for two points below the detection limit. The content of Pb was 2.04–100.00 ppm, with an average of 26.10 ppm. The content of Bi was 1.43–43.00 ppm, with an average of 12.65 ppm.

5.3. In-Situ Sulfur Isotopic Compositions

Pyrite from three boreholes was selected for in-situ sulfur isotope analysis, and the results are shown in Table 2, with a total of 24 analytical points. Overall, the $\delta^{34}S$ values of pyrite from different boreholes have a narrow range of variation (10‰–13‰), with an average pyrite $\delta^{34}S$ value of + 11.65‰, which is positively deviated from meteoric sulfur and has a small range of variation, with a high degree of sulfur homogenization, characterized by an enrichment of $\delta^{34}S$ (Figure 6).

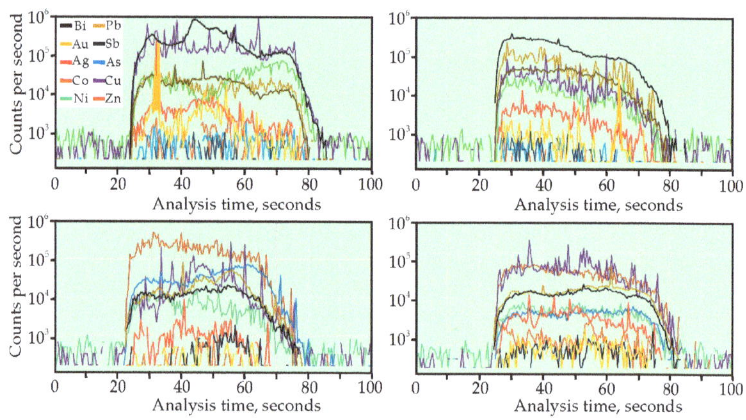

Figure 6. LA-ICP-MS output for analyses of pyrite from the gold deposit offshore of Northern Sanshandao.

Table 2. In-situ sulfur isotopic analyses from the gold deposit offshore of Northern Sanshandao.

Sample Serial Number	Mineral	$\delta^{34}S_{CDT}$ (‰)	Sample Serial Number
ZK8401-1	Pyrite	11.9	920.0 m
ZK8401-2	Pyrite	12.6	920.0 m
ZK8401-3	Pyrite	12.6	920.0 m
ZK8401-4	Pyrite	12.1	920.0 m
ZK8401-5	Pyrite	12.0	920.0 m
ZK8401-6	Pyrite	11.6	963.5 m
ZK8401-7	Pyrite	12.4	963.5 m
ZK8401-8	Pyrite	11.4	963.5 m
ZK8401-9	Pyrite	12.1	963.5 m
ZK8401-10	Pyrite	11.6	963.5 m
ZK8401-11	Pyrite	10.5	963.5 m
ZK3612-1	Pyrite	11.1	1537.0 m
ZK3612-2	Pyrite	12.0	1537.0 m
ZK3612-3	Pyrite	11.3	1537.0 m
ZK3612-4	Pyrite	12.9	1537.0 m
ZK3410-1	Pyrite	10.0	1552.0 m
ZK3410-2	Pyrite	10.3	1552.0 m
ZK3410-3	Pyrite	10.6	1552.0 m
ZK3410-4	Pyrite	11.9	1552.0 m
ZK3410-5	Pyrite	11.4	1552.0 m
ZK3410-6	Pyrite	10.8	1613.0 m
ZK3410-7	Pyrite	10.6	1613.0 m
ZK3410-8	Pyrite	13.0	1613.0 m
ZK3410-9	Pyrite	12.8	1613.0 m

5.4. Rb–Sr Age

Test samples of pyrite Rb–Sr isotopes were taken from borehole ZK3410, and the contents of the Rb–Sr and isotope ratios are shown in Table 3. The values of $^{87}Rb/^{86}Sr$ and $^{87}Sr/^{86}Sr$ were (0.797712–21.364224) and (0.713049–0.747714), respectively; the $^{87}Rb/^{86}Sr$ and $^{87}Sr/^{86}Sr$ plots showed good linear relationships. ISOPLOT software was used to calculate the isochron age, which was t = 118.50 ± 0.65 Ma, and the initial $^{87}Sr/^{86}Sr$ (Isr) value was 0.711686 ± 0.000059 (Figure 7).

Table 3. The contents and isotope ratios of Rb and Sr from the gold deposit offshore of Northern Sanshandao.

Sample Serial Number	Ore Type	$^{87}Rb/^{86}Sr$	$^{87}Sr/^{86}Sr$	StdErr (%)
ZK3410-2	Pyrite	2.466515	0.715794	0.000012
ZK3410-5	Pyrite	3.847805	0.718191	0.000022
ZK3410-7	Pyrite	21.364224	0.747714	0.000019
ZK3410-9	Pyrite	19.944731	0.745217	0.000017
ZK3410-11	Pyrite	8.088107	0.725319	0.000015
ZK3410-12	Pyrite	0.797712	0.713049	0.000024

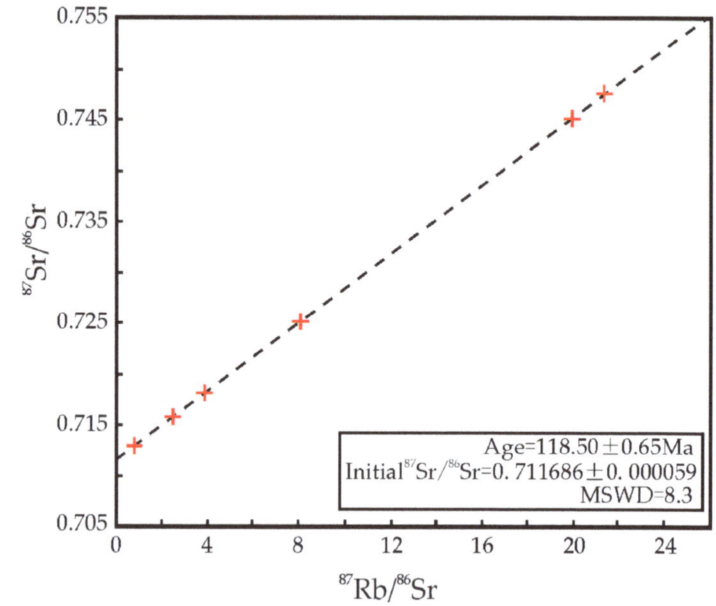

Figure 7. Rb–Sr isotope data and isochron age of pyrite from the gold deposit offshore of Northern Sanshandao.

6. Discussion

6.1. Trace Element Distribution Characteristics of Pyrite

The trace element composition of pyrite can effectively reflect the characteristics of ore-forming fluids, trace the source of ore-forming materials, and indicate the genesis of the deposit [25–33]. The form of occurrence form of trace elements in pyrite can be intuitively and effectively distinguished using a LA-ICP-MS time-resolved signal spectrum. Generally, if the signal intensity of a certain element shows a gentle straight line with the denudation time, it means that the element mainly exists in the form of a solid solution in pyrite. If the peak of an element appears on the time-resolved signal spectrum, it means that the element exists in the form of micron-sized mineral inclusions in pyrite [32]. Correlations among the trace elements in pyrite were analyzed, and a relatively flat signal area was selected for data processing so that the data could reliably and truly reflect the changes in trace elements in pyrite (Figure 6, Table 1). The Au content of pyrite in gold deposits is relatively low, and the contents of Bi, Pb, Cu, Co, and Ni are generally high. Among these, Co and Ni are strong pro-copper elements, which can replace Fe in the lattice of pyrite minerals [34]. In addition, As and Sb also have isomorphism with S or Fe. According to the trace element correlation diagram, Au and Ag, and As and Cu showed certain positive correlations, and Sb and Ag, and Zn and Cu showed positive correlations (Figure 8). The correlation between Au and Ag indicates that Au in gold-bearing pyrite may exist in the form of an electrum, while the correlation with As is more likely to indicate that As-bearing pyrite is a good gold carrier.

Among the trace elements of pyrite, As, Co, and Ni are often affected by the physical and chemical conditions in which they are precipitated. Therefore, it is of great significance to determine the genetic information of pyrite using the contents of these trace elements. It is generally believed that the content ratio of Co and Ni in pyrite of sedimentary or sedimentary transformation is generally less than 1.0, and the content ratio of Co and Ni in pyrite of magmatic origin is often close to 1.0. For pyrite related to hydrothermal genesis, the content ratio of Co and Ni is generally greater than 1 and less than 10, while for pyrite of volcanic origin, the content ratio of Co and Ni is greater than 10 [34,35]. From the test results, it can be seen that the extreme values of Co and Ni elements in pyrite are large, and the element contents jump greatly. In addition, the Co/Ni ratios of each test point are also different. The Co/Ni ratio of the gold deposit offshore of Northern Sanshandao is mainly between 0.1 and 10, and the Co/Ni ratio has a wide range of changes. Most of the casting points on the Co–Ni diagram are located in magmatic and hydrothermal genesis areas, and a few are located in sedimentary genesis areas (Figure 8f). The Mesozoic sedimentary strata are missing in the Jiaodong area. It is speculated that the pyrite of gold deposits in the northern sea area of Sanshandao is of hydrothermal genesis. The reason that most of the casting points deviate from the hydrothermal genesis area may be that the metallogenic process has experienced different forms of hydrothermal evolution or different fluid mixing.

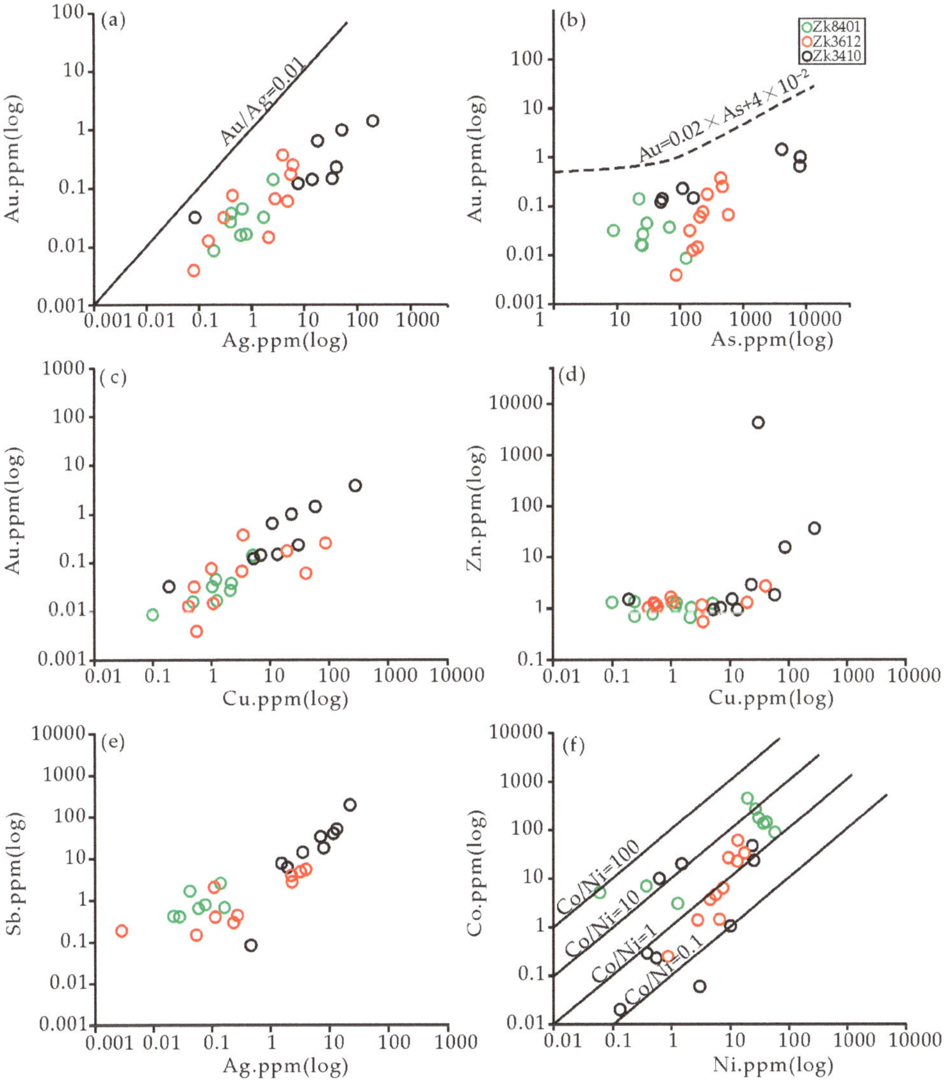

Figure 8. Binary plots of trace elements of pyrites in the mineralization stage from the gold deposit. (**a**) Au vs. Ag; (**b**) Au vs. As; (**c**) Au vs. Cu; (**d**) Zn vs. Cu; (**e**) Sb vs. Ag; (**f**) Co vs. Ni.

6.2. Occurrence of Gold

The LA-ICP-MS signal curves of pyrite from different boreholes are stable, indicating that Au mainly exists in pyrite in the form of invisible lattice gold or nanoparticles. However, there are also abnormal peaks of Au (Figure 6a,b). The inclusion of gold can also be seen under the reflected light of the microscope (Figure 4b,c), indicating that there was also a micron-sized inclusion of gold in the gold-bearing pyrite. The maximum amount of soluble Au (C_{Au}) in pyrite depends on the As content (C_{As}). According to the Au saturation line of pyrite, $C_{Au} = 0.02 \times C_{As} + 4 \times 10^{-5}$, and it can be found that below this curve, Au mainly exists in the form of invisible Au in the pyrite lattice. Above the curve, Au mainly appears in the form of natural gold or mineral inclusions, such as gold and silver, and the Au/As ratio can reflect the occurrence state of invisible gold in pyrite [36]. In the LA-ICP-MS time-resolved signal spectrum of this study, the signal intensity of Au generally shows a smooth state with the ablation time, and all test points were below the gold solubility line, indicating that invisible gold mainly exists in the form of a solid solution (Figures 6 and 8b). In the form of atoms or ions, gold enters the lattice of the medium through isomorphism and forms a solid solution with the medium. In hydrothermal fluids, gold may mainly migrate in the form of Au $(HS)^{2-}$ complexes. In the process of pyrite–sericite–quartz lithification, sulfidation makes the Au $(HS)^{2-}$ complex unstable, and it decomposes, resulting in Au precipitation and enrichment mineralization [15].

Au and As generally show a certain positive correlation (Figure 8b), indicating that As has a certain correlation with the enrichment of Au in the gold deposit offshore of Northern Sanshandao, which is not completely consistent with previous research results of the Sanshandao gold deposit [37]. There are usually two mechanisms for the simultaneous entry of As and Au in gold-bearing pyrite, that is, Au and As replace Fe as follows: $As^{3+} + yAu^{+} + (1–y)M = Fe^{2+}$, where M refers to other elements in lattice defects [38], or Au replaces Fe, and As replaces S^{-} with As^{-}. The former is usually found in high-sulfur epithermal gold deposits with high oxygen fugacity [39], while the precipitation of gold in the Jiaodong area is in a relatively reducing environment [40]. Therefore, it is speculated that As in the gold deposit offshore of Northern Sanshandao enters into pyrite in the form of As^{-} instead of S^{-}, causing its lattice defects to promote the entry of Au^{+} into gold-bearing pyrite.

6.3. Metallogenic Age

In this study, the Rb–Sr isochron age of pyrite is 118.5 $\pm$ 0.65Ma, which can represent the metallogenic age of the deposit (Figure 7). It belongs to the Early Cretaceous and is consistent with the gold mineralization age (120.0 $\pm$ 5Ma) in the Jiaodong area [41–44]. Previous studies have determined the crystallization ages of Sanshandao granite (128.0–129.0 Ma) [45,46]. The sericite Rb–Sr isochron age of the Sanshandao gold deposit is 117.6 $\pm$ 3.0 Ma [47], and the $^{40}Ar/^{39}Ar$ plateau age of sericite from the Sanshandao gold deposit is 118.35 $\pm$ 1.21 Ma [48]. The Cangshang gold deposit has a sericite $^{40}Ar/^{39}Ar$ plateau age of 121.3 $\pm$ 0.2 Ma [49]. The younger age peak is 8–10 Ma later than the crystallization ages of igneous rocks in the study area and may represent the main mineralization period in the Sanshandao gold belt. These ages suggest that gold mineralization in the Jiaodong area was completed in a short time under the same tectonic conditions.

6.4. Sources of Ore-Forming Material

The $\delta^{34}S$ values of sulfides from different sources usually have a certain range, for example, that of a meteorite or mantle source is 0.0‰–3.0‰, and that of granitic magma sulfur is −5.0‰–10.0‰. The average value of sulfur in seawater is +20.0‰, and sedimentary source reduction sulfur is <0.0‰ [50–52]. The sulfur isotope samples of the three boreholes studied in this test are all pyrite, and the $\delta^{34}S$ values are concentrated in the 10‰–13‰ range. The test results generally show high $\delta^{34}S$ values (Figure 9), which are close to the $\delta^{34}S$ value of the Sanshandao and Cangshang gold deposits [53,54], all of which have characteristics of a positive deviation from meteorite sulfur and a relatively concentrated distribution. This shows that the degree of sulfur homogenization is high, and it also shows that the possibility of sulfur from only mantle-derived magma is relatively small. The $\delta^{34}S$ values are relatively higher than those of the Precambrian metamorphic strata, Neoarchean TTG rock series, and Mesozoic granites [55–58]. Previous researchers believe that sulfur isotopes are mainly derived from ore-bearing surrounding rocks, initially from early Precambrian metamorphic complexes, and finally from late Mesozoic granitoids [59]. Previous studies have suggested that sulfur is dominated by mantle-derived magmatic sulfur and mixed with crust-derived sulfur [60,61]. It is also believed that sulfur may be derived from devolatilization during plate subduction [62]. Other studies have suggested that the value of $\delta^{34}S$ gradually increases from east to west in the Jiaodong area, which is due to sulfur in the seawater participating in mineralization [58]. However, some recent studies suggest that the high sulfur isotope composition of the Jiaodong gold deposit was related to subduction of the Neoproterozoic sedimentary strata of the Yangtze Craton under the lithospheric mantle of the North China Craton [63]. It is speculated that the sulfur isotope composition has mantle-derived components. The large-scale, deep cutting, and high degree of fragmentation of the Sanshandao fault zone are conducive to the full development of water–rock interactions and the mixing of seawater sulfur, increasing the $\delta^{34}S$ value.

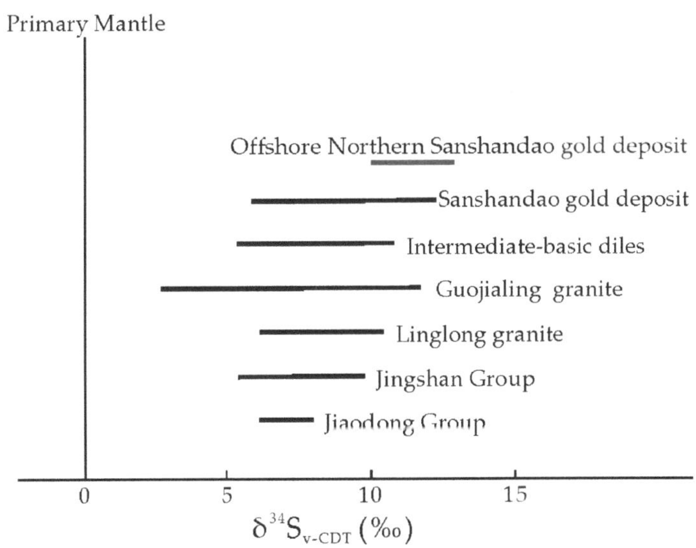

Figure 9. Sulfur isotopic composition of pyrite from the gold deposit offshore of Northern Sanshandao and regional rocks.

The initial $^{87}Sr/^{86}Sr$ value is also an essential indicator for judging the source of diagenetic and metallogenic materials. It is often used to trace the source of metallogenic materials, magmatic fluid, and deep source fluid in the geological study of deposits [64]. Generally, values of $^{87}Sr/^{86}Sr > 0.710$ are considered to reflect a crustal source, whereas values of $^{87}Sr/^{86}Sr < 0.705$ are considered to reflect a mantle source [64–67]. The initial Sr value from the gold deposit offshore of Northern Sanshandao was lower than the continental crust but significantly higher than the mantle. The Sr isotope composition also shows that the source of the minerals has the attributes of a mixed crust–mantle source. It is speculated that the volatile-containing gold-rich initial mantle fluid upwells along the deep fault. In the process of rising and intrusion, a large amount of seawater sulfur is mixed to form a mixed fluid. This fluid carries mineralized materials as it rises, and it precipitates to form gold deposits in the fracture zones with increased tectonic space.

7. Conclusions

The gold deposit offshore of Northern Sanshandao contains visible gold and invisible gold. The visible gold is mainly electrum and natural gold, and the invisible gold is mainly invisible lattice gold or nanoparticles. The gold most likely precipitated directly from the Au–As-rich fluid.

The Rb–Sr isochron age of pyrite from the main mineralization period was 118.5 ± 0.65 Ma, representing the ore formation age of the gold deposits. It belongs to the Early Cretaceous and is consistent with the gold mineralization age (120.0 ± 5 Ma) in the Jiaodong area.

The mineralization of gold deposits originated from the mixing of a large amount of seawater sulfur in the rising process of mantle magmatic fluid. The fluid underwent phase separation and sulfidation under suitable temperature and pressure conditions, and gold deposits were formed in the Early Cretaceous.

Author Contributions: J.T. and J.W. conceived and designed the research ideas; J.T., J.W., T.T. and L.W. participated in the field investigation; J.T., Y.W., X.Y., W.Z. and T.T. performed the data processing; T.R., B.S. and J.W. reviewed and edited the draft. All data were obtained from previous work performed by the project team. All authors have read and agreed to the published version of the manuscript.

Funding: This study was financially supported by the National Key Research and Development Program of China (grant 2016YFC0600100), geological and mineral survey special project from Shandong province (grants 12120113060400, LKZ2021–18, and LKB2020–1), and the Doctoral Research Foundation of Shandong Jianzhu University (grant X21005Z).

Data Availability Statement: All data are contained within the article.

Conflicts of Interest: The authors declare no conflicts of interest.

References

1. Li, J.J.; Luo, Z.K.; Liu, X.Y.; Xu, W.D.; Luo, H. Geodynamic setting for formation of large-superlarge gold deposits and Mesozoic granites in Jiaodong area. *Miner. Depos.* **2005**, *24*, 361–372. (In Chinese)
2. Deng, J.; Yang, L.Q.; Ge, L.S.; Wang, Q.F.; Zhang, J.; Gao, B.F.; Zhou, Y.F.; Jiang, S.Q. Research advances in the Mesozoic tectonic regimes during the formation of Jiaodong ore cluster area. *Prog. Nat. Sci.* **2006**, *16*, 777–784.
3. Deng, J.; Wang, Q.F. Gold mineralization in China: Metallogenic provinces, deposit types and tectonic framework. *Gondwana Res.* **2016**, *36*, 219–274. [CrossRef]
4. Li, S.R.; Santosh, M. Metallogeny and craton destruction: Records from the North China Craton. *Ore Geol. Rev.* **2014**, *56*, 376–414. [CrossRef]
5. Goldfarb, R.J.; Santosh, M. The dilemma of the Jiaodong gold deposits: Are they unique? *Geosci. Front.* **2014**, *5*, 139–153. [CrossRef]
6. Song, M.C.; Ding, Z.J.; Zhang, J.J.; Song, Y.X.; Bo, J.W.; Wang, Y.Q.; Liu, H.B.; Li, S.Y.; Li, J.J.; Li, R.X. Geology and mineralization of the Sanshandao supergiant gold deposit (1200 t) in the Jiaodong Peninsula, China: A review. *China Geol.* **2021**, *4*, 686–719. [CrossRef]
7. Qiu, K.F.; Goldfarb, R.J. *Gold Deposits of the Jiaodong Peninsula, Eastern China*; Society of Economic Geologists: Littleton, CO, USA, 2020; pp. 1–10.
8. Song, M.C.; Zhang, J.J.; Zhang, P.J.; Yang, L.Q.; Liu, D.H.; Ding, Z.J.; Song, Y.X. Discover and tectonic-magmatic background of superlarge gold deposit in offshore of northern Sanshandao, Shandong peninsula. *Acta Geol. Sin.* **2015**, *89*, 365–383. (In Chinese)
9. Liu, D.H.; Lv, G.X.; Zhang, P.J.; Ding, Z.J.; Zhang, J.J.; Lin, D.W.; Lv, C.X.; Wang, Z.Y. A study of 3D ore-controlling of the tectonic altered rocks of the Sanshandao fault in Jiaodong Peninsular and the discovery of an offshore super-large gold deposit in China. *Earth Sci. Front.* **2015**, *22*, 162–172. (In Chinese)
10. Gao, J.W.; Shen, J.F.; Li, G.W.; Wang, L.M.; Sun, N.Y.; Liu, H.D.; Chi, L.; Zhang, H.F.; Du, B.S. Characteristics of Cell Parameters and Thermoelectricity of the Pyrite in the Sanshandao North Offshore Gold Deposit, Shandong Province, China. *Bull. Mineral. Petrol. Geochem.* **2020**, *39*, 1205–1214. (In Chinese) [CrossRef]
11. Gao, J.W.; Liu, W.Q.; Deng, H.J.; Shen, J.F.; Zhao, G.C. Hydrothermal Alteration Characteristics and Migration Rules of Trace Elements in the North Sanshandao Sea Gold Deposit, Shandong, China. *Northwestern Geol.* **2023**, *56*, 245–253. (In Chinese) [CrossRef]
12. Lyu, J.Y.; Wang, J.H.; Mao, M.Q.; Zhang, W.; Wang, Y.P.; Yu, X.W.; Wang, L.G.; Zhang, G.L. The Genesis of the Haiyu Gold deposit in the northern part of Sanshandao gold metallogenic belt in the eastern Shandong: Constraints from geological characteristics and fluid inclusion study. *Acta Mineral. Sin.* **2023**, *43*, 521–532. (In Chinese)

13. Song, Y.X.; Li, S.R.; Shen, J.F.; Zhang, L.; Li, W.T.; Zeng, Y.J. Characteristics and prospecting significance of thermoluminescence patterns and cell parameters of quartz from the undersea gold deposit off northern Sanshandao, Jiaodong Peninsula. *Earth Sci. Front.* **2021**, *28*, 305–319. (In Chinese)
14. Wang, J.; Zhu, L.X.; Ma, S.M.; Tang, S.X.; Zhang, L.L.; Zhou, W.W. Application of the multi-attribute anomaly model for prospecting potential at depth: A case study of the Haiyu Au deposit in the Jiaodong Gold Province, China. *J. Geochem. Explor. J. Assoc. Explor. Geochem.* **2019**, *207*, 106359. [CrossRef]
15. Wang, J.P. Geochemical Mechanism of Gold Accumulation in Offshore of Northern Sanshandao Gold Deposit in the Jiaodong Gold Province, Eastern China. Ph.D Thesis, China University of Geosciences, Wuhan, China, 2020.
16. Zhang, C.; Huang, T.; Liu, X.D.; Liu, Y.; Wang, X.D. Hydrothermal alteration of the xincheng gold deposit, northwestern Jiaodong, China. *Acta Pet. Sin.* **2016**, *32*, 2433–2450. (In Chinese)
17. Fan, Y.C. Origins of Ore-Forming Materials and the Metallogenic Model of Gold Deposits in the Jiaoxibei Area. Ph.D Thesis, Jilin University, Changchun, China, 2022.
18. Ding, Z.J. Study on Metallogenic Regularity of Mesozoic Precious and Non-Ferrous Deposits in Jiaodong Area. Ph.D. Thesis, Jilin University, Changchun, China, 2014.
19. Hu, Z.C.; Zhang, W.; Liu, Y.S.; Gao, S.; Li, M.; Zong, K.Q.; Chen, H.H.; Hu, S.H. "Wave" Signal-Smoothing and Mercury-Removing Device for Laser Ablation Quadrupole and Multiple Collector ICPMS Analysis: Application to Lead Isotope Analysis. *Anal. Chem.* **2015**, *87*, 1152–1157. [CrossRef] [PubMed]
20. Liu, Y.S.; Hu, Z.C.; Gao, S.; Gunther, D.; Xu, J.; Gao, C.G.; Chen, H.H. In situ analysis of major and trace elements of anhydrous minerals by LA-ICP-MS without applying an internal standard. *Chem. Geol.* **2008**, *257*, 34–43. [CrossRef]
21. Fu, J.L.; Hu, Z.C.; Zhang, W.; Yang, L.; Liu, Y.S.; Li, M.; Zong, K.Q.; Gao, S.; Hu, S.H. In situ sulfur isotopes (δ34S and δ33S) analyses in sulfides and elemental sulfur using high sensitivity cones combined with the addition of nitrogen by laser ablation MC-ICP-MS. *Anal. Chim. Acta* **2016**, *911*, 14–26. [CrossRef] [PubMed]
22. Zhang, W.; Hu, Z.C.; Liu, Y.S. Iso-Compass: New freeware software for isotopic data reduction of LA-MC-ICP-MS. *J. Anal. At. Spectrom.* **2020**, *35*, 1087–1096. [CrossRef]
23. Tian, J.P. The Mesozoic Gold Polymetallic Regional Metallogeny in Qipengfu Ore Concentration Area, Jiaodong Peninsula. Ph.D. Thesis, China University of Geosciences, Beijing, China, 2020.
24. Liu, W.G.; Liu, H.B.; Li, G.Z.; Xiao, Z.B.; Tu, J.R.; Li, H.M. The Application of Ion Exchange Resins in Sr-Nd Isotopic Assay of Geological Samples. *Acta Geol. Sin.* **2017**, *91*, 2584–2592. (In Chinese)
25. Barker, S.L.; Hickey, K.A.; Cline, J.S.; Dipple, G.M.; Kilburn, M.R.; Vaughan, J.R.; Longo, A.A. Uncloaking invisible gold: Use of nanoSIMS to evaluate gold, trace elements, and sulfur isotopes in pyrite from Carlin-type gold deposits. *Econ. Geol.* **2009**, *104*, 897–904. [CrossRef]
26. Ciobanu, C.L.; Cook, N.J.; Utsunomiya, S.; Kogagwa, M.; Green, L.; Gilbert, S.; Wade, B. Gold-telluride nanoparticles revealed in arsenic-free pyrite. *Am. Mineral.* **2012**, *97*, 1515–1518. [CrossRef]
27. Ciobanu, C.L.; Cook, N.J.; Pring, A.; Brugger, J.; Danyushevsky, L.V.; Shimizu, M. 'Invisible gold' in bismuth chalcogenides. *Geochim. Et Cosmochim. Acta* **2009**, *73*, 1970–1999. [CrossRef]
28. Cook, N.J.; Ciobanu, C.L.; Meria, D.; Silcock, D.; Wade, B. Arsenopyrite-Pyrite Association in an Orogenic Gold Ore: Tracing Mineralization History from Textures and Trace Elements. *Econ. Geol. Bull. Soc. Econ. Geol.* **2013**, *108*, 1273–1283. [CrossRef]
29. Cook, N.J.; Ciobanu, C.L.; Mao, J.W. Textural control on gold distribution in As-free pyrite from the Dongping, Huangtuliang, and Hougou gold deposits, North China Craton (Hebei Province, China). *Chem. Geol.* **2009**, *264*, 101–121. [CrossRef]
30. Ulrich, T.; Long, D.G.F.; Kamber, B.S.; Whitehouse, M.J. In Situ Trace Element and Sulfur Isotope Analysis of Pyrite in a Paleoproterozoic Gold Placer Deposit, Pardo and Clement Townships, Ontario, Canada. *Econ. Geol.* **2011**, *106*, 667–686. [CrossRef]
31. Fan, H.R.; Li, X.H.; Zuo, Y.B.; Chen, L.; Liu, S.; Hu, F.F.; Feng, K. In-situ LA-(MC)-ICPMS and (Nano) SIMS trace elements and sulfur isotope analyses on sulfides and application to confine metallogenic process of ore deposit. *Acta Petrol. Sin.* **2018**, *34*, 3479–3496. (In Chinese)
32. Zhang, H.Y.; Zhao, Q.Q.; Zhao, G.; Hong, J.X.; Liu, J.J.; Zhai, D.G. In situ LA-ICP-MS trace element analysis of pyrite and its application in study of Au deposit. *Miner. Depos.* **2022**, *41*, 1182–1199. (In Chinese)
33. Zhou, L.L.; Zeng, Q.D.; Sun, G.T.; Duan, X.X.; Bonnetti, C.; Biegler, T.; Long, D.G.F.; Kamber, B. Laser Ablation-Inductively Coupled Plasma Mass Spectrometry (LA-ICP-MS) elemental mapping and its applications in ore geology. *Acta Petrol. Sin.* **2019**, *35*, 1964–1978. (In Chinese) [CrossRef]
34. Koglin, N.; Frimmel, H.E.; Lawrie Minter, W.E.; Brätz, H. Trace-element characteristics of different pyrite types in Mesoarchaean to Palaeoproterozoic placer deposits. *Miner. Depos.* **2010**, *45*, 259–280. [CrossRef]
35. Bralia, A.; Sabatini, G.; Troja, F. A revaluation of the Co/Ni ratio in pyrite as geochemical tool in ore genesis problems. *Miner. Depos.* **1979**, *14*, 353–374. [CrossRef]
36. Reich, M.; Kesler, S.E.; Utsunomiya, S.; Palenik, C.S.; Chryssoulis, S.L.; Ewing, R.C. Solubility of gold in arsenian pyrite. *Geochim. Et Cosmochim. Acta* **2005**, *69*, 2781–2796. [CrossRef]
37. Lin, Z.W.; Zhao, X.F.; Xiong, L.; Zhu, Z.X. In-situ trace element analysis characteristics of pyrite in Sanshandao Gold Deposit in Jiaodong Peninsula: Implications for ore genesis. *Adv. Earth Sci.* **2019**, *34*, 399–413. (In Chinese) [CrossRef]
38. Deditius, A.P.; Reich, M.; Kesler, S.E.; Utsunomiya, S.; Chryssoulis, S.L.; Walshe, J.; Ewing, R.C. The coupled geochemistry of Au and As in pyrite from hydrothermal ore deposits. *Geochim. Cosmochim. Acta* **2014**, *140*, 644–670. [CrossRef]

39. Simmons, S.F.; White, N.C.; John, D.A. Geological Characteristics of Epithermal Precious and Base Metal Deposits. In *One Hundredth Anniversary Volume*; Society of Economic Geologists: Littleton, CO, USA, 2005; pp. 485–522. [CrossRef]
40. Zhu, Z.X.; Zhao, X.F.; Lin, Z.W.; Zhao, S.R. In Situ Trace Elements and Sulfur Isotope Analysis of Pyrite from Jinchiling Gold Deposit in the Jiaodong Region: Implications for Ore Genesis. *Earth Sci.* **2020**, *45*, 945–959. (In Chinese) [CrossRef]
41. Zhai, M.G.; Fan, H.R.; Yang, J.H.; Miao, L.C. Large-scale cluster of gold deposits in east Shandong: Anorogenic metallogenes. *Earth Sci. Front.* **2004**, *11*, 85–98. (In Chinese)
42. Deng, J.; Qiu, K.F.; Wang, Q.F.; Goldfarb, R.; Yang, L.Q.; Zi, J.W.; Geng, J.Z.; Ma, Y. In Situ Dating of Hydrothermal Monazite and Implications for the Geodynamic Controls on Ore Formation in the Jiaodong Gold Province, Eastern China. *Econ. Geol.* **2020**, *115*, 671–685. [CrossRef]
43. Yang, K.F.; Jiang, P.; Fan, H.R.; Zuo, Y.B.; Yang, Y.H. Tectonic transition from a compressional to extensional metallogenic environment at similar to 120 Ma revealed in the Hushan gold deposit, Jiaodong, North China Craton. *J. Asian Earth Sci.* **2018**, *160*, 408–425. [CrossRef]
44. Zhang, L.; Weinberg, R.F.; Yang, L.Q.; Groves, D.I.; Sai, S.X.; Matchan, E.; Phillips, D.; Kohn, B.P.; Miggins, D.P.; Liu, Y.; et al. Mesozoic Orogenic Gold Mineralization in the Jiaodong Peninsula, China: A Focused Event at 120 +/− 2 Ma During Cooling of Pregold Granite Intrusions. *Econ Geol.* **2020**, *115*, 415–441. [CrossRef]
45. Yang, K.F.; Fan, H.R.; Santosh, M.; Hu, F.F.; Wilde, S.A.; Lan, T.G.; Lu, L.N.; Liu, Y.S. Reactivation of the Archean lower crust: Implications for zircon geochronology, elemental and Sr-Nd-Hf isotopic geochemistry of late Mesozoic granitoids from northwestern Jiaodong Terrane, the North China Craton. *Lithos* **2012**, *146*, 112–127. [CrossRef]
46. Wang, L.G.; Qiu, Y.M.; McNaughton, N.J.; Groves, D.I.; Luo, Z.K.; Huang, J.Z.; Liu, Y.K. Constraints on crustal evolution and gold metallogeny in the northwestern Jiaodong Peninsula, China, from SHRIMP U-Pb zircon studies of granitoids. *Ore Geol. Rev.* **1998**, *13*, 275–291. [CrossRef]
47. Hu, F.F.; Fan, H.R.; Jiang, X.H.; Yang, K.F. Fluid inclusions in different depths at Sanshandao gold deposit, Jiaodong Peninsula: Implication for ore genesis. *Geofluids* **2013**, *13*, 528–541. [CrossRef]
48. Yang, Z.Y.; Yu, X.W.; Zhang, W.; Wang, L.G.; Wang, Q.Y.; Guo, R.P. 40Ar/39Ar Age and Its Significance of Sericite in Pyrite Sericite in Sanshandao Gold Deposit in Northwest of Shandong Province. *Shandong Land Resour.* **2020**, *36*, 1–8. (In Chinese)
49. Zhang, X.; Cawood, P.A.; Wilde, S.A.; Liu, R.; Song, H.; Li, W.; Snee, L.W. Geology and timing of mineralization at the Cangshang gold deposit, north-western Jiaodong Peninsula, China. *Miner. Depos.* **2003**, *38*, 141–153. [CrossRef]
50. Ohmoto, H. Systematics of Sulfur and Carbon Isotopes in Hydrothermal Ore Deposits. *Econ. Geol.* **1972**, *67*, 551–578. [CrossRef]
51. Chaussidon, M.; Albarède, F.; Sheppard, S.M.F. Sulphur isotope variations in the mantle from ion microprobe analyses of micro-sulphide inclusions. *Earth Planet Sci. Lett.* **1989**, *92*, 144–156. [CrossRef]
52. Rollinson, H.R. *Using Geochemical Data: Evaluation, Presentation, Interpretation*; Routledge: London, UK, 1993; Available online: https://www.taylorfrancis.com/books/mono/10.4324/9781315845548/using-geochemical-data-hugh-rollinson (accessed on 1 October 2023).
53. Wen, B.J.; Fan, H.R.; Hu, F.F.; Liu, X.; Yang, K.F.; Sun, Z.F.; Sun, Z.F. Fluid evolution and ore genesis of the giant Sanshandao gold deposit, Jiaodong gold province, China: Constrains from geology, fluid inclusions and H-O-S-He-Ar isotopic compositions. *J. Geochem. Explor. J. Assoc. Explor. Geochem.* **2016**, *171*, 96–112. [CrossRef]
54. Liu, Y.Z. Characteristics of the Ore-Forming Fluid and Thermodynamic Simulation of Water-Rock Interaction at the Sanshandao Gold Deposit, Shandong Province, China. Ph.D. Thesis, China University of Geosciences, Beijing, China, 2019.
55. Zhu, R.X.; Fan, H.R.; Li, J.W.; Meng, Q.R.; Li, S.R.; Zeng, Q.D. Decratonic gold deposits. *Sci. China Earth Sci.* **2015**, *58*, 1523–1537. [CrossRef]
56. Deng, J.; Wang, C.M.; Bagas, L.; Carranza, E.J.M.; Lu, Y.J. Cretaceous-Cenozoic tectonic history of the Jiaojia Fault and gold mineralization in the Jiaodong Peninsula, China: Constraints from zircon U-Pb, illite K-Ar, and apatite fission track thermochronometry. *Miner. Depos.* **2015**, *50*, 987–1006. [CrossRef]
57. Wang, Y.W.; Zhu, F.S.; Gong, R.T. Tectonic isotope geochemistry—Further study on sulphur isotope of Jiaodong Gold Concentration Area. *Gold* **2002**, *23*, 1–16. (In Chinese)
58. Huang, D.Y. Sulfur isotope studies of the metallogenic series of gold deposits in Jiaodong area. *Miner. Depos.* **1994**, *13*, 75–86. (In Chinese)
59. Yang, L.Q.; Deng, J.; Wang, Z.L.; Zhang, L.; Guo, L.N.; Song, M.C.; Zheng, X.L. Mesozoic gold metallogenic system of the Jiaodong gold province, eastern China. *Acta Petrol. Sin.* **2014**, *30*, 2447–2467. (In Chinese)
60. Li, J.; Song, M.C.; Liang, J.L.; Jiang, M.Y.; Li, S.Y.; Ding, Z.J.; Su, F. Source of ore-forming fluids of the Jiaojia deeplyseated gold deposit: Evidences from trace elements and sulfur-helium-argon isotopes of pyrite. *Acta Geol. Sin.* **2020**, *36*, 297–313. (In Chinese)
61. Li, J.; An, M.Y.; Song, M.C.; Wang, M.Y.; Ding, Z.J.; Bao, Z.Y.; Wang, S.S. Sulfur isotopic composition and its source of Jiaodong gold deposit. *Geol. Bull. China* **2022**, *41*, 993–1009. (In Chinese) [CrossRef]
62. Feng, K.; Fan, H.R.; Groves, D.I.; Yang, K.F.; Hu, F.F.; Liu, X.; Cai, Y.C. Geochronological and sulfur isotopic evidence for the genesis of the post-magmatic, deeply sourced, and anomalously gold-rich Daliuhang orogenic deposit, Jiaodong, China. *Miner. Depos.* **2020**, *55*, 293–308. [CrossRef]
63. Deng, J.; Yang, L.Q.; Groves, D.I.; Zhang, L.; Qiu, K.F.; Wang, Q.F. An integrated mineral system model for the gold deposits of the giant Jiaodong province, eastern China. *Earth Sci. Rev.* **2020**, *208*, 103274. [CrossRef]

64. Hou, M.L.; Jiang, S.Y.; Jiang, Y.H.; Ling, H.F. S-Pb isotope geochemistry and Rb-Sr geochronology of the Penglai gold field in the eastern Shangdong province. *Acta Petrol. Sin.* **2006**, *22*, 2525–2533. (In Chinese)
65. Chen, Y.J.; Pirajno, F.; Lai, Y.; Li, C. Metallogenic time and tectonic setting of the Jiaodong gold province, eastern China. *Acta Petrol. Sin.* **2004**, *20*, 907–922.
66. Faure, M.; Caridroit, M.; Charvet, J. The Late Jurassic oblique collisional orogen of SW Japan: New structural data and synthesis. *Tectonics* **1986**, *5*, 1089–1114. [CrossRef]
67. Sun, X.L. Sr and O isotopic characteristics of porphyries in the Qinling molybdenum deposit belt and their implication To genetic mechanism and type. *Sci. China Ser. D Earth Sci.* **2001**, *43*, 82–94.

minerals

MDPI

Article

Occurrence of State of Gold in Crude Oil and Its Economic Significance

Zhiyong Ni [1,2,*], Wen Zhang [1,2], Jie Liu [1,2], Shengbao Shi [1,2], Xue Wang [1,2] and Yang Su [1,2]

[1] National Key Laboratory of Petroleum Resources and Engineering, China University of Petroleum (Beijing), Beijing 102249, China; 2023210028@student.cup.edu.cn (W.Z.); mifengguan@outlook.com (J.L.); shengbao@cup.edu.cn (S.S.); 2020010029@student.cup.edu.cn (X.W.); suyangcupb@163.com (Y.S.)

[2] College of Geosciences, China University of Petroleum (Beijing), Beijing 102249, China

* Correspondence: nizhy@cup.edu.cn

Abstract: Gold and petroleum are also strategic resources of great importance to national security. With the increasing demand for energy, multi-energy cooperative exploration has become an inevitable trend of resource development and utilization. Petroleum and hydrothermal gold deposits may form together, with similar evolutionary trends in their formation, migration, and enrichment. Petroleum reservoirs and gold deposits are closely coupled under certain geological conditions. The solubility of gold in crude oil and its forms of occurrence are important in determining the mechanisms of interaction between gold and petroleum and in facilitating the recovery of gold from gold-bearing petroleum. In this study, the occurrence of gold in crude oil from the Linnan Depression in the Bohai Bay Basin, China, was studied using inductively coupled plasma–mass spectrometry and X-ray photoelectron spectroscopy. Concentrations of gold in crude oil from the Linpan and Shanghe oilfields averaged 44.5 ppb, which is well above the minimum concentration required for hydrothermal gold mineralization. Gold has an affinity with carbon, oxygen, and sulfur, and its concentration in crude oil is positively correlated with total acid and sulfur contents. We speculate that gold may exist in crude oil as complexes with organic acids or thiols, with crude oil thus being a transport medium for gold.

Keywords: crude oil; gold; affinity; organic acid

Citation: Ni, Z.; Zhang, W.; Liu, J.; Shi, S.; Wang, X.; Su, Y. Occurrence of State of Gold in Crude Oil and Its Economic Significance. *Minerals* **2024**, *14*, 351. https://doi.org/10.3390/min14040351

Academic Editor: Jan Golonka

Received: 27 February 2024
Revised: 21 March 2024
Accepted: 27 March 2024
Published: 28 March 2024

1. Introduction

Petroleum and low-temperature hydrothermal Au deposits may form together and exhibit similar evolutionary trends in their formation, migration, and enrichment. Under certain circumstances, oil reservoirs and low-temperature hydrothermal Au deposits, such as Carlin-type deposits, exhibit a close coupling relationship on macro- to micro-scales [1–7] that may manifest in three ways, as follows. (i) Oil and gas reservoirs and Au deposits may have similar spatial distributions. For example, northeastern Nevada, USA, hosts abundant Carlin-type Au deposits and contains numerous oil and gas fields [8,9]; such deposits are also widely developed in the Youjiang Basin, China, together with numerous ancient and residual oil and gas reservoirs [10,11], and many low-temperature hydrothermal Hg, Sb, As, and Au deposits are distributed in the Tongren–Wanshan and Kaili–Majiang–Danzhai paleo-reservoir belts of eastern Guizhou Province, China [12]. (ii) Hydrothermal Au deposits provide records of oil and gas activities. Carlin-type deposits are a type of low-temperature hydrothermal deposit containing ore-bearing rocks that are rich in organic matter, with evidence of organic components participating in mineralization [13]. Liquid hydrocarbon inclusions were found in Carlin-type Au deposits in Nevada, and temperature analyses indicated that the capture temperature of oil-containing inclusions is below 150 °C and within the oil-generation window (80 °C–160 °C) [8,14]. In the Shuiyindong Au mine in the Youjiang Basin, the ore is rich in organic matter and bitumen, and CH_4, CO_2, and N_2 inclusions occur in ore-bearing quartz veins. The sources of these inclusions were linked to the decomposition of organic matter [15–19]. (iii) Many types of organic matter are enriched in metals. For example, lower Paleozoic dark mudstones of eastern Guizhou, China, are rich in organic C and elements such as Au, V, Ni, and Mo, with metal contents often being positively correlated with total organic C contents. These mudstones are considered "double-source beds" with both mineral and hydrocarbon sources [20,21]. There are also examples of metal-rich bitumen of varying maturity [22]. For example, the Boss deposit in Nevada has a high Au content of 1.3 mg g^{-1} in bitumen within dolomitic limestone. Scanning electron microscopy studies show that Au is present in the bitumen in the form of a Hg alloy [(Pd_{81}, Au_{13}, Pt_{01}) Hg] [23]. It is possible that metals can be transported in crude oil [24–27]. The contents of Au in the Shengli oilfield, China, are close to the industrial grade of 980 ppb [28], which is several orders of magnitude higher than the average abundance of Au in continental crust (1.8 ppb) [29]. However, there are few examples of the industrial recovery of Au from crude oil worldwide, and there have been no reports of the recovery of Au, Ag, and other precious metals in China because of the difficulty in their recovery [28].

Although many low-temperature hydrothermal Au deposits and oil and gas reservoirs are closely coupled, the mechanisms of ore-formation and oil-accumulation coupling in the interaction between Au and hydrocarbons are poorly understood. Currently, many geologists focus on the role of hydrocarbon fluid in the metallogenic process, the ability to mobilize and transport metal elements, and the metallogenic mechanism of hydrocarbon fluid during basin evolution [3,4,27]. Noble metals such as gold are thought to play a catalytic role in the process of hydrocarbon generation to improve the efficiency of hydrocarbon generation [5]. Petroleum is thought to provide reduced sulfur directly or indirectly for sulfide precipitation during mineralization [1]. It remains unclear whether crude oil can directly participate in the formation of low-temperature hydrothermal Au deposits as a migration medium, but there have been relatively few studies of the contents and forms of Au in crude oil worldwide. This study focused on the Linnan Sag in the Bohai Bay Basin to determine the Au content of crude oil and its form(s) of occurrence. Our results provide insights into the migration of Au in petroleum and its recovery from crude oil.

2. Geological Setting

The Linnan Sag is located in the southwest of Huimin Depression, Bohai Bay Basin, and contains the Linpan oil field, Shanghe oil field, Jiangjiadian oil field, Linan oil field, and so on [30]. The Linnan Sag is controlled by the Linshang and Xiakou Faults on the northern and southern edges, respectively (Figure 1). Xiakou Fault is the most important fault in the Linnan Sag, it controls the distribution of the settlement depocenter and also plays as the channel for petroleum [30].

The Linnan Sag is the main source region of the Huimin Depression, and the distributions of oil fields are mainly demarcated by faults. The Cenozoic tectonic movements of this region include five stages: the initial rifting stage (66.0–46.0 Ma), the intense rifting stage (46.0–38.0 Ma), the shrinking rifting stage (38.0–24.6 Ma), the uplift stage (24.6–14.0 Ma), and the subsiding stage (14.0–0.0 Ma) [30]. There exists an unconformity between the Paleogene and Neogene sequences due to uplift during the Dongying movement (24.6–14.0 Ma). The Cenozoic strata in the Linnan Sag comprise the Paleogene Kongdian Formation (Ek), Shahejie Formation (Es) and Dongying Formation (Ed), the Neogene Guantao Formation (Ng) and Minghuazhen Formation (Nm), and the Quaternary Pingyuan Formation (Qp) [31]. The Ek is dominated by glutenite, sandstone, and siltstone; it formed in a fluvial–lacustrine environment. The Es formation is divided into four parts, which are labeled, respectively, as Es4, Es3, Es2, and Es1 from the bottom to the top, including sandstones and mudstones (Figure 2). The Ng formed in a braided stream environment, including conglomerates, such as sandstones and mudstones. The Nm and Qp are floodplain sediment dominated by mudstones and sandstones [31].

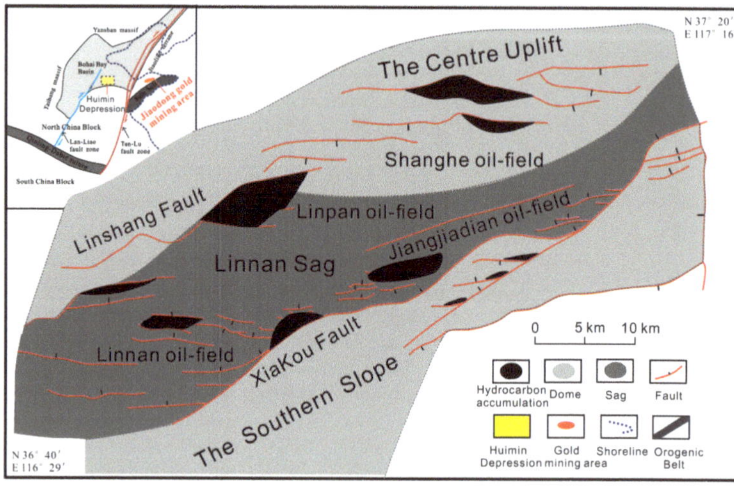

Figure 1. Simplified structural map of the Bohai Bay Basin, sketch map of Linnan Sag and the main oil fields (modified by reference [30]).

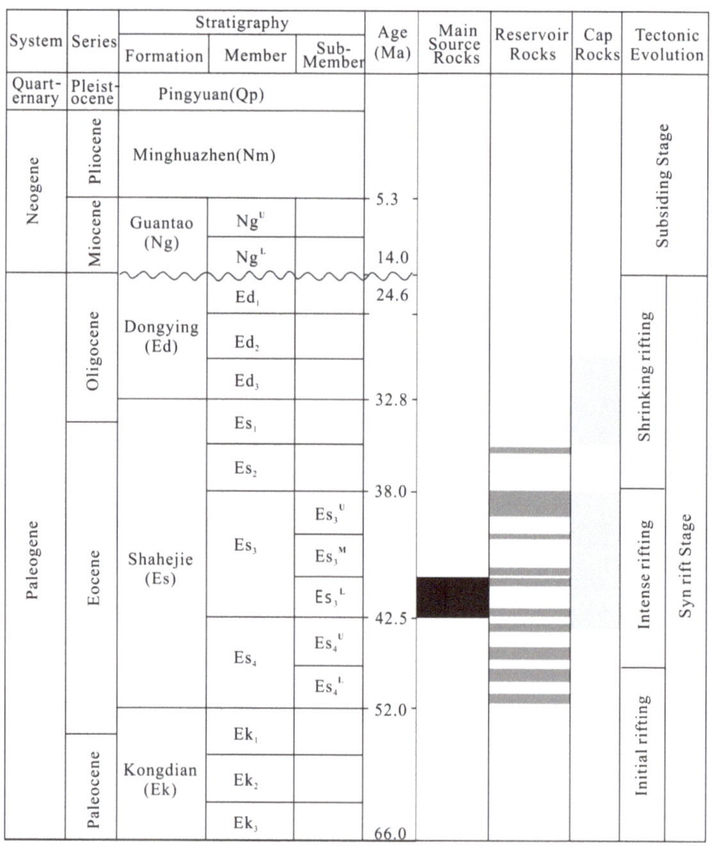

Figure 2. Schematic Cenozoic stratigraphy and tectonic evolution of the Linnan Sag.

The widely distributed lacustrine mudstones and shales in the lower part of Es3 are the dominant source rocks, with an average total organic carbon content (TOC) of 1.55% [32]. The conglomerate and sandstone in the Es4, Es3, and Es2 are main reservoirs, and Paleogene mudstones are regional cap rocks [32]. It is worth noting that China's famous Jiaodong gold mining area is distributed in the eastern part of the study area (Figure 1).

3. Materials and Methods

Crude oil samples were collected from formations in the Linpan and Shanghe oilfields of the Linnan Sag. The main production beds of crude oil were the Paleogene Shahejie and Neogene Guantao formations. The samples were black solids or viscous semi-solids at room temperature with densities of >0.86 g cm^3 (i.e., heavy oil).

3.1. Determination of Au Contents

The measurements of elemental analysis were performed at the Analytical Instrumentation Center of Peking University. Microwave digestion and inductively coupled plasma–mass spectrometry (ICP–MS) were employed to determine the Au contents of crude oil. An UltraWAVE ECR (Milestone, Milan, Italy) microwave digestion system was used: 50 mg of crude oil was weighed into a dry digestion tank, 2% HNO_3 was added (1.5 mL), and the sample was sealed and heated to 260 °C for 20 min for microwave digestion. The sample was then cooled to below 50 °C and transferred to a 30 mL volumetric flask for analysis.

A NexION 350X (PerkinElmer, Waltham, MA, USA) ICP–MS system was used for elemental analysis, with a three-cone interface (sampling, interceptor, over-interceptor cones), a triple four-pole ion deflector, and a universal cell technology with a wide linear range. Working parameters were set by automatic optimization, with a transmission power of 1250 W, carrier-gas flow rate of 0.82 L min^{-1}, peristaltic pump speed of 20.0 rpm, scanning repeated 20 times, and triplicate sample analyses. The standard-addition method was used with internal standards containing Ni, V, and other elements.

3.2. Au-Tube Thermal Simulation

The Au-tube thermal simulation equipment was developed by the State Key Laboratory of Oil and Gas Resources and Exploration, China University of Petroleum (CUP), Beijing, China. Au-tube thermal simulation treatments were undertaken at 420 °C and 40 MPa with a reaction time of 72 h. The tube was 60 mm long with an inner diameter of 5.55 mm and a wall thickness of 0.25 mm. A mudstone sample (15 mg of low-maturity source rock from the Bohai Bay Basin) and 160 mg deionized water were placed in the tube, which was sealed by welding in an Ar gas flow. The tube was opened after cooling.

3.3. XPS Analysis

Residual oil on Au-tube surfaces was analyzed by XPS (ThermoFisher, Waltham, MA, USA) at the CUP using Al Kα radiation, a spectral range of 0–1100 eV, 20 eV resolution, 0.1 eV step size, and a spot diameter of 300 μm. The instrument etched the Au tube inner surfaces with X-rays for periods of 0, 30, and 60 s, and changes in carbon (C), oxygen (O), nitrogen (N), Sulfur (S), and gold (Au) concentrations with etching depth were recorded. ORIGIN software (OriginPro 2021, 64-bit, 9.8.0.200) was used for graph plotting.

3.4. Determination of Total Acid Values of Crude Oil

The total acid values (TAN) of crude oil samples were determined by acid-base titration (GB/T 264, PRC National Standard) at the CUP [33]. Crude oil samples (2 g) were dissolved in 25 mL of neutral ethanol in a conical flask, and anhydrous ether (25 mL) was added. A blank sample containing 25 mL ethanol and 25 mL ether was also prepared. Phenolphthalide solution (1 mL) was added to each of the mixtures, titration with 0.1 mol L^{-1} KOH was undertaken until the pink endpoint, and the acid value was calculated by TAN = (V1 − V2) × C × 56.1, where V1 is the volume of KOH used in sample titration, V2 is the volume of the KOH solution used in the blank titration, and C is the concentration of the KOH solution.

3.5. The Gas Chromatography–Mass Spectrometry (GC-MS) Analysis

Saturated hydrocarbons of all the crude oils were acquired by standard column liquid chromatography. The GC-MS was performed at the CUP using an Agilent 6890 GC instrument with an Agilent Model 5975i Mass Selective Detector. HP-5 MS fused silica capillary column (30 m × 0.25 mm inner diameter, 0.25 mm thick) was used. The carrier gas was He at constant flow (1 mL/min). The initial GC oven temperature was at 50 °C (held for 1 min) and increased to 100 °C at a rate of 20 °C/min, then ramped to 310 °C at 3 °C/min (held for 10 min).

4. Results

4.1. Crude Oil Au, S, and Total Acid Contents

Twenty-five crude oil samples from the Shahejie and Guantao formations were selected for Au and S analysis; the results are shown in Table 1. Au contents ranged from 12.0 to 107.5 ppb, with an average of 44.5 ppb. S contents ranged from 0.10 to 0.56 wt.% (average 0.32 wt.%). The TAN values ranged from 0.64 to 1.75 mg g^{-1} (average 1.17 mg g^{-1}).

Table 1. Contents of Au and S and total acid values of crude oil samples.

Oilfield	Well	Fm.	Depth (m)	Density (g/cm^3)	Viscosity (mpa.s)	S (%)	TAN (mg/g)	Au (ppb)
Linpan	L2-x201	G2	1414.4–1423.6	0.945	377.75	0.51	1.63	13.4
Linpan	L23-11	G3	1642.2–1645.0	0.923	90.90	0.37	0.95	105.3
Linpan	L55	S3	2626.6–2992.8	0.866	11.07	0.12	0.92	90.7
Linpan	L7-x28	S2	1799.2–1823.6	0.909	27.46	0.10	0.64	12.0
Linpan	L75-17	S3	1859.2–1909.0	0.918	138.00	0.35	1.11	24.4
Linpan	L75-x30	S4	2474.0–2512.6	0.895	18.00	0.35	1.09	16.0
Linpan	L76-3	S4	2069.0–2094.0	0.893	41.00	0.10	0.82	32.4
Linpan	P14-6	S4	1588.6–1594.4	0.986	5951.00	0.50	1.39	36.6
Linpan	P18-x3	S3	1597.1–1630.3	0.935	119.00	0.42	1.21	38.1
Linpan	P20-6	S1	1556.0–1599.4	0.936	352.30	0.25	1.36	63.1
Linpan	P2-504	S3	1576.0–1606.4	0.965	921.00	0.56	1.21	23.2
Linpan	P40-12	G	1337.6–1354.0	0.960	384.00	0.41	1.49	47.9
Linpan	P40-17	G	1345.4–1350.4	0.951	324.00	0.38	1.33	31.4
Shanghe	S23-70	S2	1710.0–1734.2	0.897	48.94	0.25	0.90	57.2
Shanghe	S25-11	S4	2044.2–2126.8	0.899	48.84	0.14	1.28	18.9
Shanghe	S25-40	S4	1969.6–2267.8	0.897	58.23	0.44	1.21	107.5
Shanghe	S42-2	S2	1655.2–1675.4	0.932	140.30	0.44	0.70	27.3
Shanghe	S70-x5	S2	2348.5–2374.0	0.863	11.26	0.25	1.22	17.5
Shanghe	S73-13	S2	2275.8–2334.2	0.892	40.10	0.37	1.75	34.3
Shanghe	S8-29	S2	1854.2–1859.4	0.887	21.49	0.36	1.52	29.0
Shanghe	S8-300	S2	1776.0–1833.0	0.900	27.90	0.24	0.74	69.0
Shanghe	S8-33	S1	1653.0–1675.0	0.946	322.10	0.50	1.14	83.9
Shanghe	S84-4	S1	1899.0–1903.6	0.872	24.27	0.25	1.31	-
Shanghe	S847	S1	1884.6–1890.4	0.883	29.89	0.17	1.18	41.1
Shanghe	S8-52	S2	1846.0–1861.0	0.897	37.84	0.11	1.15	26.2

4.2. XPS Analyses

After the Au-tube thermal simulation, the tube was cut open to reveal the inner wall covered by generated crude oil. The tube with the most oil was analyzed by XPS, and spectra of Au, C, N, O, and S were recorded for etching times of 0, 30, and 60 s (Figure 3). The double peaks of Au ($4f_{5/2}$ and $4f_{7/2}$) were clear, with corresponding binding energies of 82 and 86 eV, respectively. Carbon (1s), N (1s), O (1s), and S ($4p_{3/2}$) had respective XPS peaks at 283, 398, 530, and 162 eV. Origin software (OriginPro 2021, 64-bit, 9.8.0.200) was used to process the data, and the results indicate that peak areas of Au, C, and O increased with etching time, whereas that of N decreased (Figure 3, Table 2).

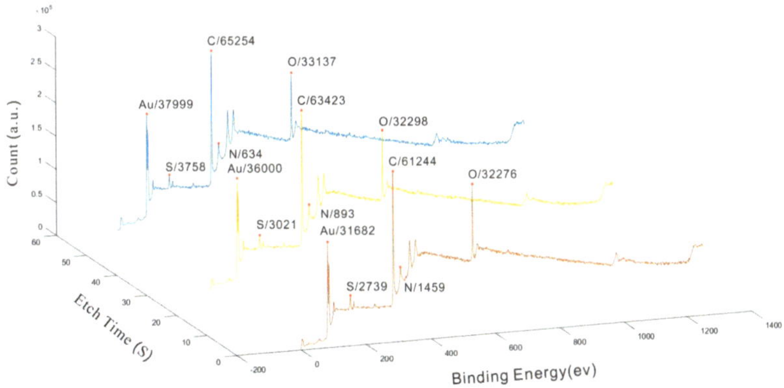

Figure 3. XPS peak areas of Au, C, N, O, and S at etching times of 0, 30, and 60 s.

Table 2. Elemental XPS spectral characteristics.

Elements	Spectral Line	Etch Time (s)	Peak Area	Peak Center (ev)	Elements	Spectral Line	Etch Time (s)	Peak Area	Peak Center (ev)
Au	4f5/2	0	31,682	82	N	1s	0	1459	398
Au	4f5/2	30	36,000	82	N	1s	30	893	398
Au	4f5/2	60	38,000	82	N	1s	60	634	398
Au	4f7/2	0	26,368	86	O	1s	0	32,276	530
Au	4f7/2	30	30,378	86	O	1s	30	32,299	530
Au	4f7/2	60	32,330	86	O	1s	60	33,137	530
C	1s	0	61,244	283	S	4p3/2	0	2739	162
C	2s	30	63,423	283	S	4p3/2	30	3021	162
C	3s	60	65,254	283	S	4p3/2	60	3758	162

5. Discussion

The average Au content of raw oil samples from the Linpan and Shanghe oilfields was 44.5 ppb in our study. However, previous work shows that the maximal Au contents in crude oil samples from the Jiyang Depression, Bohai Bay Basin, can reach 15.74 ppm [34], which is much greater than the industrial Au grade, is typically 0.3–0.5 g tonne^{-1} (300–500 ppb) [7]. Although the average Au content in our study oil samples is below industrial grade, it may have economic value if an appropriate recovery method is used. The maximal Au concentrations in our oil reached 107.5 ppb at 150 °C, which is considerably higher than the solubility of Au in hydrothermal fluids of the type that have been invoked as possible ore fluids (20 ppb) [20]. The Linpan and Shanghe oilfields are in the Linpan oil-production block, which has an oil-bearing area of 174.6 km^2 with geological reserves of 283 million tons (Mt) and recoverable reserves of 84 Mt. Based on the average crude oil Au content, the total recoverable Au would be $84 \times 10^6 \times 44.5 \times 10^{-9} = 3.74$ tons, which is equivalent to a small Au mine. Soft/hard acid–base theory [35] implies that Au, as a soft acid, would tend to bind to soft bases such as organic acids and reduced S [36–38]. Previous studies show that there is a strong correlation between the contents of N, O, S, and other heteroatoms and metal elements in crude oil [39,40]. XPS peak areas represent relative contents, and our results (Figure 3) indicate that Au and C peak areas are positively correlated with etching time (i.e., depth), implying a close relationship between Au and organic matter in crude oil. The relative content of O also increased with etching depth, indicating a further correlation between Au and oxygen contents. Trends in Au, carbon, and oxygen contents were similar, indicating that Au has a strong affinity for C- and O-containing compounds in crude oil. Most O exists as carboxylic acids in crude oil [36]; therefore, Au contents may be closely related to carboxylic acid contents. The relative abundance of S increases with etching depth, and its binding energy corresponds to the XPS peak of Au–thiolate complexes (162.4 eV; Figure 4) [41,42], implying that at least some Au in crude oil may exist in the form of Au thiolate [43,44]. The relative abundance of N decreased with increasing etching depth (Figure 3), indicating no close relationship between Au and nitrogen compounds.

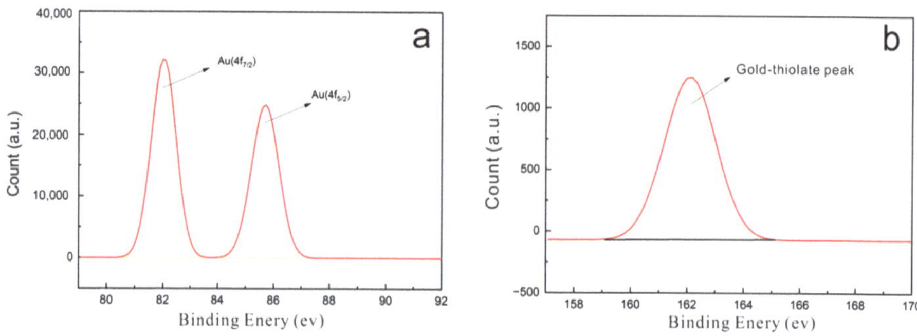

Figure 4. XPS spectra of Au and thiolate (**a**) the double peaks of Au ($4f_{5/2}$ and $4f_{7/2}$); (**b**) the peak of Au–thiolate complexes.

To further investigate the occurrence of Au in crude oil, a correlation analysis of Au contents, TAN values, and S contents of 25 crude oil samples was undertaken (Figure 5). When the TAN value was below 1.1, there was a positive correlation between Au content and TAN value; with values above 1.1, there was no obvious correlation. When the oil S content was below 0.35 wt.%, Au contents were positively correlated with S contents but not at higher S contents. The most important determinant of the TAN value of crude oil is the content of organic acids such as carboxylic acid and naphthenic acid [44–46]. Significant quantities of these acids may be released from petroleum source rocks by diffusion during early diagenesis [47]. The investigation shows levels of generation for acetic acid equivalent to 0.2%–1.2% of the TOC of the source rock and total acids at a level of 1%–2% of the TOC [48]. Organic acids are also generally formed by biodegradation as a by-product of bacterial metabolism [49,50]. Similarly, contents of N, O, S, and other heteroatoms may be released from the source rock to oil during the hydrocarbon generation and expulsion process, and the contents of these heteroatoms also increase as crude oil undergoes biodegradation [40,51,52]. Therefore, we speculate that crude oil samples with lower TAN values (<1.1) and S contents (<0.35 wt.%) appear to have a positive correlation with Au contents, which may represent the acid-releasing process from petroleum source rocks by diffusion during early diagenesis. In comparison, crude oil samples with higher TAN values (>1.1) and S contents (>0.35 wt.%) may have been affected by biodegradation, resulting in an insignificant correlation between Au contents and TAN values or S contents. The TAN values and S contents would increase through biodegradation, whereas Au contents would not. This may be due to biodegraded crude oil having a stronger ability to complex Au through its high TAN value and S content, but it would fail to bind more Au owing to its low fluidity. To further verify our speculation about the effect of biodegradation, we performed GC-MS analysis for two typical wells (Well L23-11 and Well P40-17). The total ion current (TIC) diagrams of the saturated hydrocarbons of crude oils from two wells show obviously different degrees of biodegradation. The TIC diagrams of saturated hydrocarbon show that the crude oil from Well L23-11 contains a complete distribution of n-alkane series (Figure 6), which suggests that the crude oil from Well L23-11 did not biodegrade or slightly biodegrade. By contrast, the TIC diagram of crude oil from Well P40-17 shows an intact n-alkane series and an evident 25-norhopanes, suggesting that the crude oil is a severely biodegraded oil (Figure 6). The results are totally consistent with our speculation based on Figure 5.

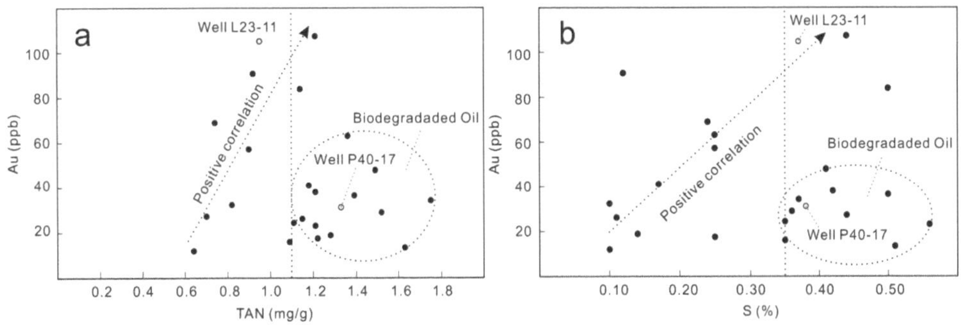

Figure 5. Relationships between crude oil Au contents and (**a**) total acid (TAN) values and (**b**) S contents. The arrows indicate the positive correlation between Au and TAN, S; the ellipse regions show that the positive correlations between Au and TAN, S are not obvious; the solid circles denote crude oil samples; the hollow circles denote crude oil samples with GC-MS data.

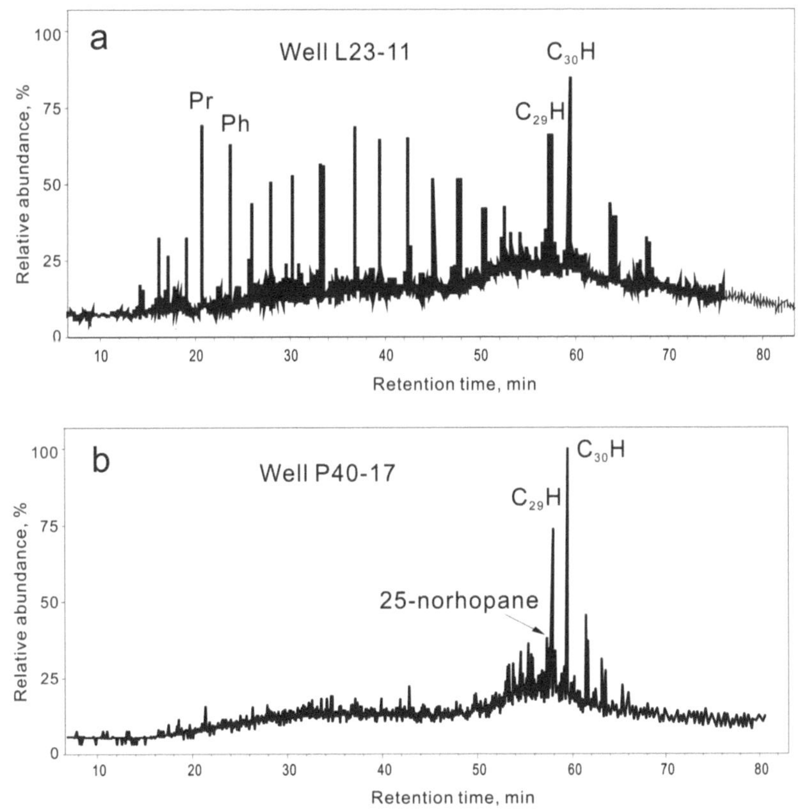

Figure 6. The total ion current diagrams of the saturated hydrocarbons of crude oils. (**a**) Well L23-11; (**b**) Well P40-17.

6. Conclusions

The analysis of gold contents of crude oil samples indicates that the gold in oil has economic value if it can be recycled effectively. The possible occurrence states of gold in crude oil are organic acid salts and thiolates. The maximal gold concentrations in our crude oil reached 10.57 ppb, which is considerably higher than the solubility of gold in potential hydrothermal ore-forming fluids. The oil with higher contents of organic acid and sulfur and a lower degree of biodegradation may function as an effective ore-forming fluid for Au deposits.

Author Contributions: Formal data analysis, Z.N., W.Z., J.L., S.S., X.W. and Y.S.; project administration, Z.N.; writing—original draft, Z.N. All authors have read and agreed to the published version of the manuscript.

Funding: This research was funded by the National Natural Science Foundation of China (Grant No. 42273069).

Data Availability Statement: The data presented in this study are available upon request from the corresponding author. The data are not publicly available due to privacy.

Acknowledgments: The help from Wu Jia, Fang Peng and Liang Yun of CUP and Liu Jiahui of Peking University in experiments were acknowledged.

Conflicts of Interest: The authors declare no conflicts of interest.

References

1. Gu, X.X.; Zhang, Y.M.; Li, B.H.; Xue, C.J.; Dong, S.Y.; Fu, S.H.; Cheng, W.B.; Liu, L.; Wu, C.Y. The coupling relationship between metallization and hydrocarbon accumulation in sedimentary basins. *Earth Sci. Front.* **2010**, *17*, 83–105.
2. Clark, J.R.; Williams-Jones, A.E. Analogues of epithermal gold–silver deposition in geothermal well scales. *Nature* **1990**, *346*, 644–645. [CrossRef]
3. Williams-Jones, A.E.; Migdisov, A.A. An experimental study of the solubility of gold in crude oil: Implications for ore genesis. *Geochim. Cosmochim. Acta* **2006**, *70*, A703. [CrossRef]
4. Emsbo, P.; Williams-Jones, A.; Koenig, A.E.; Wilson, S.A. Petroleum as an agent of metal transport: Metallogenic and exploration implications. In Proceedings of the Tenth Biennial SGA Meeting, Townsville, Australia, 17–20 August 2009; pp. 99–101.
5. Hu, S.-Y.; Plet, C. Introduction to a special issue on organic- and microbe-metal interactions in mineral systems. *Ore Geol. Rev.* **2022**, *146*, 104938. [CrossRef]
6. Gu, X.X.; Zhang, Y.M.; Li, B.H.; Dong, S.Y.; Xue, C.J.; Fu, S.H. Hydrocarbon- and ore-bearing basinal fluids: A possible link between gold mineralization and hydrocarbon accumulation in the Youjiang basin, South China. *Miner. Depos.* **2012**, *47*, 663–682. [CrossRef]
7. Large, R.R.; Bull, S.W.; Maslennikov, V.V. A Carbonaceous Sedimentary Source-Rock Model for Carlin-Type and Orogenic Gold Deposits. *Econ. Geol.* **2011**, *106*, 331–358. [CrossRef]
8. Hulen, J.B.; Collister, J.W.; Stout, B.; Curtiss, D.K.; Dahdah, N.F. The exhumed "Carlin-type" fossil oil reservoir at Yankee Basin. *JOM* **1998**, *50*, 30–34. [CrossRef]
9. Collister, J.W.; Hulen, J.B. The oil-bearing, carlin-type gold deposits of Yankee Basin, Alligator Ridge District, Nevada. *Econ. Geol.* **1999**, *94*, 1029–1049.
10. Wang, G.Z.; Hu, R.Z.; Su, W.C.; Zhu, L.M. Fluid and mineralization of Youjiang Basin in the Yunnan-Guizhou-Guangxi area, China. *Sci. China (Ser. D Earth Sci.)* **2002**, *46*, 99–109.
11. Hu, Y.; Liu, W.; Wang, J.; Zhang, G.; Zhou, Z.; Han, R. Basin-scale structure control of Carlin-style gold deposits in central Southwestern Guizhou, China: Insights from seismic reflection profiles and gravity data. *Ore Geol. Rev.* **2017**, *91*, 444–462. [CrossRef]
12. Hu, Y.Z.; Han, R.S.; Mao, X.X. Relationship between metal mineralization and accumulation oil and gas in the eastern Guizhou. *Geol. Prospect.* **2007**, *43*, 51–56.
13. Arehart, G.B. Characteristics and origin of sediment-hosted disseminated gold deposits: A review. *Ore Geol. Rev.* **1996**, *11*, 383–403. [CrossRef]
14. Peters, K.E.; Walters, C.C.; Moldowan, J.M. *The Biomarker Guide: Volume 2, Biomarkers and Isotopes in Petroleum Systems and Earth History*; Cambridge University Press: Cambridge, UK, 2007.
15. Su, W.C. Geochemistry of the Ore-Forming Fluids of the Carlin-Type Gold Deposits in the Southwestern Margin of theYangtze Block. Doctoral Dissertation, Institute of Geochemistry. Chinese Academy of Scicnces, Guiyang, China, 2002; pp. 1–120.
16. Xia, Y.; Chen, Q.N.; Zhou, W.C. The simulating experiments of gold mineralization in low temperature and open system. *J. Nanjing Univ. (Nat. Sci.)* **1996**, *32*, 634–639.
17. Liu, J.Z. Ore characteristics and gold occurrence of the Shuiyindong gold deposit, guizhou. *Guizhou Geol.* **2003**, *20*, 30–34.

18. Liu, J.Z.; Deng, Y.M.; Liu, C.Q.; Zhang, X.C.; Xia, Y. Metallogenic conditions and model of the superlarge Shuiyindong stratabound gold deposit in Zhenfeng County, Guizhou Province. *Geol. China* **2006**, *33*, 169–177.
19. Liu, J.Z.; Xia, Y.; Zhang, X.C.; Deng, Y.M.; Su, W.C.; Tao, Y. Model of strata Karlin-type gold deposit: The Shuiyindong super-scale gold deposit. *Gold Sci. Technol.* **2008**, *16*, 1–15.
20. Tu, G.Z. *Gechemisty of Strata-Bound Deposits in China (Volume III)*; Science Press: Beijing, China, 1988.
21. Fu, J.M.; Peng, P.A.; Lin, Q.; Liu, D.H.; Jia, R.F.; Shi, J.X.; Lu, J.L. Several problems in the study of organic gechemisty of strata-bound Deposits. *Prog. Earth Sci.* **1990**, *5*, 43.
22. Parnell, J. Metal enrichments in solid bitumens: A review. *Miner. Depos.* **1988**, *23*, 191–199. [CrossRef]
23. Jedwab, J.; Badaut, D.; Beaunier, P. Discovery of a palladium-platinum-gold-mercury bitumen in the Boss Mine, Clark County, Nevada. *Econ. Geol.* **1999**, *94*, 1163–1172. [CrossRef]
24. Baker, W.E. The role of humic acid in the transport of gold. *Geochim. Et Cosmochim. Acta* **1978**, *42*, 645–649. [CrossRef]
25. Liu, J.Z.; Fu, J.M.; Lu, J.L. Experimental study on the role of organic matter in the mineralization of Sedimentary-reformed gold deposits. *Sci. China (Ser. B)* **1993**, *9*, 993–1000.
26. Zhang, J.R.; Lu, J.J.; Yang, F.; Wu, J.W.; Zhu, F.H. An experimental study on gold solubility in amino acid solution and its geological signifcance. *Geochimica* **1996**, *25*, 190–195.
27. Crede, L.-S.; Liu, W.; Evans, K.; Rempel, K.; Testemale, D.; Brugger, J. Crude oils as ore fluids: An experimental in-situ XAS study of gold partitioning between brine and organic fluid from 25 to 250 °C. *Geochim. Et Cosmochim. Acta* **2019**, *244*, 352–365. [CrossRef]
28. Xu, Z.H.; Zhan, Y.L.; Zhu, J.H.; Li, X. The research progress in the utilization of metal resources in crude oils. *Multipurp. Util. Miner. Resour.* **1998**, *3*, 22–26.
29. Taylor, S.R.; Mclennan, S.M. *The Continental Crust: Its Composition and Evolution*; Blackwell Scientific Publications: Oxford, UK, 1985.
30. Wang, D.; Wu, Z.P.; Yang, L.L.; Liu, H.; Zhang, Y. Growth and linkage of the Xiakou fault in the Linnan Sag, Jiyang Depression, Eastern China: Formation mechanism and sedimentation response. *Mar. Pet. Geol.* **2020**, *116*, 104319. [CrossRef]
31. Li, C.; Zhang, L.K.; Luo, X.R.; Wang, B.; Lei, Y.H.; Cheng, M.; Luo, H.M.; Wang, C.J.; Yu, L. Modeling of overpressure generation-evolution of the Paleogene source rock and implications for the Linnan Sag, Eastern China. *Front. Earth Sci.* **2022**, *10*, 829322. [CrossRef]
32. Guo, X.L.; Xiong, M.; Zhou, Q.; Tian, H.; Xiao, X.M. Petroleum generation and expulsion kinetics: A sase study of the Shahejie Formation source rocks from Linnan Sag of Huimin Depression. *ACTA Sedmentol. SINICA* **2009**, *27*, 723–731.
33. *GB/T 264-1983*; Petroleum Products—Determination of Acid Number. National Bureau of Standards: Beijing, China, 1983.
34. Sun, Y.Z.; Jiao, W.W.; Zhang, S.C.; Qin, S.J. Gold enrichment mechanism in crude oils and source rocks in jiyang depression. *Energy Explor. Exploit.* **2009**, *27*, 133–142. [CrossRef]
35. Pearson, R.G. Hard and Soft Acids and Bases. *J. Am. Chem. Soc.* **1963**, *85*, 3533–3539. [CrossRef]
36. Seward, T.M. Thio complexes of gold and the transport of gold in hydrothermal ore solutions. *Geochim. Cosmochim. Acta* **1973**, *37*, 379–399. [CrossRef]
37. Giordano, T.H. Evaluation of Organic Ligands and Metal-Organic Complexing in Mississippi Valley-Type Ore Solutions. *Econ. Geol.* **1985**, *80*, 96–106. [CrossRef]
38. Vlassopoulos, D.; Wood, S.A.; Mucci, A. Gold speciation in natural waters: II. The importance of organic complexing- Experiments with some simple model ligands. *Geochim. Cosmochim. Acta* **1990**, *54*, 1575–1586. [CrossRef]
39. Sanz-Robinson, J.; Williams-Jones, A.E. The solubility of Nickel (Ni) in crude oil at 150, 200 and 250°C and its application to ore genesis. *Chem. Geol.* **2020**, *533*, 119443. [CrossRef]
40. Hughey, C.A.; Rodgers, R.P.; Marshall, A.G.; Walters, C.C.; Qian, K.; Mankiewicz, P. Acidic and neutral polar NSO compounds in Smackover oils of different thermal maturity revealed by electrospray high field Fourier transform ion cyclotron resonance mass spectrometry. *Org. Geochem.* **2004**, *35*, 863–880. [CrossRef]
41. Castner, D.G.; Hinds, K.; Grainger, D.W. X-ray Photoelectron Spectroscopy Sulfur 2p Study of Organic Thiol and Disulfide Binding Interactions with Gold Surfaces. *Langmuir* **1996**, *12*, 5083–5086. [CrossRef]
42. Sanz-Robinson, J.; Sugiyama, I.; Williams-Jones, A.E. The solubility of palladium (Pd) in crude oil at 150, 200 and 250 °C and its application to ore genesis. *Chem. Geol.* **2020**, *531*, 119320. [CrossRef]
43. Lewan, M.D. Factors controlling the proportionality of vanadium to nickel in crude oils. *Geochim. Cosmochim. Acta* **1984**, *48*, 2231–2238. [CrossRef]
44. Giordano, T.H. Metal Transport in Ore Fluids by Organic Ligand Complexation. In *Organic Acids in Geological Processes*; Pittman, E.D., Lewan, M.D., Eds.; Springer: Berlin/Heidelberg, Germany, 1994; pp. 319–354.
45. Speight, J.G. *Handbook of Petroleum Analysis*; John Wiley & Sons Inc: Hoboken, NJ, USA, 2001.
46. Park, L.K.E.; Liu, J.; Yiacoumi, S.; Borole, A.P.; Tsouris, C. Contribution of acidic components to the total acid number (TAN) of bio-oil. *Fuel* **2017**, *200*, 171–181. [CrossRef]
47. Lewan, M.D.; Fisher, J.B. Organic Acids from Petroleum Source Rocks. In *Organic Acids in Geological Processes*; Pittman, E.D., Lewan, M.D., Eds.; Springer: Berlin/Heidelberg, Germany, 1994.
48. Barth, T.; Bjørlykke, K. Organic acids from source rock maturation: Generation potentials, transport mechanisms and relevance for mineral diagenesis. *Appl. Geochem.* **1993**, *8*, 325–337. [CrossRef]

49. Kawamurat, K.; Tannenbaum, E.L.I.; Huizinga, B.J.; Kaplan, I.R. Long-chain carboxylic acids in pyrolysates of Green River kerogen. *Adv. Org. Geochem.* **1986**, *10*, 1059–1065. [CrossRef]
50. Shock, E.L. Organic acid metastability in sedimentary basins. *Geology* **1988**, *16*, 886–890. [CrossRef]
51. Head, I.M.; Jones, D.M.; Larter, S.R. Biological activity in the deep subsurface and the origin of heavy oil. *Nature* **2003**, *426*, 344–352. [CrossRef]
52. Wenger, L.M.; Davis, C.L.; Isaksen, G.H. Multiple Controls on Petroleum Biodegradation and Impact on Oil Quality. *SPE Reserv. Eval. Eng.* **2002**, *5*, 375–383. [CrossRef]

minerals

MDPI

Article

Present-Day Geothermal Regime and Thermal Evolution of the Fukang Sag in the Junggar Basin, Northwest China

Huajun Guo [1], Chenxing Li [2,3], Bo Peng [1], Xiang Shan [1], Jiabo Xu [2,3], Ze Zhang [2,3] and Jian Chang [2,3,*]

1 PetroChina Hangzhou Research Institute of Geology, Hangzhou 310023, China; guohj_hz@petrochina.com.cn (H.G.)
2 National Key Laboratory of Petroleum Resources and Engineering, China University of Petroleum, Beijing 102249, China; lichenxing2020@163.com (C.L.)
3 College of Geosciences, China University of Petroleum, Beijing 102249, China
* Correspondence: changjian@cup.edu.cn; Tel.: +86-0108-9733-332

Abstract: The Fukang Sag in the Junggar Basin is an important petroleum exploration and exploitation region. However, the geothermal regime and tectono-thermal evolution of the Fukang Sag, which control its hydrocarbon generation and conservation, are still controversial. This study involved a systematic analysis of the present-day geothermal gradient, heat flow, and thermal history of the Fukang Sag for better further exploration. According to the well log data and well-testing temperature data, we calculated that the geothermal gradient of the Fukang Sag ranges from 16.6 °C/km to 29.6 °C/km, with an average of 20.8 °C/km, and the heat flow ranges from 34.6 mWm^{-2} to 64.3 mWm^{-2}, with an average of 44.6 mWm^{-2}. Due to the basement relief, they decrease from northeast to southwest. The weight averages of the single-grain apatite (U-Th)/He ages of the core samples are 1.3–85.2 Ma, and their apatite fission track ages range from 50.9 Ma to 193.8 Ma. The thermal modeling results revealed that the Fukang Sag experienced late Permian, late Jurassic, and late Cretaceous cooling events (although the timing and magnitude of these events varied among the samples), which were related to the continuous compression of the Junggar Basin. In addition, basin modeling indicated that the heat flow of the Fukang Sag decreased from 80 mWm^{-2} in the Carboniferous to the current value of 44.6 mWm^{-2}. The Fukang Sag's edge exhibits prolific hydrocarbon generation in the Carboniferous–Permian source rocks, while the Jurassic source rocks within the sag also undergo abundant hydrocarbon generation. This study provides new insights into the present-day geothermal field and tectono-thermal evolutionary history of the Fukang Sag, which are significant in terms of regional tectonic evolution and oil and gas resource assessment.

Keywords: Fukang Sag; geothermal gradient; heat flow; thermochronometry; thermal history; tectono-thermal evolution; maturity evolution

check for updates

Citation: Guo, H.; Li, C.; Peng, B.; Shan, X.; Xu, J.; Zhang, Z.; Chang, J. Present-Day Geothermal Regime and Thermal Evolution of the Fukang Sag in the Junggar Basin, Northwest China. *Minerals* **2024**, *14*, 260. https://doi.org/10.3390/min14030260

Academic Editor: Luca Aldega

Received: 6 December 2023
Revised: 23 February 2024
Accepted: 27 February 2024
Published: 29 February 2024

1. Introduction

The Junggar Basin, which has an area of 38 × 10^4 km^2, is located in Northwest China (Figure 1a) and characterized by abundant oil and gas resources. The previously discovered oil and gas fields are mainly distributed in the uplift areas in the Junggar Basin [1]. In recent years, some new oil and gas fields, which are characterized by lithologic reservoirs [2], were found in the depression areas of the Junggar Basin [3], such as the Mahu Sag, Fukang Sag, Shawan Sag, etc. The Fukang Sag in the central Junggar Basin represents a good exploration prospect along with the discovery of the KT1 oil well (Figure 1c) [4].

The geothermal regime and tectono-thermal evolution in the sedimentary basins are the crucial factors that affect generation and accumulation [5–7]. They can be studied based on borehole temperature data, rock thermal property parameters, thermal indicators, and thermodynamic models [8–10]. The Junggar Basin was considered to be a 'cool' basin with a current mean geothermal gradient of ~21 °C/km [11,12]. However, due to insufficient borehole temperature data, the present-day geothermal characteristics of the Fukang Sag are still controversial. Previous studies discussed the tectono-thermal evolution of the secondary tectonic units of the Junggar Basin, including the Central Depression, Luliang Uplift, Southern Overthrust Belt, etc. [13–18], but there is little research on the tectono-thermal evolution of the Fukang Sag with the combination of apatite fission track (AFT), apatite (U-Th)/He (AHe), and vitrinite reflectance (Ro) data [9].

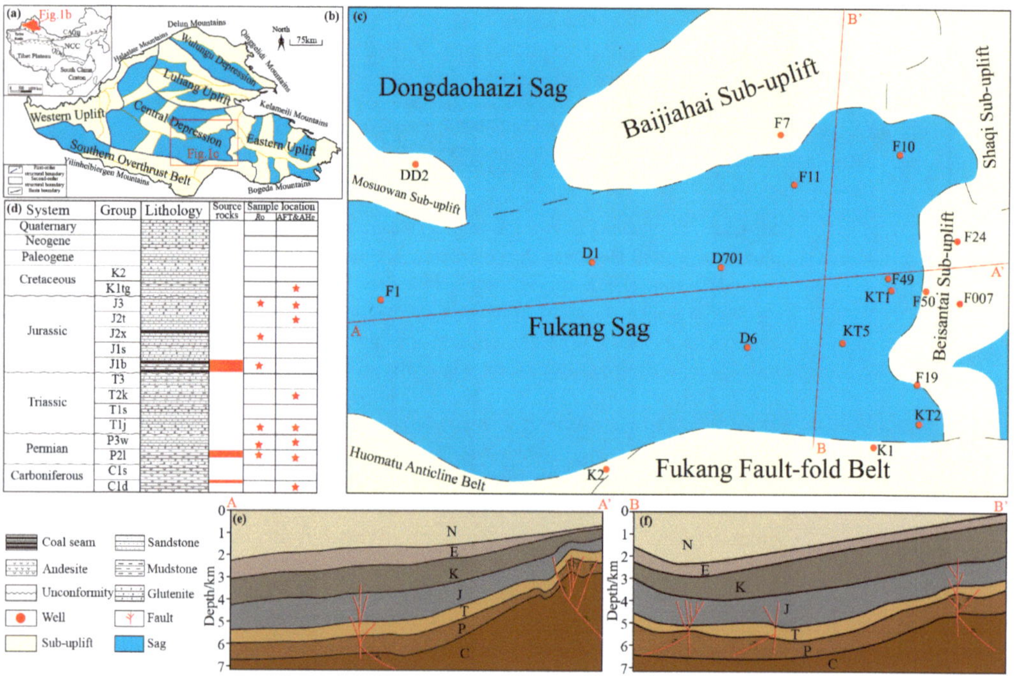

Figure 1. (**a**) Tectonic framework of China illustrating the location of the Junggar Basin [19]; (**b**) sketch of the tectonic unit divisions in the Junggar Basin [20]; (**c**) tectonic units of the Fukang Sag; (**d**) generalized stratigraphic column and the sample location (the red asterisk symbol) of the Fukang Sag [4]; (**e**,**f**) geologic cross sections of the Fukang Sag (the letters N, E, K, J, T, P, and C in the figure are the Neogene, Paleogene, Cretaceous, Jurassic, Triassic, Permian, and Carboniferous formations, respectively) [21], and the AA′ and BB′ represent the locations of geologic cross sections in the figure (**c**).

The purpose of this study is to analyze the geothermal regime, reconstruct the thermal history, and further investigate the maturity evolution of source rocks in the Fukang Sag. Firstly, we calculated the present-day geothermal gradients and heat flow using a heat conduction equation and further analyzed the distribution characteristics of the geo-thermal regime. Secondly, the tectono-thermal history of the Fukang Sag was reconstructed with the AFT, AHe, and Ro as thermal indicators. Finally, this study analyzed the maturation evolutionary history of the Carboniferous, Permian, and Jurassic source rocks in the Fukang Sag and its surrounding areas and discussed its effect on hydrocarbon generation and conservation. This study provides new insights into the Fukang Sag's present-day geothermal regime and tectono-thermal history and is of significance to understanding the tectonic evolution and the evaluation of resources in the Fukang Sag within the Junggar Basin.

2. Geological Setting

The Junggar Basin, located in Northwestern China, is a superimposed basin formed during the Carboniferous [20]. It is bounded by the Halaalate mountains to the west, the

Qinggelidi and Delun Mountains to the north, the Kelameili and Bogeda mountains to the east, and the Yilinheibiergen Mountains to the south. The Junggar Basin can be divided into six secondary tectonic units, including the Central Depression, Wulungu Uplift, Luliang Uplift, Eastern Uplift, Western Uplift, and Southern Overthrust Belt (Figure 1b). The Fukang Sag, as the studied area, is a part of the Central Depression bordered by the Southern Overthrust Belt to the south, the Baijiahai Sub-uplift to the north, the Beisantai Sub-uplift to the east, and the Mosuowan Sub-uplift to the west (Figure 1c) [1,3]. Due to multiple tectonic movements, the Fukang Sag developed several unconformities (Figure 1d). The thickness of sedimentary strata in the Fukang Sag is variable. It is thin from west (~7000 m thickness) to east (~3000 m thickness) (Figure 1e). The central Fukang Sag has a thickness of ~6500 m, while the northern Fukang Sag has a thickness of 5000 m (Figure 1f) [21–23].

Due to the variation in the sedimentary environment, the lithology in the Fukang Sag is unique for each stratum. The Carboniferous strata are predominantly composed of volcanic rocks such as andesite [22]. During the Permian, the Fukang Sag developed a lake–river delta and deposited sandstone, mudstone, and glutenite. The Triassic stratum in the Fukang Sag consists of mudstone and sandstone. During the Jurassic, the Fukang Sag developed a riverine and a lacustrine braided river delta, and deposited sandstone and mudstone with coal-bearing layers [24]. Since the Cretaceous, the Fukang Sag has predominantly deposited sandstone and glutenite. The Carboniferous Dishuiquan Formation (C_1d), Permian Lucaogou Formation (P_2l), and Jurassic Badaowan Formation (J_1b) are good source rocks [2] (Figure 1d), which provided massive amounts of hydrocarbon for the discovered oil and gas fields in the Fukang Sag and its surrounding areas [23].

3. Present-Day Geothermal Regime

The present geothermal regime represents the final state of thermal evolution in a sedimentary basin, so it is a necessary constraint for simulating tectono-thermal history [8,25]. The present-day geothermal regime of one sedimentary basin can be studied based on measured borehole temperature data [26,27]. In this study, we collected the well log data of 13 wells and well-testing temperature data of >50 wells. The well log data, which were obtained by measuring the temperature at intervals of 200 m from top to bottom for one borehole after the completion of drilling and recovery of formation temperatures, reflect the variations in the geothermal gradient with depth (Figure 2a). The well-testing temperature is equal to the burial temperature of the oil-producing formations (Figure 2b).

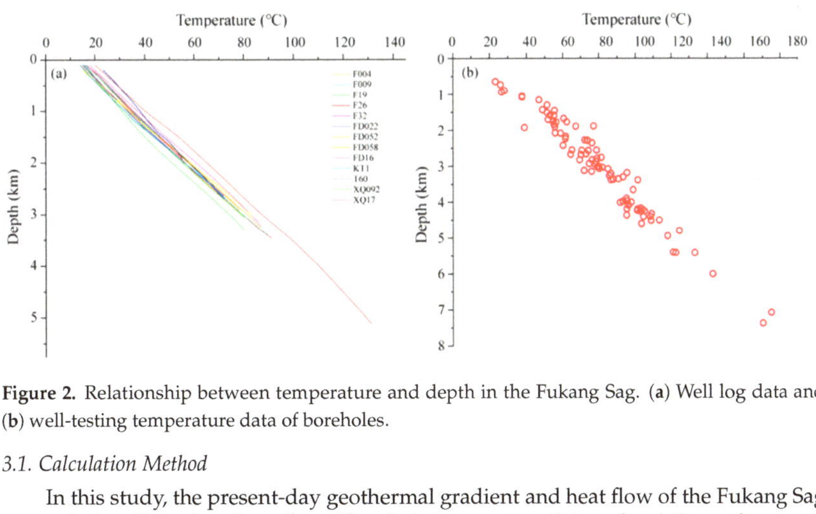

Figure 2. Relationship between temperature and depth in the Fukang Sag. (**a**) Well log data and (**b**) well-testing temperature data of boreholes.

3.1. Calculation Method

In this study, the present-day geothermal gradient and heat flow of the Fukang Sag were calculated based on the collected borehole temperature data and rock thermal property parameters. Meanwhile, their spatial distribution characteristics and relevant influencing factors were analyzed.

Due to the differential rock thermal conductivity, the deep geothermal gradient in sedimentary basins is usually smaller than the shallow geothermal gradient. Therefore, the variation in the temperature depths across different boreholes can make it difficult to correctly reveal the spatial distribution of the actual geothermal gradients for a basin. To resolve this problem, this study referred to the unified depth method provided by Chang et al. to study the geothermal gradient of the Fukang Sag [27]. First, we calculated the burial temperature at the depths of 4000 m for all the wells in the Fukang Sag with a one-dimensional steady-state heat conduction equation (Equation (1)):

$$T_i^b = T_i^t + \frac{(q_i^t \times Z_i)}{K_i} - \frac{\left(A_i \times Z_i^2\right)}{(2 \times K_i)} \tag{1}$$

where i is the structural layer number; T_i^b and T_i^t are the temperatures of the bottom and top surfaces in the structural layer i, respectively; Z_i is the thickness of structural layer (in km); A_i is the rock heat production rate (μWm^{-3}); K_i is the rock thermal conductivity (W/(m·K)) [11]; and q_i^t is the heat flow of the top of the structural layer i (mWm^{-2}).

Then, the geothermal gradient at unified depths of 0–4000 m could be calculated with Equation (2):

$$G_{4000} = \frac{Tz - T_0}{Z - Z_0} \tag{2}$$

where G_{4000} is the average geothermal gradient (°C/km), T_Z is the temperature at depths of 4000 m, T_0 is the constant temperature near the surface (14 °C), Z is the depth of the temperature measurement (in km), and Z_0 is the depth of the constant temperature zone (0.02 km).

The surface heat flow represents the energy flux through the Earth's surface. It is equal to the rock thermal conductivity times the geothermal gradient (Equation (3)):

$$q_s = -K_t \times \frac{dT}{dZ} \tag{3}$$

where q_s is the surface heat flow (mWm^{-2}), Kt is the harmonic mean thermal conductivity (W/(m·K)), and dT/dZ is the geothermal gradient (°C/km).

When the geothermal gradient and heat flow of each well in the Fukang Sag were obtained, we could analyze their spatial distribution characteristics with the Kriging method [28].

3.2. Present-Day Geothermal Gradient and Heat Flow

The geothermal gradients in the Fukang Sag and its surrounding areas at depths 0–4000 m range from 16.6 to 29.6 °C/km, with an average of 20.8 °C/km, indicating obvious variation among the different tectonic units (Figure 3). In the Fukang Sag, the geothermal gradient ranges from 17.2 to 27.6 °C/km, with an average geothermal gradient of 20.3 °C/km, showing that it gradually increases from southwest to northeast. The average geothermal gradients of the Baijiahai Sub-uplift and Beisantai Sub-uplift in the northeast of the Fukang Sag are 23.0 °C/km and 22.7 °C/km, respectively, and exhibit similar spatial distribution characteristics. Generally, the geothermal gradients in the uplift areas are larger than those in the depression areas. In addition, the geothermal gradients in the Fukang Fault-fold Belt just range from 16.6 °C/km to 21.3 °C/km, with an average value of 18.8 °C/km, which is very low and related to low thermal conductivity of the sedimentary cover [11,29–31].

The heat flow in the Fukang Sag ranges from 34.6 mWm^{-2} to 64.3 mWm^{-2}, with an average value of 44.6 mWm^{-2}, indicating an obvious decrease from northeast to southwest (Figure 4). The Shaqi Sub-uplift exhibits the highest heat flow, which ranges from

51.6 mWm^{-2} to 64.3 mWm^{-2}, with an average value of 54.1 mWm^{-2}. The Baijiahai and Beisantai Sub-uplifts also show relatively high heat flow, with average values of 47.8 mWm^{-2} and 47.9 mWm^{-2}, respectively. The heat flow in the Fukang Sag is lower than in the uplift areas and ranges from 38.27 mWm^{-2} to 55.2 mWm^{-2}, with an average of 44.13 mWm^{-2}. The Fukang Fault-fold Belt shows the lowest heat flow, with an average value of only 40.3 mWm^{-2}. Generally, the heat flow spatial distribution in the Fukang Sag and its surrounding areas is related to the basement relief (Figure 1e,f).

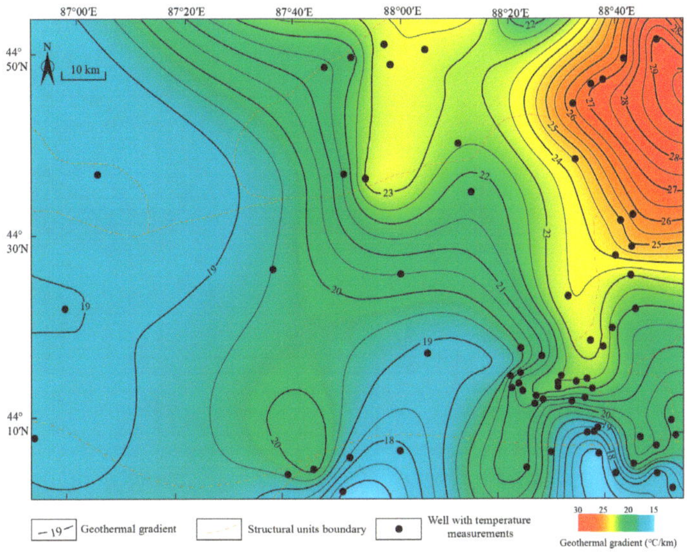

Figure 3. Distribution pattern of present-day geothermal gradient of the Fukang Sag with uniform depths of 0–4000 m.

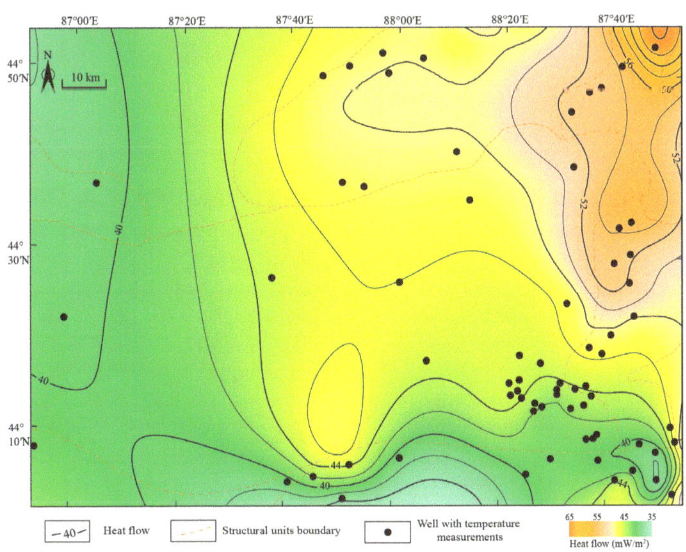

Figure 4. Distribution pattern of present-day heat flow of the Fukang Sag.

4. Thermal History

4.1. Sampling Strategy and Results

In this study, we collected 20 samples from the drilling wells F19, F49, F50, KT2, KT5, and K2 for studying the tectono-thermal history of the Fukang Sag (Tables 1 and 2; Figure 1c,d). The apatite grains for AFT and AHe dating were separated from the sandstone samples using conventional mineral separation techniques. The measurement procedures for the AFT and AHe ages were introduced in reference [32]. The vitrinite reflectance (Ro) for the mudstone samples was measured under an optical microscope equipped with a spectrophotometer [33].

Table 1. The apatite fission track data of the samples in the Fukang Sag.

Sample	Depth (m) (Fm.)	No. Grains	ρs (10^5/cm^2) (Ns [1])	U (ppm) [2]	P (χ^2) [3]	Central Age $\pm 2\sigma$ (Ma)	Mean Track Length $\pm 2\sigma$ (μm) (n [4])	Dpar [5] (μm)
F191	1865 (K1tg)	27	6.84 (763)	9.48	0.12	143.3 ± 13.4	13.2 ± 1.9 (42)	2.12 (1.25~3.07)
F192	2674 (T1j)	33	7.76 (1499)	7.71	0.2	193.8 ± 12.2	13.1 ± 1.5 (53)	2.24 (1.75~3.32)
F193	2981 (C)	6	3.52 (60)	10.46	0.09	95.7 ± 45.8	13.9 (1)	2.11 (1.36~2.52)
F491	5020 (T1j)	26	1.20 (163)	4.37	0.14	50.9 ± 9.4	—	2.11 (1.57~2.77)
F501	4288 (P3w)	11	2.50 (121)	3.20	0.31	156.0 ± 37.5	—	2.34 (1.63~2.89)
F502	4378 (P2l)	25	1.95 (190)	4.60	0.84	82.8 ± 12.0	12.2 ± 0.9 (8)	2.31 (1.73~2.86)
K21	2593 (J3)	18	8.80 (568)	10.06	0.2	139.2 ± 16.8	11.4 ± 1.6 (20)	2.30 (1.79~2.75)
K22	3250 (J2t)	3	2.21 (17)	3.38	0.28	129.4 ± 73.1	—	2.71 (2.36~2.96)
KT21	3676 (T2k)	26	6.89 (1023)	6.88	0.17	189.4 ± 14.4	11.9 ± 1.9 (24)	2.20 (1.68~2.74)
KT22	4292 (P3w)	12	5.02 (336)	6.65	0.35	161.6 ± 18.2	11.7 ± 0.8 (11)	2.58 (1.75~3.48)

[1] Ns: sum of spontaneous fission tracks for one sample. [2] U (ppm): uranium content value obtained from the LA-ICP-MS measurement. [3] P (χ^2): chi-square probability. [4] n: the number of confined fission tracks. [5] Dpar: the mean maximum diameter of etch figures parallel to the c-axis, which was used as a kinetic parameter.

Table 2. The vitrinite reflectance data of the wells in the Fukang Sag.

Well Number	Sample Depth (m)	Formation	Vitrinite Reflectance (%)
KT2	4602.32	P3w	0.74
	4867.11	P2l	0.83
KT5	6189.46	P2l	1.16
	6246.90	P2l	1.21
F50	4288.85	P3w	0.84
	4326.72	P3w	0.97
	4374.84	P2l	0.75
F49	5623.97	P2l	0.91
	5626.64	P2l	1.01
	5629.21	P2l	1.03

The samples from the wells F19, F49, F50, K2, and KT2 show the variant AFT and AHe ages (Tables 1 and 3, Figures 5 and 6), recording the differential thermal histories in the Fukang Sag. The measured central AFT ages range from 50.9 Ma to 193.8 Ma, younger than their corresponding stratigraphic ages. There is a negative correlation between the AFT ages and the depths, except for the samples F191 and F192 from the Well F19 (Table 1), which indicated that a higher temperature results in much more annealing. The AHe weighted mean ages range from 1.29 Ma to 85.2 Ma. Most of the single-grain AHe ages are younger than their corresponding stratigraphic ages (Table 3). There is a negative correlation between single-grain AHe ages and effective uranium concentration (Figure 6b) and no correlation between single-grain AHe ages and radius (Figure 6a), indicating that the dispersion of the single-grain AHe ages is related to radiation damage. Despite the intense radiation damage, some single-grain AHe ages are still older than the corresponding stratigraphic age (for the sample F191). The measured Ro data range from 0.74% to 1.21% (Table 2), indicating that the source rocks has reached the middle–late mature stage.

Table 3. The apatite (U-Th)/He data of the samples in the Fukang Sag.

Sample	Grain No.	Rs [2] (μm)	Mass (μg)	He (ncc)	U (ppm)	Th (ppm)	Th/U	eU [1] (ppm)	Raw Age (Ma)	FT [3]	Corrected Age ± 1σ (Ma)	Weight Mean Age ± 1σ (Ma) [4]
F191	1	92.8	19.31	0.2908	0.60	1.60	2.66	0.98	126.32	0.84	149.7 ± 4.04	85.2 ± 0.71
	2	50.3	3.16	0.0357	0.61	4.93	8.07	1.79	52.41	0.72	72.5 ± 0.77	
	3	59.8	5.83	0.2151	3.02	6.80	2.25	4.66	65.61	0.76	86.0 ± 11.31	
	4	43.3	3.01	0.2203	1.59	14.74	9.29	5.13	118.52	0.68	175.3 ± 2.29	
F192	1	71.2	10.94	0.4150	1.31	6.33	4.84	2.83	111.14	0.80	139.1 ± 3.09	2.52 ± 0.19
	2	66.6	7.80	0.0120	8.71	8.60	0.99	10.78	1.19	0.82	1.4 ± 0.19	
	3	71.7	11.01	0.0500	0.73	3.49	4.75	1.57	24.02	0.83	29.0 ± 1.38	
F193	1	61.1	4.84	0.0226	0.55	4.17	7.62	1.55	25.11	0.77	32.8 ± 0.54	38.26 ± 0.37
	2	53.4	3.29	0.0107	0.22	2.75	12.41	0.88	30.91	0.74	42.0 ± 0.52	
	3	120.3	42.19	0.4664	0.62	1.88	3.02	1.08	85.02	0.88	96.8 ± 7.36	
	4	71.0	8.46	0.3446	1.09	4.00	3.68	2.05	163.9	0.80	205.9 ± 6.46	
F501	1	45.6	2.37	0.0160	2.27	13.01	5.74	5.39	10.46	0.70	15.0 ± 0.50	6.29 ± 0.05
	2	63.0	6.38	0.0229	3.01	11.14	3.70	5.68	5.25	0.77	6.8 ± 0.28	
	3	67.8	7.47	0.0051	0.44	3.82	8.71	1.35	4.18	0.79	5.3 ± 0.09	
	4	51.8	3.96	0.0058	0.51	8.29	16.37	2.50	4.89	0.73	6.7 ± 0.07	
F491	1	53.0	3.96	0.0039	0.72	4.58	6.40	1.82	4.56	0.74	6.2 ± 0.05	3.86 ± 0.01
	2	62.3	6.49	0.0055	0.99	4.94	4.97	2.18	3.26	0.77	4.2 ± 0.03	
	3	98.0	22.13	0.0134	0.97	3.45	3.58	1.79	2.82	0.85	3.3 ± 0.02	
	4	56.1	4.22	0.0059	0.52	4.47	8.67	1.59	7.32	0.75	9.8 ± 0.16	
K22	1	51.4	3.32	0.0060	1.56	9.58	6.15	3.86	3.91	0.78	5.0 ± 1.25	8.44 ± 0.89
	2	49.1	3.25	0.0850	8.31	4.22	0.51	9.32	23.23	0.76	30.6 ± 2.32	
	3	45.5	2.32	0.0040	2.56	9.08	3.54	4.74	3.03	0.75	4.0 ± 1.51	
K21	1	53.5	4.25	0.0183	0.53	5.01	9.46	1.73	20.73	0.74	28.1 ± 0.25	1.65 ± 0.11
	2	75.4	10.47	0.0152	9.62	14.45	1.50	13.09	0.92	0.81	1.1 ± 0.12	
	3	72.2	8.70	0.0810	1.81	5.86	3.23	3.22	24.07	0.80	30.0 ± 1.32	
	4	68.1	7.94	0.0601	12.38	32.70	2.64	20.23	3.11	0.79	3.9 ± 0.34	
KT21	1	64.3	7.08	0.1554	1.69	10.41	6.15	4.19	43.56	0.78	56.0 ± 1.03	1.29 ± 0.04
	2	53.0	4.11	0.2969	2.98	16.35	5.49	6.90	86.84	0.73	118.3 ± 2.37	
	3	56.8	5.30	0.3676	2.29	16.45	7.19	6.24	92.36	0.75	123.1 ± 3.45	
	4	54.4	4.03	0.0091	14.01	32.06	2.29	21.70	0.86	0.74	1.2 ± 0.04	
KT22	1	81.8	12.56	0.0030	0.46	2.40	5.22	1.03	1.92	0.87	2.2 ± 0.74	15.18 ± 0.14
	2	64.5	6.16	0.0030	0.39	2.67	6.93	1.03	3.96	0.78	5.1 ± 1.69	
	3	55.0	4.41	0.0075	0.30	3.79	12.72	1.21	11.74	0.74	15.8 ± 0.15	

[1] eU: effective uranium content, eU = U + 0.235 × Th [34]. [2] Rs: radius of a sphere with an equivalent surface area-to-volume ratio to the cylindrical crystals [35]. [3] FT: the α-ejection correction [32]. [4] Weighted mean age with 95% confidence is calculated in the IsoplotR [36].

4.2. Thermal History Inversion

The partial annealing zone (PAZ) of AFT dating is considered to be between 60 °C and 125 °C [37]. The AHe partial retention zone (PRZ) is 40~80 °C [38]. By integrating AFT and AHe data in the thermal history simulation process, it is possible to reveal a thermal history of 40–125 °C. The thermal history modeling in this study was carried out by the HeFTy v1.9.1 software with Monte Carlo models [39–41]. When modeling the time–temperature paths, some geological constraints need to be inputted. We set the start time of each sample to be the corresponding stratigraphic ages and the starting temperature (at surface) to be 14 °C [42]. The present-day burial temperature was calculated based on the present-day geothermal gradient and sample depth in the Fukang Sag. The temperature changes in the thermal history were related to the tectonic evolution of the Fukang Sag. Therefore, we set a series of constraints for the possible timing of tectonic movements based on the unconformities (Figure 7).

The AFT and AHe data were modeled by using a multi-kinetic annealing model and an apatite radiation damage accumulation and annealing model, respectively [39,41]. The green region in the figures represents an 'acceptable' path range (0.05 ≤ GOF < 0.5), the purple region represents a 'good' path range (0.5 ≤ GOF ≤ 1.0) (Figure 7), and the thick black line represents the best path.

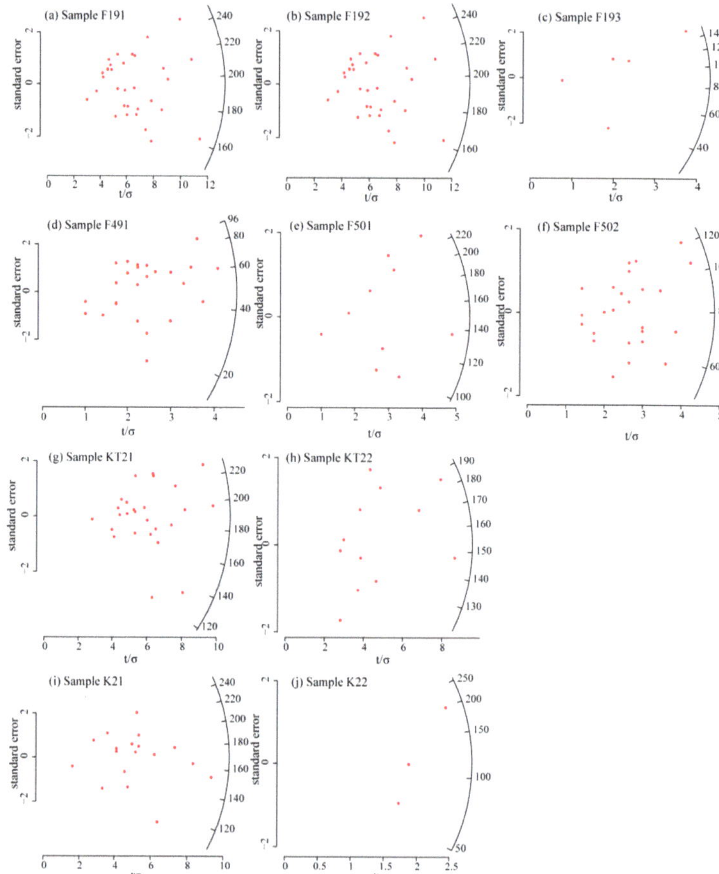

Figure 5. Radial plots of the AFT ages. The red dots represent the single-grain age of the samples.

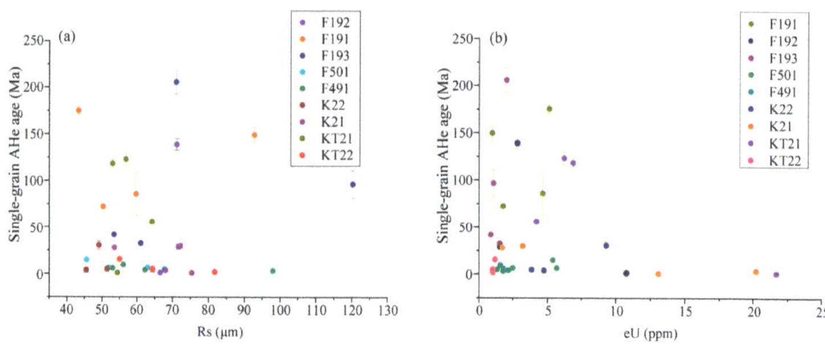

Figure 6. Correlation of single-grain AHe age with Rs (**a**) and eU (**b**) for the samples from the Fukang Sag.

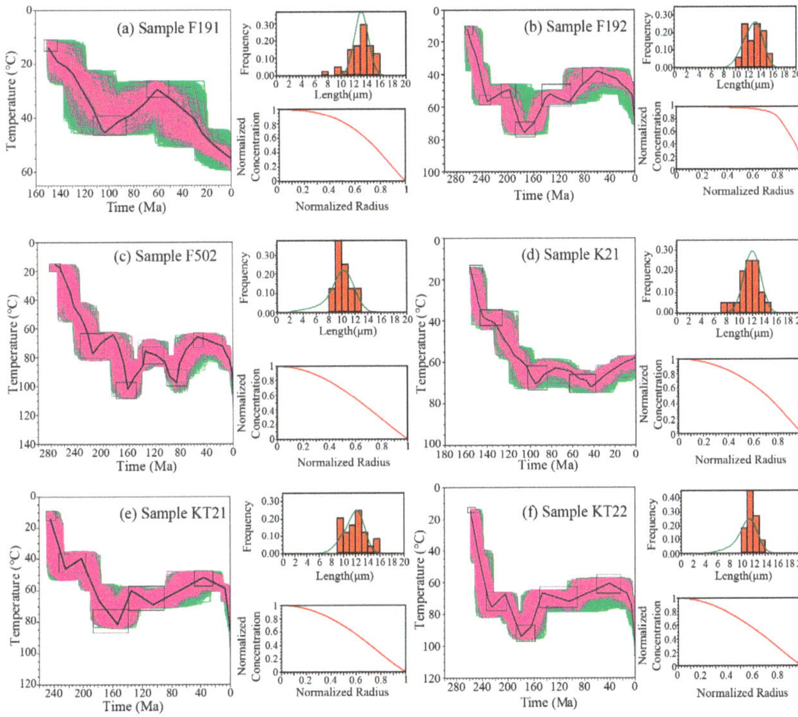

Figure 7. Thermal histories of the samples F191, F192, F502, K21, KT21, and KT22. The green lines are 'acceptable' time–temperature paths. The purple lines are 'good' time–temperature paths. And the black line is the best time–temperature path. The figures show the simulated and measured c-axis projected fission track length data and AHe data with the best time–temperature path.

The thermal history of the Permian sample F192 revealed that the Beisantai Sub-uplift has experienced three cooling events of 240 Ma–200 Ma, 165 Ma–140 Ma, and 100 Ma–60 Ma since the late Permian, while that of the lower Cretaceous sample F191 just revealed a late Cretaceous (100 Ma–60 Ma) cooling event (Figure 7a,b). The thermal histories of the sample F502 indicated that the northeast margin of the Fukang Sag experienced three cooling events, which occurred in 220 Ma–190 Ma, 160 Ma–140 Ma, and 90 Ma–40 Ma (Figure 7c). And the thermal histories of the samples KT21 and KT22 in the southeast margin of the Fukang Sag also recorded three cooling events of ~230 Ma–200 Ma, 165 Ma–150 Ma, and 100 Ma–40 Ma (Figure 7e,f). The sample K21 from well K2 in the southern margin of the Fukang Sag just record two cooling events of 100 Ma–80 Ma and 40 Ma–0 Ma (Figure 7d).

4.3. Paleo-Heat Flow Simulation

The paleo-heat flow evolutionary histories of the Fukang Sag can be modeled by combining the burial histories of the studied wells, the measured Ro, and the t-T paths of the samples mentioned above. The 1D burial and thermal histories of the studied wells were reconstructed based on the stratigraphic thickness and the erosion amounts using Basinmod 1D software [43] with the Easy% Ro model [33]. The erosion amounts were equal to the temperature difference for one cooling event revealed by the time–temperature paths of the samples divided by the geothermal gradient (Figure 7). After the burial history was rebuilt, the thermal modeling constraints, including present-day heat flow, rock thermal conductivity, and Ro data, were input. Then, we iteratively modified the paleo-heat flow to ensure that the simulated thermal history in Basinmod was consistent with the time–

temperature paths obtained from the samples of the AFT and AHe simulations, and the simulated Ro path overlapped with the measured Ro data (Figure 8).

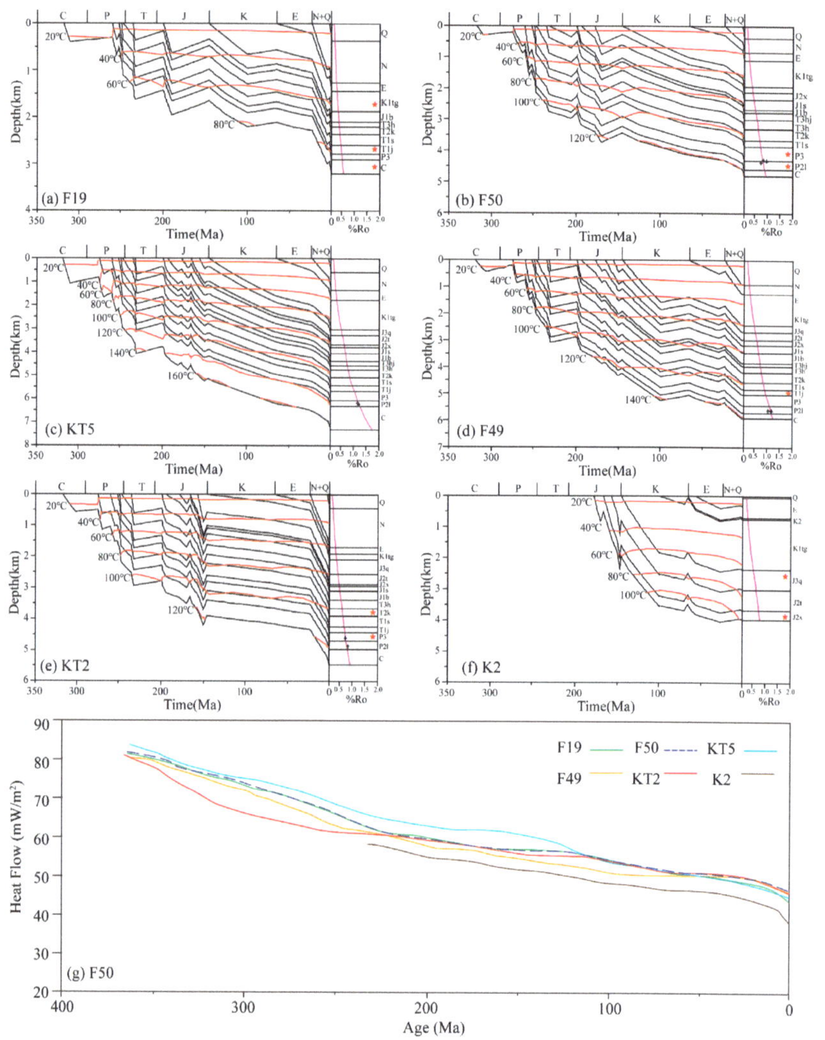

Figure 8. The burial–thermal history (**a–f**) and the paleo-heat flow evolution (**g**) in wells F19, F49, F50, KT2, KT5, and K2. The purple curve represents the simulated %Ro data. The black daggers are the measured %Ro data, and the asterisk symbol indicate the samples' layers with AFT and AHe data.

The paleo-heat flow of the Fukang Sag has gradually decreased since the Carboniferous (Figure 8). The paleo-heat flow was 75–85 mWm^{-2} in the early Carboniferous. And it decreased to 60–70 mWm^{-2} from the Carboniferous to the Permian. Due to the stress field transition from extension to compression, the paleo-heat flow began slowly deceasing during the Triassic and Jurassic and was 55–65 mWm^{-2} by the end of the Jurassic. Due to the differential deposition in the early Cretaceous [44], the paleo-heat flow among the studied wells in the Fukang Sag experienced different evolutionary histories. During the late Cretaceous, the paleo-heat flow in the Fukang Sag decreased to 50–60 mWm^{-2}. The rapid sedimentation during the Cenozoic caused the paleo-heat flow of the Fukang Sag to decrease to ~45 mWm^{-2} [12].

5. Discussion

5.1. Geothermal Controlling Factors in the Fukang Sag

The heat flow in the Fukang Sag and its surrounding area ranges from 34.55 to 64.31 mWm^{-2}, which is lower than that in sedimentary basins in Central and Eastern China and offshore areas [45–47], as well as the average heat flow in continental regions [48]. The present-day geothermal regime in the Fukang Sag should be related to the long-term stable tectonic environment [49]. A previous study indicated that the crustal structure, tectonic evolution, basement relief, subsurface fluid activity, and rock heat production play an important effect on the present-day geothermal regime of basins [50]. In this study, we discuss how the basement relief and tectonic evolution affect the geothermal regime of the Fukang Sag.

The basement of the Junggar Basin consists of lower Carboniferous strata [51]. The Beisantai, Shaqi, and Baijiahai Sub-uplifts show very shallow basements with burial depths of 3–5 km, while the burial depths of the basement in the Fukang Sag exceeds 6.5 km (Figure 1e,f). This basement relief is in accordance with the distribution of the geothermal regime in the Fukang Sag and its surrounding area. In addition to the basement relief, the present-day geothermal regime of the Fukang Sag is also related to regional tectonic evolution. During the Permian, The Fukang Sag and Beisantai, Shaqi, and Baijiahai Sub-uplifts formed due to N-S striking compression [20,52]. These areas always kept a shallow basement with thin sedimentary layers. Before the Jurassic, the areas in the southern part of the Fukang Sag were at the deposition center and deposited relatively thick sediments [44,52]. Along with the uplifting of the Tianshan from the late Jurassic to early Cretaceous, the deposition center migrated northward to the Fukang Sag [23]. During the Cenozoic, a thrust-and-fold belt formed in the south of the Fukang Sag with intense N-S striking compression, causing uplift and erosion [53]. Although the southern part of the Fukang Sag stays in a structurally elevated position right now, the existing thick sediments still make it relatively cool.

The geothermal gradient and heat flow in the shallow basement areas are usually higher those that in the deep basement areas. This phenomenon is attributed to the "thermal refraction" effect caused by the differential rock thermal conductivities between the bedrock and sedimentary rocks [54], which resulted in heat accumulation in the uplift areas. It also confirms that basins developed in a compressional regime, which had thicker sedimentary rocks, exhibit lower heat flow than basins developed in a stretching regime [55].

5.2. Differential Maturation Evolution of Source Rocks

Exploration indicates that the Jurassic and overlying formations in the Beisantai, Baijiahai, and Shaqi uplifts are rich in oil and gas resources [56], which came from the Carboniferous, Permian, and Jurassic source rocks in the Fukang Sag [4,56]. For a better understanding the hydrocarbon generation and accumulation process, this study discussed the maturation evolutionary process of the source rocks in the Fukang Sag, including the Carboniferous Dishuiquan Formation (C_1d), Permian Lucaogou Formation (P_2l), and Jurassic Badaowan Formation (J_1b) (Figure 9).

Due to the very low burial temperature throughout geological time, the Jurassic source rock in the well F19 area is currently immature, with a Ro value of ~0.3%, and the Carboniferous source rock in the Well F19 area just reached the early maturity stage in the late Cretaceous and probably began generating oil. In the well F50 area, the Jurassic source rock is also currently immature, while the Carboniferous and Permian source rocks reached the early and middle maturity stages in the early Triassic and the early Jurassic, respectively. In the well KT2 area, the Jurassic source rock entered the early maturity stage during the late Cenozoic, and the Carboniferous and Permian source rocks reached the early and middle maturity stages in the early Triassic and the late Jurassic, respectively. The

Jurassic source rock in the Fukang Sag (well F49 and KT5 areas) entered the early maturity stage in the early Cretaceous, while the Carboniferous and Permian source rocks reached the early, middle, and late maturity stages in the early Triassic, late Triassic, and Cretaceous, respectively. Due to the thick sediments developed by the long-term deposition, the current Ro values of the Carboniferous and Permian source rocks in the F49 and KT5 wells range from 1.1% to 1.3%. Generally, the differential maturity of the source rocks in the Fukang Sag is related to burial depth.

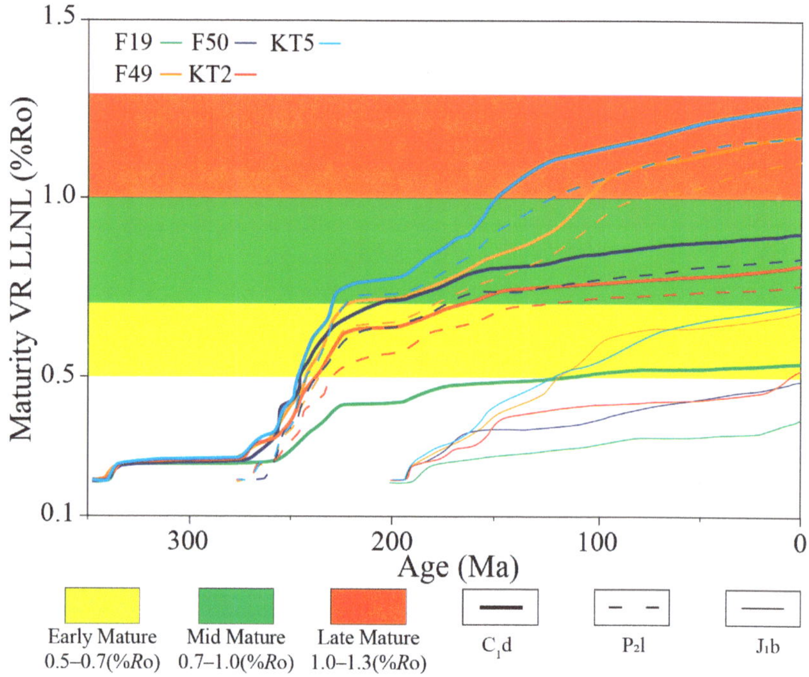

Figure 9. The maturation evolutionary histories of the source rocks in the Fukang Sag. The thicker solid line, dashed line, and thinner solid line represent C_1d, P_2l, and J_1b source rocks, respectively.

The Carboniferous and Permian source rocks along the east margin of the Fukang Sag began generating oil in the late Triassic, which should provide massive amounts of oil for the peripheral oil fields. This suggests that Carboniferous and Permian source rocks in the central region of Fukang Sag had higher maturity, possibly resulting in gas generation. According to the geochemical data, the oil and gas reservoirs in the uplift areas of the Fukang Sag belonged to a mixed source [57]. Our study supports that liquid oil primarily originated from the margin of the Fukang Sag, while gas originated from the center of the Fukang Sag [58].

6. Conclusions

In this study, we comprehensively used borehole temperature data, AFT data, AHe data, and Ro data to investigate the geothermal regime and tectono-thermal history of the Fukang Sag in the Junggar Basin. The present-day average geothermal gradient and heat flow in the Fukang Sag are 20.8 °C/km and 44.6 mWm^{-2}, respectively. Due to the basement relief, they decrease from northeast to southwest. The AFT and AHe ages range between 50.9 Ma–193.8 Ma and 1.3 Ma–85.2 Ma, respectively. Thermal history simulation revealed three cooling events in the late Permian, late Jurassic, and late Cretaceous periods. In addition, the basin modeling results also indicated that the paleo-heat flow of the

Fukang Sag decreased from 80 mWm^{-2} in the Carboniferous to the current value of 44.6 mWm^{-2}. The Carboniferous–Permian source rocks in the eastern Fukang Sag entered into the middle–late maturation stage during the late Triassic, showing great potential for oil generation.

Author Contributions: Conceptualization, H.G. and J.C.; methodology, C.L. and B.P.; software, C.L. and J.X.; validation, B.P., X.S. and C.L.; formal analysis, J.C.; investigation, C.L.; resources, B.P. and X.S.; data curation, Z.Z.; writing—original draft preparation, C.L.; writing—review and editing, J.C. and H.G.; visualization, B.P.; project administration, J.C.; funding acquisition, H.G. All authors have read and agreed to the published version of the manuscript.

Funding: This research was funded by the China National Petroleum Corporation's "Fourteenth Five Year Plan" Basic Prospective Major Scientific and Technological Projects (No. 2021DJ0202 and No. 2021DJ0108) and the China National Petroleum Corporation's Important Science and Technology Project (No. 2023ZZ24-01).

Data Availability Statement: The data are contained within the article.

Conflicts of Interest: Huajun Guo, Bo Peng, and Xiang Shan are employees of PetroChina Hangzhou Research Institute of Geology. The paper reflects the views of the scientists and not the company.

References

1. He, W.; Fei, L.; Ablimiti, Y.; Yang, H.; Lan, W.; Ding, J.; Bao, H.; Guo, W. Accumulation conditions of deep hydrocarbon and exploration potential analysis in Junggar Basin, NW China. *Earth Sci. Front.* **2019**, *26*, 189–201. [CrossRef]
2. Tao, S.; Yuan, X.; Hou, L.; Zhang, G.; Yang, F.; Tao, X.; Wei, Y.; Li, X.; Zhang, C.; Wang, L.; et al. Formation and distribution of large lithologic-stratigraphic oil & gas fields (provinces). *J. Nat. Gas Geosci.* **2018**, *3*, 51–65. [CrossRef]
3. Yang, Z.; Zou, C. "Exploring petroleum inside source kitchen": Connotation and prospects of source rock oil and gas. *Pet. Explor. Dev.* **2019**, *46*, 181–193. [CrossRef]
4. He, H.Q.; Zhi, D.M.; Tang, Y.; Liu, C.W.; Chen, H.; Guo, X.G.; Wang, Z. A great discovery of Well Kangtan 1 in Fukang Sag in the Junggar Basin and its significance. *China Pet. Explor.* **2021**, *26*, 1–11. [CrossRef]
5. Hu, S.B.; Wang, J.Y. Principles and progresses on thermal regime of sedimentary basins—An overview. *Earth Sci. Front.* **1995**, *2*, 171–180.
6. Zhang, G. Co-control of source and heat: The generation and distribution of hydrocarbons controlled by source rocks and heat. *Acta Pet. Sin.* **2012**, *33*, 723–738. [CrossRef]
7. Ren, Z.; Cui, J.; Qi, K.; Yang, G.; Chen, Z.; Yang, P.; Wang, K. Control effects of temperature and thermal evolution history of deep and ultra-deep layers on hydrocarbon phase state and hydrocarbon generation history. *Nat. Gas Ind. B* **2020**, *7*, 453–461. [CrossRef]
8. Qiu, N.; Chang, J.; Zhu, C.; Liu, W.; Zuo, Y.; Xu, W.; Li, D. Thermal regime of sedimentary basins in the Tarim, upper Yangtze and north China cratons, China. *Earth-Sci. Rev.* **2022**, *224*, 103884. [CrossRef]
9. Qiu, N.; He, L.; Chang, J.; Zhu, C. Research progress and challenges of thermal history reconstruction in sedimentary basins. *Pet. Geol. Exp.* **2020**, *42*, 790–802. [CrossRef]
10. Ren, Z.; Cui, J.; Qi, K.; Yang, P.; Liu, X.; Zhang, C.; Yang, G.; Gao, Y.; Zhang, Y.; Xing, G. New progress in research on theories and methods for reconstruction of deep and ultra—Deep thermal evolution history in superimposed basins. *J. Northwest Univ. (Nat. Sci. Ed.)* **2022**, *52*, 910–929.
11. Rao, S.; Hu, S.B.; Zhu, C.Q.; Tang, X.Y.; Li, W.W.; Wang, J.Y. The characteristics of heat flow and lithospheric thermal structure in Junggar Basin, northwest China. *Chin. J. Geophys.* **2013**, *56*, 2760–2770. [CrossRef]
12. Qiu, Z.; Zhang, Z.; Xu, E. Geothermal regime and Jurassic source rock maturity of the Junggar Basin, northwest China. *J. Asian Earth Sci.* **2008**, *31*, 464–478. [CrossRef]
13. Li, D.; He, D.; Santosh, M.; Ma, D.; Tang, J. Tectonic framework of the northern Junggar Basin part I: The eastern Luliang Uplift and its link with the East Junggar terrane. *Gondwana Res.* **2015**, *27*, 1089–1109. [CrossRef]
14. He, D. Multi-cycle superimposed sedimentary basins in China: Formation, evolution, geologic framework and hydrocarbon occurrence. *Earth Sci. Front.* **2022**, *29*, 24–59. [CrossRef]
15. Zhou, Z.; Pan, C.; Fan, S.; Han, L. Geothermal characteristics of the Junggar Basin and its significance to oil searching. *Xinjiang Pet. Geol.* **1989**, *10*, 67–74.
16. Liu, G.; Zhang, H. The Hydrocarbon Occurrence and Characteristics of Geothermal field in Junggar Basin. *Xinjiang Pet. Geol.* **1992**, *13*, 100–107.
17. Pan, C.; Zhou, Z.; Wang, Q. Study of thermal history of source beds in Junggar basin by apatite fission track. *Oil Gas Geol.* **1989**, *10*, 35–39.
18. Rao, S.; Zhu, Y.K.; Hu, S.B.; Wang, Q. The thermal history of Junggar Basin: Constraints on the tectonic attribute of the Early-Middle Permian basin. *Acta Geol. Sin.* **2018**, *92*, 1176–1195.

19. Zhao, G.C.; Sun, M.; Wilde, S.A.; Li, S.Z. Late Archean to Paleoproterozoic evolution of the North China Craton: Key issues revisited. *Precambrian Res.* **2005**, *136*, 177–202. [CrossRef]
20. He, D.; Zhai, G.; Kuang, J.; Zhang, Y.; Shi, X. Distribution and tectonic features of paleo-uplifts in the Junggar Basin. *Chin. J. Geol.* **2005**, *40*, 248–261.
21. Ma, C.; Wu, K.Y.; Pei, Y.W.; Li, T.R. Quantitative analysis of tectonic characteristics and evolution in the eastern Junggar Basin. *J. Geomech* **2019**, *25*, 313–323.
22. He, D.F.; Zhang, L.; Wu, S.T.; Li, D.; Zhen, Y. Tectonic evolution stages and features of the Junggar Basin. *Oil Gas Geol.* **2018**, *39*, 845–861.
23. Fang, S.; Guo, Z.; Jia, C.; Zhang, Z.; Wang, X.; Wang, M. Meso-Cenozoic Heavy Minerals' Assemblages in the Southern Junggar Basin and Its Implocations for Basin-orogen Patter. *Chin. J. Geol.* **2006**, *41*, 648–662.
24. Zhu, M.; Yuan, B.; Liang, Z.; Yang, D.; Tang, X. Fault properties and evolution in the periphery of Junggar Basin. *Acta Pet. Sin.* **2021**, *42*, 1163–1173. [CrossRef]
25. Chang, J.; Qiu, N.; Xu, W. Thermal regime of the Tarim Basin, Northwest China: A review. *Int. Geol. Rev.* **2017**, *59*, 45–61. [CrossRef]
26. Förster, A. Analysis of borehole temperature data in the Northeast German Basin: Continuous logs versus bottom-hole temperatures. *Pet. Geosci.* **2001**, *7*, 241–254. [CrossRef]
27. Chang, J.; Qiu, N.-S.; Zhao, X.-Z.; Xu, W.; Xu, Q.-C.; Jin, F.-M.; Han, C.-Y.; Ma, X.-F.; Dong, X.-Y.; Liang, X.-J. Present-day geothermal regime of the Jizhong depression in Bohai Bay basin, East China. *Chin. J. Geophys.* **2016**, *59*, 1003–1016. [CrossRef]
28. Oliver, M.A.; Webster, R. A tutorial guide to geostatistics: Computing and modelling variograms and kriging. *Catena* **2014**, *113*, 56–69. [CrossRef]
29. Bullard, E.C. The time necessary for a bore hole to attain temperature equilibrium. *Geophys. J. Int.* **1947**, *5*, 127–130. [CrossRef]
30. Wang, S.J.; Hu, S.B.; Wang, J.Y. The characteristics of heat flow and geothermal field in Junggar Basin. *Chin. J. Geophys.* **2000**, *43*, 816–824. [CrossRef]
31. Jin, J.; Liu, M.; Liu, Y.; Wang, J.; Gao, C.; Luo, Z.; Lu, F.; Ren, Y. Present-day temperature-pressure field and its controlling factors of the lower composite reservoir in the southern margin of Junggar Basin. *Chin. J. Geol.* **2021**, *56*, 28–43. [CrossRef]
32. Chang, J.; Yang, L.; Li, C.; Qiu, N.; Zhang, H.; Wang, X. Tectono-thermal evolution of the northern Tarim Basin, Central Asia: New insights from apatite low-temperature thermochronometers. *J. Asian Earth Sci.* **2004**, *259*, 105919. [CrossRef]
33. Sweeney, J.J.; Burnham, A.K. Evaluation of a simple model of vitrinite reflectance based on chemical kinetics. *AAPG Bull.* **1990**, *74*, 1559–1570. [CrossRef]
34. Farley, K.A.; Wolf, R.A.; Silver, L.T. The effects of long alpha-stopping distances on (U-Th)/He ages. *Geochim. Cosmochim. Acta* **1996**, *60*, 4223–4229. [CrossRef]
35. Shuster, D.L.; Flowers, R.M.; Farley, K.A. The influence of natural radiation damage on helium diffusion kinetics in apatite. *Earth Planet. Sci. Lett.* **2006**, *249*, 148–161. [CrossRef]
36. Meesters, A.G.C.A.; Dunai, T.J. Solving the production–diffusion equation for finite diffusion domains of various shapes: Part II. Application to cases with α-ejection and nonhomogeneous distribution of the source. *Source. Chem. Geol.* **2002**, *186*, 57–73. [CrossRef]
37. Vermeesch, P. IsoplotR: A free and open toolbox for geochronology. *Geosci. Front.* **2018**, *9*, 1479–1493. [CrossRef]
38. Green, P.F.; Duddy, I.R.; Gleadow, A.J.; Tingate, P.R.; Laslett, G.M. Thermal annealing of fission tracks in apatite: 1. A qualitative description. *Chem. Geol. Isot. Geosci. Sect.* **1986**, *59*, 237–253. [CrossRef]
39. Flowers, R.M.; Ketcham, R.A.; Shuster, D.L.; Farley, K.A. Apatite (U–Th)/He thermochronometry using a radiation damage accumulation and annealing model. *Geochim. Cosmochim. Acta* **2009**, *73*, 2347–2365. [CrossRef]
40. Nielsen, S.B.; Balling, N. Joint inversion of temperature, vitrinite reflectance and fission tracks in apatite with examples from the eastern North Sea area—Response to discussion. *Basin Res.* **2023**, *35*, 2045–2048. [CrossRef]
41. Ketcham, R.A.; Carter, A.; Donelick, R.A.; Barbarand, J.; Hurford, A.J. Improved modeling of fission-track annealing in apatite. *Am. Mineral.* **2007**, *92*, 799–810. [CrossRef]
42. Qiu, N.S.; Ming, Z.; Wang, X.L.; Yang, H.B. Tectono-thermal evolution of the Junggar Basin, NW China: Constraints from Ro and apatite fission track modelling. *Pet. Geosci.* **2005**, *11*, 361–372.
43. Vincent, C.J.; Merriman, R.J. *Thermal Modelling of the Cheshire Basin Using BasinModTM*; British Geological Survey: Nottingham, UK, 2004.
44. Cao, S.; He, D. Evolution of the Caiwopu basin, Xinjiang. *Geotecton. Metallog.* **1997**, *21*, 262–269.
45. Fang, S.H.; Guo, Z.J.; Song, Y.; Wu, C.D.; Zhang, Z.C.; Wang, M.N.; Fan, R.D. Sedimentary facies evolution and basin pattern of the Jurassic in southern margin area of Junggar Basin. *J. Palaeogeogr.* **2005**, *7*, 347–356.
46. Qiu, N.; Kang, Y.; Jing, Z. Temperature and pressure field in the Tertiary succession of the western Qaidam basin, northeast Qinghai-Tibet Plateau, China. *Mar. Pet. Geol.* **2003**, *20*, 493–507. [CrossRef]
47. Ren, Z.; Zhang, S.; Gao, S.; Cui, J.; Xiao, Y.; Xiao, H. Tectonic thermal history and its significance on the formation of oil and gas accumulation and mineral deposit in Ordos Basin. *Sci. China Ser. D Earth Sci.* **2007**, *50*, 27–38. [CrossRef]
48. He, L.; Wang, K.; Xiong, L.; Wang, J. Heat flow and thermal history of the South China Sea. *Phys. Earth Planet. Inter.* **2001**, *126*, 211–220. [CrossRef]

49. Jiang, G.-Z.; Gao, P.; Rao, S.; Zhang, L.-Y.; Tang, X.-Y.; Huang, F.; Zhao, P.; Pang, Z.-H.; He, L.-J.; Hu, S.-B.; et al. Compilation of heat flow data in the continental area of China. *Chin. J. Geophys.* **2016**, *59*, 2892–2910. [CrossRef]
50. Popov, Y.A.; Pimenov, V.P.; Pevzner, L.A.; Romushkevich, R.A.; Popov, E.Y. Geothermal characteristics of the Vorotilovo deep borehole drilled into the Puchezh-Katunk impact structure. *Tectonophysics* **1998**, *291*, 205–223. [CrossRef]
51. Wang, X.; Song, Y.; Bian, B.; Zhao, J.; Zhang, H.; Yan, M.; Pei, S.; Xu, Q.; Wang, S.; Liu, H.; et al. Basement structure of the Junggar Basin. *Earth Sci. Front.* **2021**, *28*, 235–255. [CrossRef]
52. Zheng, M.; Fan, X.; He, W.; Yang, T.; Tang, Y.; Ding, J.; Wu, H.; Chen, L.; Guo, J. Superposition of deep geological structural evolution and hydrocarbon accumulation in the Junggar Basin. *Earth Sci. Front.* **2019**, *26*, 22–32. [CrossRef]
53. Hendrix, M.S.; Graham, S.A.; Carroll, A.R.; Sobel, E.R.; Mcknight, C.L.; Schulein, B.J.; Wang, Z. Sedimentary record and climatic implications of recurrent deformation in the Tian Shan: Evidence from Mesozoic strata of the north Tarim, south Junggar, and Turpan basins, northwest China. *Geol. Soc. Am. Bull.* **1992**, *104*, 53–79. [CrossRef]
54. Liu, S.; Lei, X.; Feng, C.; Li, X. Heat flow, deep formation temperature and thermal structure of the Tarim Basin, Northwest China. *Earth Sci. Front.* **2017**, *24*, 41–55.
55. Mckenzie, D. Some remarks on the development of sedimentary basins. *Earth Planet. Sci. Lett.* **1978**, *40*, 25–32. [CrossRef]
56. Li, Z.; Peng, S. U-Pb geochronological records and provenance system analysis of the Mesozoic-Cenozoic sandstone detrital zircons in the northern and southern piedmonts of Tianshan, Northwest China: Responses to intracontinental basin-range evolution. *Acta Petrol. Sin.* **2013**, *29*, 739–755.
57. Zheng, J.; Xiang, C.; Wang, X.; Lei, D.; Yang, N.; A, B.; Zhang, L.; Zhang, Y.; Wang, L. Characteristics of the Jurassic source rocks and their shale gas exploration potential in the Fukang Sag of the Junggar Basin. *Geol. Rev.* **2015**, *61*, 217–226.
58. Yang, F.; Song, Y.; Chen, H.; Gong, D.; Bian, B.; Liu, H. Evaluation of Carboniferous Songkaersu Formation source rocks and gas-source correlation in the Fukang Sag of eastern Junggar Basin. *Nat. Gas Geosci* **2019**, *30*, 1018–1026.

minerals

Article

Mineralogical Characteristics and Genetic Types of Pyrite with Different Occurrence: Constraints from Spectroscopy, Geochemistry and $\delta^{34}S$ Stable Isotopes

Shiyu Ma [1], Miao Shi [1,2,3,4,*], Cun Zhang [5] and Qinyuan Cao [1]

1 School of Geology and Material Science, Hebei GEO University, Shijiazhuang 050031, China; masy0107@163.com (S.M.); maxcaoqinyuan@163.com (Q.C.)
2 Hebei Key Laboratory of Green Development of Rock Mineral Materials, Hebei GEO University, Shijiazhuang 050031, China
3 Engineering Research Center for Silicate Solid Waste Resource Utilization of Hebei Province, Hebei GEO University, Shijiazhuang 050031, China
4 State Key Laboratory of Geological Processes and Mineral Resources, China University of Geosciences, Wuhan 430074, China
5 School of Materials Science and Engineering, Qilu University of Technology (Shandong Academy of Sciences), Jinan 250353, China; gemzhangc@qlu.edu.cn
* Correspondence: miaoer727@126.com

Abstract: Pyrite is widely distributed in the Earth's crust and oceanic systems, and it generally occurs as pentagonal dodecahedra, cubic, octahedral, idiomorphic crystals, or dense, massive, granular, and nodular aggregates. In this study, representative pyrite samples from Hunan, Fujian, Jiangxi, Anhui in China and from Peru were collected. By utilizing a range of analytical techniques, including petrography, X-ray diffraction, infrared spectrum, Raman spectrum, major and trace element analysis, as well as sulfur stable isotope analysis, we comprehensively depict the mineralogical and spectroscopic characteristics of pyrite, and the evolution processes of the physical and chemical conditions of mineralization can be qualitatively constrained. The spectroscopic results indicate all samples show a relatively narrow absorption band with weak to moderate intensity in the vicinity of 343 cm^{-1}, which represents the bending vibration of the Fe-$[S_2]^{2-}$ molecular bond. The Co content of pyrite exhibits the characteristics of a positive correlation with temperature and a negative correlation with oxygen fugacity, respectively. The $\delta^{34}S$ isotopic compositions of colloidal pyrite are in the range of 0.03 to 0.67, which are close to meteoric sulfur and mantle sulfur compositions, while the $\delta^{34}S$ values of nodular pyrite fall within the range of granite, indicating the characteristics of mixtures of sulfur sources are mainly related to magmatic activity. Our results provide insight into the formation mechanisms of pyrite in different environments, its mineralization, and the ore genesis of deposits. Moreover, the integrated analytical methods for pyrite are provided, which can define theoretical guidance for the exploration and development of mineral resources.

Keywords: pyrite; spectroscopic characteristics; geochemical characteristics; $\delta^{34}S$ stable isotopes

check for updates

Citation: Ma, S.; Shi, M.; Zhang, C.; Cao, Q. Mineralogical Characteristics and Genetic Types of Pyrite with Different Occurrence: Constraints from Spectroscopy, Geochemistry and $\delta^{34}S$ Stable Isotopes. *Minerals* **2024**, *14*, 52. https://doi.org/10.3390/min14010052

Academic Editor: Jiří Sejkora

Received: 10 November 2023
Revised: 5 December 2023
Accepted: 21 December 2023
Published: 30 December 2023

1. Introduction

Pyrite (FeS_2) belongs to the equiaxial crystal system, with an NaCl-type crystal structure, and is the most common sulfide in the Earth's crust [1]. The different morphology and grain size characteristics of pyrite could reflect its different growth environments and processes. Autocrystalline pyrite is generally produced in shale cracks or matrix, and the idiomorphic pyrite in the matrix is either dispersed or aggregated into lenticular bands, but the amount of development is relatively small [2].

A previous investigation has shown pyrite generated in hydrothermal deposits was associated with other sulfide, oxide, quartz, and other minerals, and sometimes huge accumulations of pyrite deposits could be formed [3]. The crystal morphology of hydrothermal pyrite showed the crystallization habits have changed continuously with the evolution of mineralization. Generally, when the sulfur fugacity in the hydrothermal fluid is high, pyrite is prone to form "e" or "o" crystal shape, accompanied by complex polymorphic crystals. However, under high oxygen fugacity and low sulfur fugacity conditions, this is conducive to the growth of an "a" single crystal [4]. In metamorphic rocks, pyrite is often formed as a newborn product of metamorphism, while in sedimentary rocks, coal strata and sedimentary deposits, pyrite is often produced in massive, nodules or lenticular form, and its formation is inextricably linked to the organic matter. Pyrite nodules and agglomerates were dominant during the sedimentary period, and a total of three types of framboids pyrite, autoclasts and pyrite-accounted filling organisms were observed microscopically [1].

Meanwhile, the crystal shape and grain pattern of pyrite also play a great role in the study of the pyrite formation environment. The appearance of striated and nodular pyrite indicates the transformation of physical and chemical conditions of the sedimentary environment into an acidic sulfide-reduced environment, which is conducive to the preservation and enrichment of organic matter [5]. While, the formation of cubic crystals implies pyrite formed under the conditions of low supersaturation, low sulfur fugacity, and insufficient supply of material is favored to cubic growth, and the span of pyrite formation temperatures for cubic forms is relatively large [6].

Ornamental pyrite crystals with economic value have been reported in many regions of China, such as Hunan, Hubei, Zhejiang, Anhui, and Guangdong Provinces, as well as Spain, the United States, Mexico, Peru, and so on. The representative pyrite samples studied in this paper were collected from Anhui, Fujian, Jiangxi, Hunan, and Peru, including different sedimentary and hydrothermal genesis types. Most of the previous studies have focused on specific origin or type of genesis of pyrite, and the data made by the previous researchers were still incomplete, especially for the analysis of pyrite specimens with multiple origins and genesis. Therefore, this investigation carried out spectroscopy, major and trace element analysis etc., on specimens of different genesis types, and then synthesized the characteristics of its deposits, aiming to provide implications for the research of pyrite for the future. The study of trace elements in pyrite is of great significance to the understanding of the genesis of metallic ore deposits such as gold deposits, and guiding the search for exploration. Moreover, it can also be used to predict the enrichment pattern of shale gas and provide theoretical guidance for the exploration and development of mineral resources [7–9].

2. Materials and Methods

2.1. Sample Description

Five typical samples were collected from Anhui, Fujian, Jiangxi, Hunan Provinces, China and Peru, respectively (Figure 1). Sample HTK-3 (Figure 1a) from Ruijin of Jiangxi Province has no obvious crystal shape, and its pyrite product shows the characteristics of multi-stage output. The ore types are mainly divided into pyrite-quartz vein type and pyrite-silicified rubble breccia type, respectively. The whole sample presents a copper-yellow color with the size of about 3 × 5 cm, spherical, nodular, and dense internal crystal structures, and the concentric ring structure can be observed on the longitudinal section. Sample HTK-4 (Figure 1b) was collected from Shangbao in Hunan Province, which was situated in the center of the South China fold system. This specimen exhibits a prominent copper-yellow hue and measures a substantial size of approximately 6.7 × 4 cm. It is intricately linked with quartz, displaying a fully intact cubic form. Notably, the cubic grain size demonstrates a considerable range of variation. Additionally, the surface of the sample bears visible longitudinal striations, adding to its distinctive characteristics. The surface is relatively clean and smooth, and the development of cubic {100} single crystals can be found. Sample HTK-2 (Figure 1c) from Fujian shows a brown color with a size of about 7.4 × 6 cm, and the pyrite occurs in association with quartz and calcite. It has small particle size with a tiny deviation range. Overall, pyrite is cubic, granular, and scattered in the crystal clusters. Sample HTK-1 (Figure 1d) originates from Anhui, which is a prototypical skarn-type copper-gold deposit. This region generates not only exemplary skarn deposits but also an array of porphyry deposits. Remarkably, in certain locations, both skarn and porphyry mineralizations converge, leading to the coexistence of these distinct geological features. Furthermore, the area is distinguished by the presence of numerous layered skarn orebodies and, notably, layered massive sulfide orebodies, highlighting the rich diversity and complexity of the mineral deposits in this locale. The sample shows typical characteristics of a size of 2.1 × 1 cm, copper-yellow color, and symbiosis with calcite. The crystals are pentagonal dodecahedral, layered structure, with a certain gradient of flat and straight growth striations on the surface. The grain size is medium, and the range of variation is narrow. Sample HTK-5 (Figure 1e) is symbiotic with sphalerite from a high-sulfidation epithermal deposit in Peru, shows a copper-yellow color, and the pyrite is octahedral. The sphalerite can be observed in a red fluorescence crystal shape, the grain size is medium, and the range of variation is relatively large.

Figure 1. Macroscopic characteristics of pyrite samples: (**a**) HTK-3 occurs as concretions, and a concentric ring structure can be found on the longitudinal section, possibly having undergone synsedimentation. (**b**) HTK-4 occurs as a cube particle, and the pyrite occurs in association with quartz. (**c**) Sample HTK-2 is cubic and granular, with pyrite grains scattered in the crystal clusters. (**d**) Sample HTK-1 shows a pyritohedron shape, coexisting with calcite. (**e**) Sample HTK-5 is octahedral, in symbiosis with sphalerite which can be observed in red fluorescence.

2.2. Methods

A German Leica DW27009 polarizing microscope from the Jewelry Center Laboratory of Hebei GEO University was used to observe the internal structure and crystallization state of the minerals, then we circled the specific locations required for subsequent experiments; all the samples have been pretreated.

The reflectance method was used to detect the samples through a r1110 Nicolet Fourier infrared spectrometer instrument (Thermo Fisher Scientific, Waltham, MA, USA) in the Jewelry Center of Hebei GEO University. The wavelength range was set to 400~4000 cm^{-1} with the scanning times and rates being 32 and 4 cm^{-1}, respectively. The laboratory temperature was 25 °C, and the humidity was 25% RH.

Raman spectroscopy of pyrite samples was performed at the Jewelry Center in Hebei GEO University, using a RENISHAW-Invia micro confocal laser Raman spectrometer from Renishaw, London, UK. The wavelength used was 532 nm. We used a scanning time of 20 s of spectrum in an extending mode spanning 200~3000 cm^{-1}. The laboratory temperature was 25 °C, and the humidity was 25% RH.

The X-ray fluorescence spectrometer model X-3680 from Tsushima Company, Nagasaki of Japan, was used in the Jewelry Center of Hebei GEO University, using an EDX-7000 energy dispersive X-ray fluorescence spectrometer. Each sample was tested using the Al-U channel and Na-Sc channel, respectively. And, the testing conditions were a tube voltage of 50 kV and 15 kV, respectively, while the tube current was set to 767 μA and 1000 μA, respectively. The time was 30 mm.

The X-ray powder crystal diffraction instrument was adopted from the Smart Lab Rigaku of Nihon, Hebei GEO University, China, with the following specific test conditions: copper target (Cu) test, emission and scattering slit of 1°, receiving slit of 0.3 mm, operating voltage of 48 kV, current of 1000 mA, scanning speed of 6° (2θ) min, and scanning range of 2.6°~70°.

The measurement of S isotope composition of the five samples was carried out at Wuhan Spectrum Analytical Technology Co., Ltd. (Wuhan, China). The instrument used is a multi-collector mass spectrometer model Neptune Plus MC-ICP-MS (Thermo Fisher Scientific, Waltham, MA, USA) and a coherent 193 nm excimer laser deposition system (Geo Las HD) to determine the material sources of the mineral, with a beam spot of 44 μm, an energy intensity of 6 MJ/cm^2, a frequency of 2 Hz, and a carrier gas of 500 mL/min.

3. Results and Discussion

3.1. Pyrite Occurrence

The structural features of the samples could be clearly observed as in Figure 2.

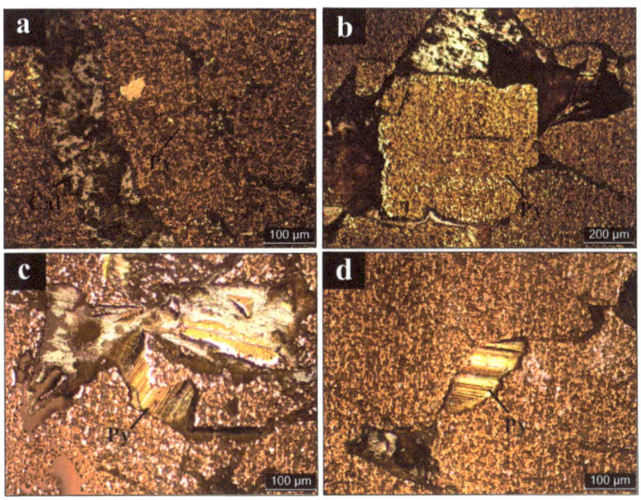

Figure 2. Representative photomicrographs showing different forms of pyrite. (**a**) Nodular pyrite in sample HTK-3. (**b**) Cubic pyrite in sample HTK-4. (**c**) Euhedral grains in sample HTK-2, and crystal face striations are visible. (**d**) Heteromorphic pyrite in sample HTK-1, and crystal face striations are also visible.

The whole sample HTK-3 (Figure 2a) presents an interwoven structure, while the crystals cannot be observed clearly. Generally, the structure is relatively dense, and the concentric ring structure can be observed on the longitudinal section, indicating its formation process is characterized by multiple phases [10]. The white minerals are distributed in stripes in the fissures, presumed to be calcite. The colloidal pyrite in Ruijin of Jiangxi Province was closely related to the activity of anaerobic microorganisms and the induced changes of sedimentary environment and pH [11]. Thus, the appearance of laminar and nodular pyrite represented the physical and chemical conditions of the depositional environment have been changed into an acidic sulfide-reduced environment, which was conducive to the preservation and enrichment of organic matter, corresponding to previous studies [12]. The surface of the pyrite crystals is relatively rough due to the abrasion and carving effects during the process of deposition, transportation, and diagenesis [13,14]. In sample HTK-4, a large amount of authigenic pyrite can be observed, the symbiotic minerals which are presumed to be the chlorite, can be found to be in different shades of blue mixed with pyrite under a single polarized light microscope (Figure 2b). The parallel crystal surface streaks were clearly evident. Combined with the hand specimen, it can be inferred the pyrite is mostly in the shape of a cube, which reflects the ore-forming fluids may be in the low sulfur fugacity, high oxygen fugacity, high temperature, or middle temperature environments. In sample HTK-1 (Figure 2d), pyrite is observed in two different colors of copper-yellow and brown-yellow, with light brown as its tarnish. Combined with the hand specimen, the sample is a pentagonal dodecahedral agglomerate, indicating the mineralized fluid may have been in a medium-temperature environment with a high sulfur concentration. The rest of the samples have obvious crystalline grains, higher crystallinity, and are exhibited in granular and massive structures. Besides, crystalline grains can be observed in both light and dark basal parts clearly, and the relatively brighter position is the area where the crystalline grains are more concentrated when observed under polarized light.

Meanwhile, through the observation of backscattering images (BSE), it was found the sample HTK-2 contained colloidal pyrite (Figure 3). Chen et al. (1989) studied the mineralogy of pyrite genesis in the Jiaodong gold deposits and concluded the pyrite precipitated in the low-temperature late stage of hydrothermal mineralization, suggested a higher probability of occurrence of pyrite in the form of cubes, and was mostly produced in the low-temperature and surface conditions [13]. Xie et al. (2021) initially investigated the microscopic structure of colloidal pyrite, and concluded there were hydrothermal and sedimentary colloidal pyrite in the Baochun deposits, with the former produced in the form of veins and the latter produced in the form of calcite grains [14]. The pyrite with fine particles in the sample HTK-2 is distributed in cubic shape on the crystal clusters, it can be discovered the crystal clusters are more complete, and the calcite particles are larger, so it is assumed the formation environment is more stable, and it is justified to be a low-temperature, surface environment [15,16].

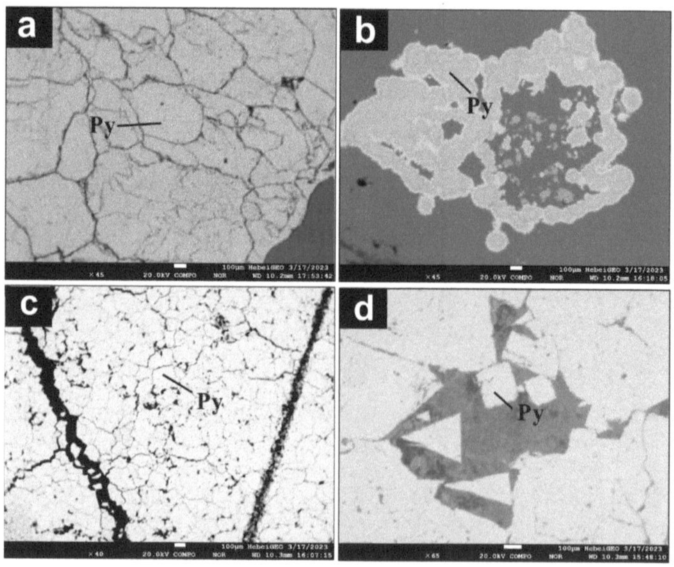

Figure 3. Representative photomicrographs of Backscatter Electron images (BSE), showing different growth habits of pyrite samples. (**a**) Euhedral grains in sample HTK-1. (**b**) Colloform pyrite in sample HTK-2. (**c**) Concretions in sample HTK-3. (**d**) Clearly cubic grains in sample HTK-4.

3.2. X-ray Diffraction Analysis

The five samples were analyzed by X-ray diffraction, and the diffraction peaks of pyrite, crystal, and other minerals can be detected. The distinction is the intensity of the diffraction peaks of the same minerals is distinguished in different samples, which indicates the relative contents of the minerals in the samples differ from each other (Figure 4).

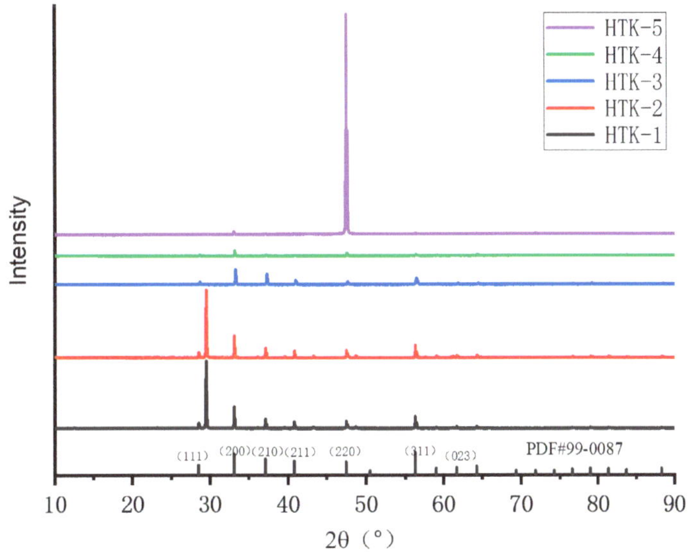

Figure 4. XRD diffraction pattern of samples. The diffraction peaks of 28.62°, 33.13°, and 40.86° can be observed in the five samples, which are basically consistent with the standard pyrite card (PDF#71-2219). Moreover, the diffraction peaks of phlogopite with different degrees exist around 47.44 and 56.36°.

Compared with the spectrum peaks with the standard database of pyrite card (PDF#99-0087), the spectrums of samples are basically inconsistent with other coexisting minerals [17,18]. Peaks of different intensities of pyrite {200} crystal plane and {311} crystal plane are visible in all five samples [19]. Diffraction peaks of sphalerite (PDF#99-0097) were detected at 29.49° in addition to pyrite in sample HTK-3 and sample HTK-4. Besides, the structure of the belt in the sample HTK-3 is obvious, and the dark belt contains more marcasite. The relative intensity of the diffraction peaks in sample HTK-2 is weak, indicating characteristics of poor crystallinity. Almost no other peaks are visible except for the diffraction peaks in the standard card of pyrite, indicating its high purity and no other co-occurring minerals during the growth process [19]. The higher intensity of sample HTK-5 on crystal surface of {220}, the spectrum of the peak is sharper, revealing the pyrite sample crystallization degree is more intact. It shows a minor portion of the diffraction peaks displacement phenomenon, which is caused by other impurity ions into the lattice resulted from lattice defects and changes in the degree of crystallization caused by the mixing, respectively.

3.3. Infrared Spectrum Analysis

The infrared absorption peaks of pyrite are mainly caused by the stretching and bending vibration of Fe-S. The samples were tested for IR spectra separately, and the IR absorption spectra were shown in Figure 5 [19]. Compared with the absorption peak at 407 cm^{-1} of the standard spectrum, the spectrum wave number of pyrite in the samples was shifted to the low-frequency region, where was the stretching and bending vibration of Fe^{2+}-$[S_2]^{2-}$ in the fingerprint region. This shift may be caused by the contained impurities. Once the Fe or S is replaced by impurity elements in pyrite, it changes the intensity and frequency of the resonance absorption, thus changing the characteristics of the infrared absorption spectrum. The intensity of the absorption peak at 440 cm^{-1} which has shift to the low-frequency region is larger for sample HTK-3. Combined with trace element test results, it is indicated the substitution of Co and Ni for Fe is larger. And, the absorption peaks of sample HTK-3, sample HTK-1, and sample HTK-5 in this region are more intense, suggesting they contain more symbiotic minerals. While the spectrum of sample HTK-4 is generally smoother, indicating the sample contains less impurity minerals.

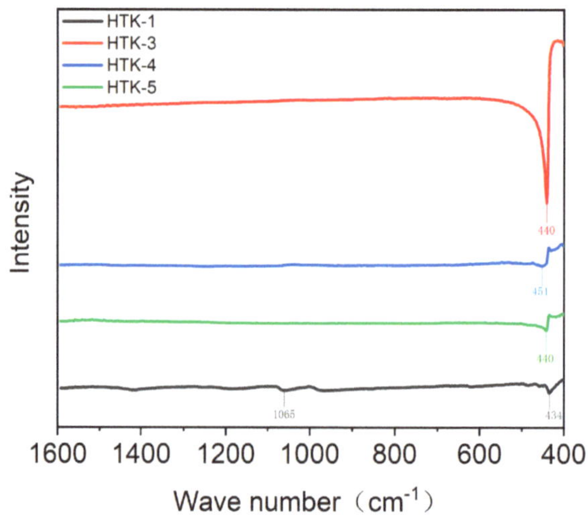

Figure 5. Infrared absorption spectra of each sample. The absorption peaks of different intensities were found at 440 cm^{-1}, and the absorption intensity of HTK-3 was stronger. A weak absorption peak at 1065 cm^{-1} can be found in sample HTK-1, and the absorption curves of HTK-4 and HTK-5 are relatively gentle.

3.4. Raman Spectrum Analysis

The results of Raman spectroscopy show three dominant characteristic peaks of pyrite with a strong peak at 341 cm^{-1}, a strong peak at 376 cm^{-1}, and a weak peak at 430 cm^{-1}, which is attributed to the S-S stretching vibration peak [20]. By comparing the sample data with the standard spectrum, we can observe the lower half-height width (FWHM) of peaks of pyrite indicates a higher degree of crystallinity and order, and the lower formation temperature [20].

The results of Raman spectra are plotted in Figure 6. The standard Raman spectrum of pyrite mainly consists of the three following characteristic peaks. In the vicinity of 343 cm^{-1}, all the samples show relatively narrow absorption bands of weak to moderate intensity, which represents the bending vibration of the Fe-$[S_2]^{2-}$ molecular bond (Eg). In the vicinity of 376 cm^{-1}, all the samples show narrow absorption bands of moderate to strong intensity, suggesting the stretching vibration of Fe$[S_2]^{2-}$ molecular bonds (Ag) in pyrite. According to a previous study, the characteristic Raman shift of pyrite in the vicinity of 379 cm^{-1} can effectively reveal the crystallinity of pyrite and its degree of ordering, and if the FWHM is smaller, it means the pyrite exhibits a higher degree of ordering and crystallinity [20]. However, near 430 cm^{-1}, some samples show weaker intensity and narrower absorption band, indicating the stretching vibration of the S-S bond (Tg) inside the $[S_2]^{2-}$ anion cluster in complex sulfides.

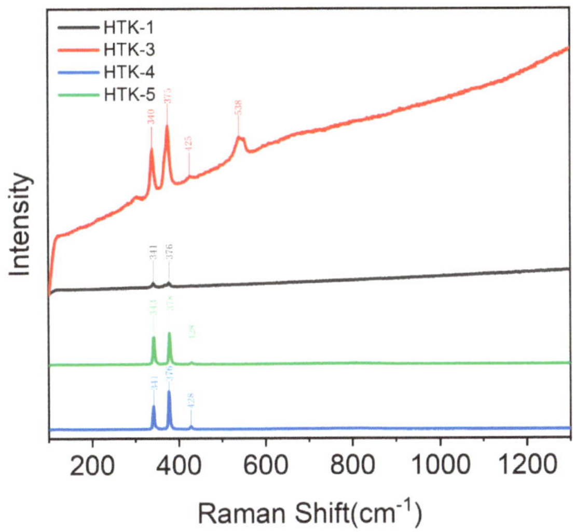

Figure 6. Raman spectra of each sample. Samples show different intensity absorption peaks around 341 cm^{-1} and 376 cm^{-1}, among which the HTK-3 sample also shows a strong absorption peak at 538 cm^{-1}.

Therefore, combined with the probe analysis results, it can be inferred the shifts of Raman spectra peaks of different crystal morphology may be mainly related to the growth environment. The morphology of pyrite can reflect the change of physical and chemical properties of ore-forming fluid well, and the crystal morphology of pyrite in different ore-forming stages is obviously different. Generally, the sample HTK-1 which is pentadodecahedral polygonal shows the weakest absorption intensity and is missing the absorption peak near 430 cm^{-1}, while the lower FWHM indicates poorer ordering and crystallinity and a higher formation temperature at this location. The Sample HTK-3 which is a nodular structure presents the strongest absorption intensity and broader FWHM, suggesting the degree of crystallinity and order is low. The intensity and FWHM of samples HTK-4 and HTK-5 are quite similar, so it can be inferred the formation temperatures and pressures of pyrite from Shangbao of Hunan Province and Peru may have similar characteristics. The sample HTK-5 presents almost no change in the position of the intensity peak at 343 cm^{-1}, indicating the Fe-$[S_2]^{2-}$ groups are more stable and less likely to be replaced by other anion groups.

All samples show characteristic peaks near 341 cm^{-1}, 376 cm^{-1}, and 430 cm^{-1} in the characteristic fingerprint area, which indicates the overall samples are pyrite mineral phases. Raman spectra of the pyrite in this area reflects the changes of the bond lengths and angles, and in turn, these changes in chemical bonds can also imply the temperature and pressure conditions of its formation [8]. In general, the lithostatic pressure is the main force on the mineralizing fluid. With the increase of depth, the lithostatic pressure of the overlying rock layer increases, the pressure of the crystallization of minerals in the mineralizing fluid increases, while the bond length of the crystal atoms is shortened accordingly, and the force constants increase. Thus, the Raman spectrum peaks of the minerals are shifted to higher frequency. The peaks of all samples present a shift to the left to a certain extent, while the left shift of the sample HTK-3 is more obvious. Combined with the test results of Co and Ni elements below, we could infer it was formed in a deeper depth and higher pressure environment. Finally, it could be concluded the sample HTK-3 was formed in a high-temperature sedimentary environment. The peak of pyrite spectrum at 324 cm^{-1} can be observed in the Raman spectrum of sample HTK-3, which is consistent with the results of the XRD test, suggesting the formation of colloidal pyrite is related to marcasite, and an acidic environment is more likely to occur here.

3.5. Major and Trace Elements Analysis

3.5.1. Major Elements Analysis

According to the contents of Fe and S elements of pyrite, the morphology can be determined. The pyrite samples were analyzed by electron microprobe analysis (EMPA) for the main quantitative elements, with w(Fe) 45.78% and w(S) 53.12% in the sample HTK-1, w(Fe) 45.99% and w(S) 53.43% in the sample HTK-2, and w(Fe) 46.16% and w(S) 53.59% in the sample HTK-4, w(Fe) 45.81% and w(S) 53.97% in the sample HTK-5, respectively. The theoretical value of w(Fe) is 46.55%, and the content of w(S) is 53.45% [21]. When the ratio of S/Fe is less than 2, it is called a sulfur deficit, while the ratio of S/Fe is greater than 2, it is called an iron deficit. The two samples are attributed to an iron deficit of comparable degree, which is consistent with the hydrothermal gold deposit of pyrite in the hydrothermal type [21]. The sample HTK-3 contained w(Fe) 46.56%, w(S) 51.88% which was attributed to being Fe-rich and S-poor compared with the sulfur and iron contents in the metal ores formed by sedimentation and the theoretical values [w(Fe) = 46.16%, w(S) = 53.84%], indicating the depositional environment was one of oxygenated bottom water overlying oxygen-poor sediments, with bacterial sulfate reduction occurring below the sediment-water interface, and possibly formed in a higher temperature environment.

3.5.2. Trace Elements Analysis

As one of the important gold-carrying minerals, pyrite is rich in a variety of trace elements, such as Au, Ag, Pb, Zn, Cu, Mo, Co, Ni, As, Sb, Se, Te, Bi, Tl, and Hg, which are closely related to the genesis of gold mineralization [22]. They are mainly presented in pyrite in the form of solid solutions, invisible inclusions at the nanometer scale, and visible inclusions at the micrometer scale endowed in the pyrite lattice or in pyrite.

Generally, the pyrite generated from magmatic hydrothermal gold deposits is enriched in high-, medium-, and low-temperature elements such as Ti, Cr, Mo, and Hg, and the pyrite from volcanic hydrothermal gold deposits is enriched in medium-temperature elements such as Mo and Sn [23]. The pyrite from metamorphic hydrothermal gold deposits is enriched in elements such as Ti and Cr, whereas pyrite from Carlin-type gold deposits is enriched in low-temperature elements such as Tl, Zn, and Hg [24]. Moreover, elements such as Se and Te are mainly enriched in volcanic and magmatic hydrothermal gold deposits, and elements such as Sb and Bi mainly occur in metamorphic hydrothermal gold deposits [25].

In our present study, microscopic observation shows sphalerite and galena inclusions are often observed in pyrite during the metallogenic period, while relatively small pyrite inclusions are also discovered in large-grained sphalerite, and Pb, Cu, and Zn in pyrite are mainly in the form of galena, sphalerite, and chalcopyrite.

The trace elements analytical results of pyrite are listed in Table 1, and the analysis results show pyrite from different origins shows distinct trace elements characteristics.

The results show the contents of trace elements in the sample HTK-3 were high except Co, Ni, Pb, and Cu, and the other elements were low. The contents of Co and Ni in pyrite are related to the high concentration of components in the forming medium and the physical and chemical conditions of their formation, respectively. It has been shown the Co content in pyrite showed a positive correlation with temperature [26], and a negative correlation with oxygen fugacity, so it could be inferred the sample HTK-3 was formed at higher temperature and lower oxygen fugacity. In addition, the sample HTK-3 with nodular and concentric ring structure can be discovered in the longitudinal profile, which suggests it has undergone the same deposition process. Combined with the characteristics of trace elements, it was inferred sample HTK-3 has undergone two stages of sedimentation and diagenesis. The presence of elements such as Au and Ag in pyrite may depend on the mineral inclusions, which is sometimes finely dispersed.

Table 1. Analytical data of major and trace elements in pyrite ($\times 10^{-6}$). Three test points were selected for each sample.

Number	HTK-1-1	HTK-1-2	HTK-1-3	HTK-3-1	HTK-3-2	HTK-3-3	HTK-4-1	HTK-4-2	HTK-4-3
S	313,282	182,869	1436.13	1,916,226	282,969	140,255	38,785,600	39,412,100	4,149,543
Cr	1.25	1.21	115.67	2.71	2.23	8.28	3054.49	1319	1622
Fe	1,452,510	1,531,390	13,285	1,408,446	1,397,150	1,272,798	10,446,837	10,422,333	11,208,000
Co	0.27	1.55	22.42	114.09	3.83	92.27	2948	52,811	134,578
Ni	2.25	11.03	73.14	68.06	3.15	115.47	1902	84.49	71.91
Cu	0.10	0.14	2.39	67.91	6.03	20.38	255.9	191.20	81.71
Zn	1.13	1.23	34.99	6.23	4.61	7.85	194.4	33.50	53.06
As	1926	1992	38.27	7.51	6.22	10.88	5415	6680	23,496
Se	−0.14	−0.04	4.05	0.81	0.86	0.72	31.15	32.51	61.23
Ag	0.00	0.00	6.35	1.40	0.03	0.35	63.62	127.4	−1.78
Sb	20.24	21.77	10.91	0.55	0.11	0.69	108.8	285.15	249.4
Te	0.00	0.00	−0.30	−0.01	−0.01	−0.01	70.66	338.9	1892
Au	0.00	0.00	−0.24	0.00	0.00	0.00	6.00	6.39	−3.91
Hg	-	-	-	-	-	-	122.5	102.3	114.5
Tl	0.28	0.88	1.32	0.15	0.06	0.09	−3.42	7.96	3.50
Pb	0.01	0.04	69.37	52.36	15.64	51.72	1807	8037	478.8
Bi	0.00	0.00	10.28	0.01	0.00	0.03	27,174	46,509	830.9

Both the sample HTK-1 and HTK-4 are associated with quartz and produced in automorphic to semi-automorphic form. The contents of As, Se, and Tl are relatively high in both samples (Table 1). The sample HTK-1 presents a pentagonal dodecahedron with high content of low-temperature elements such as Tl and Zn, suggesting it exhibits Carlin hydrothermal pyrite characteristics. The sample HTK-4 with cubic shape shows a high degree of automorphism. It can be observed pyrite is large particle with a cube structure, and the content of Cr, Hg, and other low-temperature elements is relatively high. It is speculated it should be attributed to magmatic hydrothermal pyrite type.

Generally, the volcanic-hydrothermal pyrite is rich in Mo and Sn medium-temperature elements, and the magmatic-hydrothermal pyrite is rich in Ti, Cr, Mo, Hg, and other high- and low-temperature elements. While the metamorphic-hydrothermal pyrite is rich in high-temperature elements such as Ti and Cr [25]. Besides, As, Ni, Co, Sb, Se, Mo, and other elements are generally uniformly distributed in pyrite, directly into the crystal structure or mixed at nanometer level. Elements such as Bi, Pb, Ag, Au, Te, Cu, and other elements are mixed with pyrite at low density, but can enter the pyrite lattice at high concentration. Zn and Cd elements are in sphalerite microcrystals of pyrite, and their contents vary greatly [27]. The sample HTK-1 shows high As and Sb contents and low Co content. The lattice distortion caused by the addition of As greatly improves the efficiency of Au adsorption by pyrite and may determine the dissolution limit of Au in pyrite. Therefore, the Au in pyrite is usually positively correlated with the As content for most hydrothermal gold deposits [25]. The Co content is affected by the temperature and the oxygen fugacity of the fluids, which suggests it may be formed at lower temperature, higher oxygen fugacity environments. In the sample HTK-1, the trace element contents are generally higher than other samples, especially Co, As, and Bi elements, and the present study shows the Au and Sb contents of pyrite show significant positive correlation [28], which indicates the trace element contents in pyrite crystallized in high-temperature environment are usually higher and richer in variety. While the pyrite crystallized in low-temperature environment implies fewer varieties of trace elements and lower content.

4. S Stable Isotope

Based on S stable isotope characteristics, four main sources of sulfur are proposed: (1) primitive sulfur, which comes from sulfur directly released by magma in the mantle or sulfur leaching from magma sulfide, with $\delta^{34}S$ value close to 0‰, indicating sulfur in the metallogenic fluids exists in single source; (2) seawater and oceanic evaporites, with a

$\delta^{34}S$ value of about 20‰; (3) mixture of sulfur, with a $\delta^{34}S$ value of about 5‰~15‰, which is a mixed source of mantle sulfur and seawater sulfate; (4) organic (bacterial) reduced genesis sulfur, when the $\delta^{34}S$ value is negative, at this time, pyrite under open depositional conditions and belongs to sedimentary type [29].

In this study, $\delta^{34}S$ stable isotopes were carried out on different pyrite production forms, including framboids, colloid, and euhedral grains, and the data are shown in Table 2 and plotted in Figure 7. The sulfides in the samples were all pyrite, and no other sulfate minerals formed under high oxygen fugacity were found, so the sulfur in the hydrothermal fluids mainly existed in the form of S^{2-}, H_2S, and crystallized under the low-oxygen reduction conditions, and the $\delta^{34}S$ values of pyrite could approximately reflect the total sulfur in the ore-forming hydrothermal fluid [30,31].

The $\delta^{34}S$ value of colloidal pyrite in the sample HTK-2 is 0.15‰ to 2.37‰ with an extremely constrained range of variations, and the $\delta^{34}S$ values are very close to the meteoric sulfur range and the mantle sulfur range (0‰ ± 3‰) [31], indicating its $\delta^{34}S$ stable isotopic compositions are relatively stable and the sulfur sources are homogeneous. The $\delta^{34}S$ value of framboidal pyrite shows wide range of variation between 6.21‰ and 30.86‰, and the sample HTK-3 (concretion) has $\delta^{34}S$ values of 2.35‰ to 4.58‰, which mainly falls within the granite range, suggesting its sulfur source is mainly derived from magmatic activity [32,33]. Among them, the $\delta^{34}S$ values of the sample HTK-1 (euhedral grains) are among 16.32‰ to 18.73‰, and the $\delta^{34}S$ stable isotope characteristics are very similar to those in modern seawater, probably suggesting the magma obtaining heavy sulfur from seawater sulfate during underwater spate. Notably, the well-crystallized sample HTK-3 shows the heaviest $\delta^{34}S$ isotopic composition. The $\delta^{34}S$ values of euhedral pyrite in sedimentary rock are relatively high, with $\delta^{34}S$ values ranging from 29.84‰ to 30.01‰, which mainly falls in the range of sedimentary rocks (−50‰~35‰), the high $\delta^{34}S$ value can be considered magma degassing (H_2S relative deficit ^{34}S, H_2S degassing can lead to enriching ^{34}S in melt), or partly due to the sources of seawater jet hydrothermal.

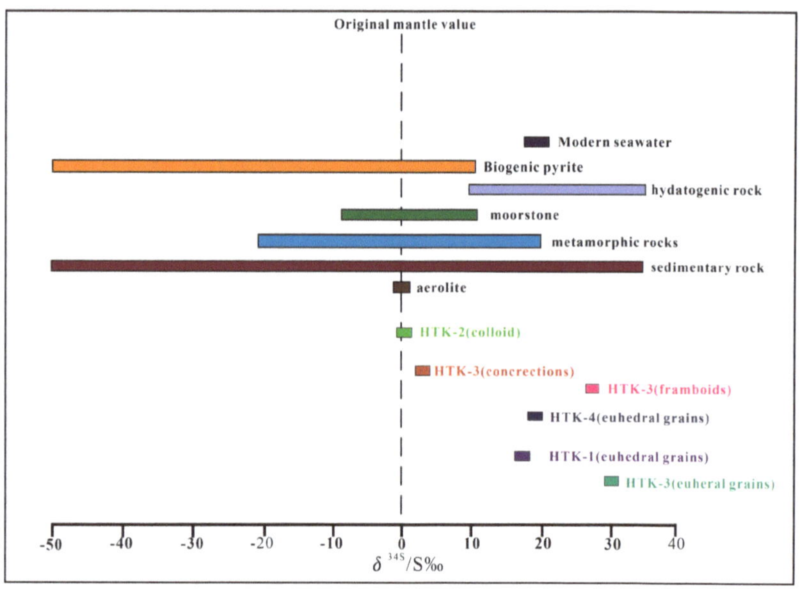

Figure 7. Diagram of $\delta^{34}S$ stable isotope distribution in pyrite (modified after Li, 2016 [33]).

Table 2. $\delta^{34}S$ values of pyrite with different occurrences in the present study.

Samples	Lithology	Mineral	Form	$\delta^{34}S$‰
HTK-3	sedimentary	Pyrite	euhedral grains	29.15~31.01
HTK-3	sedimentary	Pyrite	framboids	5.32~7.01
HTK-3	sedimentary	Pyrite	framboids	29.96~30.86
HTK-2	hydrothermal	Pyrite	colloid	0.15~2.37
HTK-3	sedimentary	Pyrite	concretions	2.35~4.58
HTK-4	hydrothermal	Pyrite	euhedral grains	18.34~20.92
HTK-1	hydrothermal	Pyrite	euhedral grains	16.32~18.73

5. Conclusions

(1) The presence of nodular pyrite formed during sedimentary processes indicated the sedimentary environment has been transitioned into an acidic sulfide reduction environment, with deeper formation depths, higher environmental pressure, and higher temperatures. Spectral and trace element analyses show this type of pyrite exhibits significant substitution of Fe by Co and Ni elements, and the pyrite crystallized in this environment shows higher and more diverse trace element contents.

(2) The cubic pyrite in hydrothermal genesis has high crystallinity and is predominated by coarse-grained euhedral pyrite in the early stages and fine-grained pyrite in the late stages of mineralization, respectively. The ore-forming fluids may be in low-sulfur and high-oxygen fugacity environment, with high As and Cr contents in pyrite. Pentagonal decahedral polycrystalline pyrite grown in layered structure with high-temperature morphological characteristics, so the ore-forming fluids could be in the high-sulfur-concentration medium-temperature environment with high As, Bi, Pb, and Cu contents. Meanwhile, the octahedral pyrite presents good crystallinity and fewer spectral peaks in its spectral characteristics, and magmatic activity controls the formation of the deposit.

(3) The sulfur isotope compositions suggest the sulfur sources in sedimentary pyrite are mainly granite and sedimentary rocks, and the sulfate supply can be divided into two cases: SO_4^{2-} open system and SO_4^{2-} closed system, respectively. The sulfur sources of hydrothermal pyrite exhibit mixed sulfur characteristics, indicate a dominant mantle sulfur signature with mixture of seawater sulfate sources.

Author Contributions: Conceptualization, S.M. and M.S.; methodology, S.M. and M.S.; software, S.M.; validation, S.M., M.S. and C.Z.; formal analysis, S.M., M.S. and Q.C.; investigation, S.M. and Q.C.; resources, M.S.; data curation, S.M. and M.S.; writing—original draft preparation, S.M.; writing—review and editing, S.M., M.S. and C.Z.; supervision, Q.C.; project administration, S.M. and M.S.; funding acquisition, S.M. and Q.C. All authors have read and agreed to the published version of the manuscript.

Funding: This research was jointly funded by National Natural Science Foundation of China (Grant No.42002156), Natural Science Foundation of Hebei Province of China (Grant No.D2021403015), Science and Technology Project of Hebei Education Department (Grant No.QN2021027), The 19th student Science and Technology Fund research project of Hebei University of Geosciences (Grant No.KAY202311), and Hebei Province Key Laboratory of strategic key mineral resources open fund (Grant No.HGU-SCMR2316), and Hebei GEO University National Pre-research Project Funding(Grant No.KY2024YB01). The APC was funded by [National Natural Science Foundation of China], grant number [42002156], [Natural Science Foundation of Hebei Province of China], grant number [D2021403015], and [Science and Technology Project of Hebei Education Department], grant number [QN2021027].

Data Availability Statement: All data are contained within the work.

Acknowledgments: We are grateful to the Editor-in-Chief and the Academic Editors, as well as two anonymous reviewers for their constructive comments and suggestions that greatly improve this manuscript.

Conflicts of Interest: The authors declare they have no known competing financial interests or personal relationships that could have appeared to influence the work reported in this paper.

References

1. Bragg, W.L. The Analysis of Crystals by the X-ray Spectrometer. *J. Proc. R. Soc. Lond.* **1914**, *613*, 468–489.
2. Cook, N.J. Mineralogy of the sulphide deposits at Sulitjelma, northern Norway. *J. Ore Geol. Rev.* **1996**, *5*, 303–338. [CrossRef]
3. Liu, J.; Li, S.; Li, Z.; Liu, Q.; Guo, W.; Zhou, X.; Ma, X. Development characteristics and geological significance of shale pyrite in Chang 7_3 submember of Ordos Basin. *Nat. Gas Geosci.* **2021**, *12*, 1830–1838.
4. Chu, F.Y.; Hu, D.Q.; Yu, H.L.; Yao, J. Significance of pyrite crystal morphology speciation in gold ore evaluation. *J. Jilin Univ. Earth Sci. Ed.* **2004**, *4*, 531–535.
5. Zhang, G.; Nie, H.; Tang, X.; Du, W.; Sun, C.; Chen, S. Types of pyrite in shales and their effects on shale gas enrichment-A case study of Wufeng Formation-Longmaxi Formation shales in the Sichuan Basin and its periphery. *Exp. Pet. Geol.* **2020**, *3*, 459–466.
6. Russo, R.E.; Mao, X.L.; Borisov, O.V.; Liu, H. Influence of wavelength on fractionation in laser ablation ICP-MS. *J. Anal. At. Spectrom.* **2000**, *9*, 1115–1120. [CrossRef]
7. Sack, P.J.; Large, R.R.; Gregory, D.D. Geochemistry of shale and sedimentary pyrite as a proxy for gold fertility in the Selwyn basin area, Yukon. *J. Miner. Depos.* **2018**, *7*, 997–1018. [CrossRef]
8. Shaheen, M.E.; Gagnon, J.E.; Fryer, B.J. Femtosecond (fs) la sers coupled with modern ICP-MS instruments provide new and improved potential for in situ elemental and isotopic analyses in the geosciences. *J. Chem. Geol.* **2012**, *330*, 260–273. [CrossRef]
9. Reich, M.; Kesler, S.E.; Utsunomiya, S.; Palenik, C.S.; Chryssoulis, S.L.; Ewing, R.C. Solubility of gold in arsenian pyrite. *J. Geochi Mica Cosmochim. Acta* **2005**, *11*, 2781–2796. [CrossRef]
10. Norris, A.; Danyushevsky, L. *Towards Estimating the Complete Uncertainty Budget of Quantified Results Measured by LA-ICP MS*; J. Goldschmidt: Boston, MA, USA, 2018.
11. Peterson, E.C.; Mavrogenes, J.A. Linking high-grade gold min eralization to earthquake-induced fault-valve processes in the Porgera gold deposit, Papua New Guinea. *J. Geol.* **2014**, *5*, 383–386. [CrossRef]
12. Chen, G.Y.; Shao, W.; Sun, D.S. *Genetic Mineralogy of Gold Deposits in Jiaodong Region with Emphasis on Gold Prospecting*; Chongqing Publishing House: Chongqing, China, 1989.
13. Xie, Q.; Sun, S.; Ma, Z.; Xu, L.; Wang, J.; Zhou, Y.; Chen, T.; Xu, X. Mineralogical and metallogenic analysis of colloidal pyrite in the Baocungold (Cu) deposit, Tongling. *Acta Geol. Sin.* **2021**, *5*, 1481–1494.
14. Norman, M.; Robinson, P.; Clark, D. Major-and trace-element analysis of sulfide ores by laser-ablation ICP-MS, solution ICP MS, and XRF: New data on international reference materials. *J. Can. Mineral.* **2003**, *2*, 293–305. [CrossRef]
15. Müller, W.; Shelley, M.; Miller, P.; Broude, S. Initial performance metrics of a new custom-designed ArF excimer LA-ICPMS system coupled to a two-volume laser-ablation cell. *J. Anal. At. Spectrom.* **2009**, *2*, 209–214. [CrossRef]
16. Rickard, D.; Luther, G.W. Chemistry of iron sulfides. *J. Chem. Rev.* **2007**, *2*, 514–562. [CrossRef]
17. Mavrogonatos, C.; Voudouris, P.; Zaccarini, F. Multi-stage introduction of precious and critical metals in pyrite: A case study from the Ko nos Hill and Pagoni Rachi porphyry/epithermal prospects, NE Greece. *J. Miner.* **2020**, *9*, 784. [CrossRef]
18. Yao, C.; Song, H.; Li, Q. Raman spectrum characterization of pyrite in the Jiaojia gold deposit in Jiaodong and its diagenetic significance. *Spectrosc. Spectr. Anal.* **2020**, *8*, 2479–2483.
19. Han, Q.M. Current status of pyrite physicochemical speciation characterization. *China Met. Bull.* **2019**, *6*, 188–190.
20. Huston, D.L.; Sie, S.H.; Suter, G.F.; Cooke, D.R.; Both, R.A. Trace elements in sulfide minerals from eastern Australian volcani-chosted massive sulfide deposits; Part I, Proton microprobe analyses of pyrite, chalcopyrite, and sphalerite, and Part II, Selenium levels in pyrite; comparison with delta ^{34}S values and implications for the source of sulfur in volcanogenic hydrothermal systems. *Econ. Geol.* **1995**, *5*, 1167–1196.
21. Belousov, I.; Large, R.; Meffre, S.; Danyushevsky, L.; Steadman, J.; Beardsmore, T. Pyrite compositions from VHMS and orogenic Au deposits in the Yilgarn Craton, Western Australia: Implications for gold and copper exploration. *J. Ore Geol. Rev.* **2016**, *79*, 474–499. [CrossRef]
22. Zhang, H.; Zhao, Q.; Zhao, G.; Hong, J.; Liu, J.; Zhai, D. Technical method for in situ microzonation analysis of pyrite trace elements by LA-ICP-MS and its application in the study of gold deposits. *Geol. Ore Depos.* **2023**, *6*, 1–18.
23. Yan, Y.; Li, S.; Jia, B.; Zhang, N.; Yan, L. Characteristics of metallogenic speciation of pyrite composition during the metallogenic period of gold deposits of different genesis types. *GOLD* **2012**, *3*, 11–16.
24. Román, N.; Reich, M.; Leisen, M.; Morata, D.; Barra, F.; Deditius, A.P. Geochemical and micro-textural fingerprints of boiling in pyrite. *Geochim. Cosmochim. Acta* **2019**, *246*, 60–85. [CrossRef]
25. Gregory, D.D.; Large, R.R.; Halpin, J.A.; Baturina, E.L.; Lyons, T.W.; Wu, S.; Danyushevsky, L.; Sack, P.J.; Chappaz, A.; Maslennikov, V.V.; et al. Trace element content of sedimentary pyrite in black shales. *J. Econ. Geol.* **2015**, *6*, 1389–1410. [CrossRef]
26. Li, W.; Cook, N.J.; Xie, G.Q.; Mao, J.W.; Ciobanu, C.L.; Li, J.W.; Zhang, Z.Y. Textures and trace element signatures of pyrite and arsenopyrite from the Gutaishan Au-Sb deposit, South China. *Miner. Depos.* **2019**, *4*, 591–610. [CrossRef]
27. Liu, Y.; Xie, H.; Xu, L.; Ni, Z.; Lu, F.; Tan, H. Sulfur isotope geochemical characteristics of Ziziping lead-zinc deposit in Jiulong County, Sichuan Province. *Geol. Bull. China* **2020**, *12*, 8.

28. Large, R.R.; Halpin, J.A.; Danyushevsky, L.V.; Maslennikov, V.V.; Bull, S.W.; Long, J.A.; Gregory, D.D.; Lounejeva, E.; Lyons, T.W.; Sack, P.J.; et al. Trace element content of sedimentary pyrite as a new proxy for deep-time ocean–atmosphere evolution. *J. Earth Planet. Sci. Lett.* **2014**, *389*, 209–220. [CrossRef]
29. Ohmoto, H. Systematics of sulfur and carbon isotopes in hydrothermal ore deposits. *Econ. Geol.* **1972**, *5*, 551–578. [CrossRef]
30. Keith, M.; Haase, K.M.; Klemd, R.; Krumm, S.; Strauss, H. Systematic variations of trace element and sulfur isotope compositions in pyrite with stratigraphic depth in the Skouriotissa volcanic-hosted massive sulfide deposit. Troodos ophiolite, Cyprus. *J. Chem. Geol.* **2016**, *423*, 7–18. [CrossRef]
31. Raymond, O.L. Pyrite composition and ore genesis in the Prince Lyell copper deposit, MtLyell mineral field, western Tasmania, Australia. *Ore Geol. Rev.* **1996**, *3*, 231–250. [CrossRef]
32. Chaussidon, M.; Lorand, J.P. Sulphur isotope composition of orogenic spinel lherzolite massifs from Ariege (North-Eastern Pyrenees, France): An ion microprobe study. *Geochim. Cosmochim. Acta* **1990**, *10*, 2835–2846. [CrossRef]
33. Li, J. *Typomorphic Characteristics and Genesis of Pyrite in the Gangcha-Kemo Area of Gansu Province*; Geological Publishing House: Beijing, China, 2016.

minerals

MDPI

Article

Mineralogy and Geochemical Characteristics of Scheelite Deposit at Xuebaoding in Pingwu, Sichuan Province, China

Qinyuan Cao [1], Miao Shi [1,2,3,*], Ye Yuan [4], Shiyu Ma [1] and Haoyu Lu [1]

1 School of Earth Sciences, Hebei GEO University, Shijiazhuang 050031, China; maxcaoqinyuan@163.com (Q.C.); masy0107@163.com (S.M.); l18866545749@163.com (H.L.)
2 Hebei Key Laboratory of Green Development of Rock Mineral Materials, Hebei GEO University, Shijiazhuang 050031, China
3 Engineering Research Center for Silicate Solid Waste Resource Utilization of Hebei Province, Hebei GEO University, Shijiazhuang 050031, China
4 School of Gemmology, China University of Geosciences, Beijing 100083, China; yuenyeah@163.com
* Correspondence: miaoer727@126.com; Tel.: +86-0311-8720-8485

Abstract: Featuring subtle lithological alterations in the host rocks and containing colossal gemstone crystals, the scheelite deposit at Xuebaoding in the Pingwu region of Sichuan Province exhibits characteristics typical of a vein-like hydrothermal-type deposit. The scheelite from the Xuebaoding region is renowned for its high saturation of color, perfect crystal shape, and pure color. In this study, its crystal structure and mineralogical, geochemical, and in situ Sr-Nd isotope characteristics are all systematically characterized. Our objective is to determine the source of ore-forming materials, the timing of the mineralization, and the chemical composition of scheelite, including major elements, trace elements, and rare earths elements (REE). The scheelite samples were analyzed with a variety of methods such as polarizing microscopy, X-ray powder diffraction (XRD), X-ray fluorescence spectrometry (XRF), electron probing, and laser ablation inductively coupled plasma mass spectrometry (LA-ICP-MS). In addition, conventional gemological tests were performed using instruments including gemstone microscopes, GI-UVB ultraviolet fluorescent lamps, grating spectroscopy, etc. The results demonstrate that scheelite exhibits a high refractive index, excellent crystallinity, and a granular structure. Clear color bands and ring structures are observed within the minerals, accompanied by interference colors of light blue, blue, and yellow. Additionally, the mineral components are relatively concentrated, with muscovite and illite serving as accessory minerals. Furthermore, the chemical composition of scheelite reveals a WO_3/CaO mass ratio that approaches or exceeds the ideal value. Moreover, it exhibits a wide range of variations in total rare earth element (∑REE) content, which is characterized by an enrichment of light rare earths (LREE), significant negative Eu anomalies, and insignificant Ce anomalies. In addition, the metallogenic formation of scheelite can be estimated to have occurred during the Toarcian stage in the Lower Jurassic Epoch period, approximately 183 Ma. The study further revealed that A-type granite serves as the genesis type of scheelite, with most of the ore-forming materials originating from the upper crust and a few derived from younger crustal sources.

Keywords: Xuebaoding; scheelite; mineralogy characteristics; in situ elements analysis; Sr-Nd isotopes

check for updates

Citation: Cao, Q.; Shi, M.; Yuan, Y.; Ma, S.; Lu, H. Mineralogy and Geochemical Characteristics of Scheelite Deposit at Xuebaoding in Pingwu, Sichuan Province, China. *Minerals* **2024**, *14*, 38. https://doi.org/10.3390/min14010038

Academic Editor: Anthony Bell

Received: 16 November 2023
Revised: 17 December 2023
Accepted: 19 December 2023
Published: 28 December 2023

1. Introduction

Scheelite is a commonly occurring tungstate mineral and an anhydrous compound with the chemical formula Ca[WO_4], belonging to the tetragonal crystal system. Key characteristics of scheelite include an island structure, a greasy luster, and transparency to translucency. Scheelite occurs in skarn, pegmatites, and hydrothermal veins with high to moderate temperatures [1]. Notably, scheelite is also an important mineral for understanding varieties of geologically diverse ore deposit setting, including prophyries, vein/stockworks, greisens and orogenic Au deposits [2–5]. As a significant tungsten producer, China possesses over 60% of the world's tungsten resources. Among these resources, the reserves and mining output of scheelite consistently ranked first globally [6].

The tungsten–tin–beryllium (W-Sn-Be) deposit in Xuebaoding is located on the northern margin of the Western Longmen Mountains, representing the sole deposit discovered to date. Within the Songpan Ganzi and Qinling orogenic belts, it constitutes a hydrothermal vein deposit formed at high to medium temperatures with W, Sn and Be as its primary elements found in the ore. The deposit is predominantly located within tensile fractures in marble positioned between Pankou and Pukouling granites. The ore veins are composed of beryl, cassiterite, and muscovite within the granite. Meanwhile, the veins are surrounded by rocks composed of marble, which contain beryl, scheelite, cassiterite, and quartz. The scheelite produced from this deposit exhibits relatively large crystals, a beautiful appearance, and reaches gemstone grade [7,8].

Extensive research has been conducted by previous scholars on the scheelite from Xuebaoding. In terms of gemological and mineralogical characteristics [9], study has focused on orange toned scheelite, which exhibits translucency, a greasy luster, and strong blue and white fluorescence observed at short wavelengths (S.W.), as well as good crystallinity. Vibration spectroscopy has been employed to investigate the color genesis of orange–red, orange–yellow, and colorless scheelite [10]. According to the crystal field theory, the color genesis of scheelite from Xuebaoding may be related to the smaller microstructure within the crystal structure, as well as trace elements and REEs. In addition, most research efforts have primarily concentrated on the W-Sn-Be deposit in Xuebaoding [7,11], encompassing investigations into the individual mineral characteristics, mineral zonation features, metallogenic age determination, analysis of metallogenic properties, and exploration of the formation mechanism of the Xuebaoding deposit. Nevertheless, there has been relatively limited research on the scheelite with different color tones in this region.

The present study reveals the gemological and mineralogical characteristics, rare-earth patterns, genesis types, and the timing of the mineralization of scheelite. These findings provided the scientific basis for the identification and prospecting of scheelite from Xuebaoding, thereby enriching the geochemistry and geochronological research content concerning scheelite.

2. Regional Geological Setting

Xuebaoding peak, located at 103.8° E and 32.7° N in Songpan County, Aba Tibetan Autonomous Prefecture, reaches an impressive altitude of 5588 m. As the highest summit in the Minshan mountain range, it stands in the southern section, which extends from north to south. The region below 5000 m elevation is primarily composed of rocky terrain, with the primary peak constituted by carboniferous limestone. The W-Sn-Be gemstone deposit in Pingwu, Sichuan Province, is produced in Moziping, the east-west complex tectonic belt of the Motianling, and is the secondary structure of the upper nano-complex syncline core located in the Zibaishan dome tectonics. The predominantly exposed strata within the mining area encompass the Lower Triassic Bocigou Formation (T_1b), the Middle Triassic Zanaishan Formation (T_2z), and the Upper Triassic Jurassic Formation (T_3zh). Among them, the Zanaishan Formation represents a sequence of metamorphic clastic rock mixed with carbonate rocks, which serve as the main ore-bearing layer of the W-Sn-Be deposit in Xuebaoding [12]. The exposed magmatic rocks in this region predominantly consist of granite with a high content of $WO_3{}^{2-}$, and certain veins are directly situated within the granite mass, which indicates a genetic connection between the late Indosinian to early Yanshan magmatic activity and W-Sn-Be mineralization [13]. In addition, the Upper Triassic Jurassic Formation is mainly composed of slate, schist, and marble, which serve as one of the main perimeter rocks of the W-Sn-Be ore body in Xuebaoding [14]. The lower section of the Zagunao Formation in the Middle Triassic can be divided into eight layers from bottom to top, consisting of phyllite and marble from the shallow metamorphic rock series. Furthermore, the magmatic rocks in the region are a group of granite bodies intruded in the late Indosinian period, which are well developed and occur as rock stocks and veins. Among the rock bodies closely associated with W-Sn-Be mineralization, the most prominent one is Zibaishan rock, which exhibits excellent exposure [8]. The Xuebaoding deposit primarily occurs within a monocline stratum formation at the contact between granite and marble in the Pankou and Pukouling Mountains [7].

3. Materials and Methods

3.1. Materials

In the study, five scheelite samples were tested, including three raw samples with a light yellow to orange tone, a nearly colorless sample with quartz as the accessory mineral, and a light yellow tone sample with muscovite as the accessory mineral (Figure 1). The samples were categorized into two groups: the first group was named raw scheelite (R−Sch), with samples labeled Sch−1, Sch−2, and Sch−3, while the second group was named accessory minerals, with scheelite (A−Sch) samples labeled as Sch−4 and Sch−5.

Figure 1. Samples of scheelite from Xuebaoding. R−Sch samples: Sch−1, Sch−2, and Sch−3 (light yellow to orange tone). A−Sch samples as Sch−4 (transparency and nearly colorless), and Sch−5 (translucent and light yellow tone). R−Sch: raw scheelite A−Sch: accessory minerals with scheelite Sch: scheelite Qtz: quartz Ms: muscovite.

3.2. Methods

A conventional gemstone mineralogy testing experiment was conducted at the Hebei Key Laboratory of Green Development of Rock Mineral Materials. Experimental instruments used included gemstone microscope, polarizing microscope, electronic balance, GI-UVB ultraviolet fluorescent lamp, polar scope, and grating spectroscope. The experiments were carried out under conditions of room temperature at 27.2 °C and an air humidity level of 10%.

X-ray powder diffraction (XRD) experimental testing of the A−Sch samples was completed in the Hebei Key Laboratory of Green Development of Rock Mineral Materials. The experimental instrument used was a Rigaku 9 kW X-ray diffraction analyzer from the Nippon Institute of Science. Experimental conditions with the sample were ground to 200 mesh and the experiment employed a Cu target, K α radiation, a working voltage of 45 kV, and a working current of 200 mA. Scanning was performed within the range of 3°–90° (2θ), with a scanning speed set at 10° (2θ/min, continuous scanning).

X-ray fluorescence spectroscopy (XRF) experiment was conducted at the Gemology Experimental Teaching Center of China University of Geosciences (Beijing), where non-destructive testing and analysis were performed on R−Sch samples. The experimental instrument used was the SHIMADZU EDX-7000 energy dispersive X-ray fluorescence spectrometer from Japan. Experimental conditions included Rh target, 5 mm collimator, a testing element range of 11Na–92U, testing voltages of 15 kV (Na-SC) and 50 kV (Al-U), and a testing current of 1000 μA. The basic parameter method (FP method) was adopted, with the testing environment maintained in a vacuum atmosphere.

An electron probe experiment was conducted at the Hebei Key Laboratory of Green Development of Rock Mineral Materials. Two groups of samples were tested for 15 points and precise data points were obtained for subsequent analysis. The experimental instrument used was the JEOL-JXA-8230 electronic probe manufactured in Japan. Experimental conditions included an acceleration voltage of 15 kV, a working current of 20 nA, and a beam spot diameter of 5 μm. Regarding integration time, the main elements (content of more than 1 wt.%) were measured for 10 s with background measurements taken for 5 s, the trace elements (content of less than 1 wt.%) took 20 s, and the background takes 10 s.

LA-ICP-MS tests and experimental analysis were completed in the Hebei Key Laboratory of Strategic Critical Mineral Resources. Two groups of samples were analyzed at 30 different points, yielding precise data for further analysis. The experimental setup involved the utilization of the Thermo iCAP RQ ICP-MS inductively coupled plasma mass spectrometer in conjunction with a 193 nm ArF laser sourced from the United States. Experimental parameters included a laser beam diameter of 29 mm and a repetition frequency of 6 Hz. NIST 610 was employed as an external standard, while Ca served as an internal standard element.

An in situ Sr-Nd isotope testing experiment was successfully conducted at the Hebei Key Laboratory of Strategic Critical Mineral Resources. Two groups of samples were analyzed at 15 different points, and precise data points were obtained for further analysis. The experimental instrument employed was an LA-MC-ICP-MS equipped with a 193 nm laser. Experimental conditions included a beam spot size of 43 μ, an energy level of 6 J, and a frequency of 5 Hz.

4. Results

4.1. *Gemological Mineralogy Characteristics of Scheelite from Xuebaoding*

4.1.1. Gemological Characteristics

Two groups of scheelite samples were subjected to magnified observations, revealing clear color bands could be observed inside the R−Sch samples with light yellow to orange tone, with some samples exhibiting relatively clean interiors (Figure 2a–c). The A−Sch samples displayed growth morphology, dark inclusions, and ring structures, while the accessory mineral muscovite was visible to the naked eye (Figure 2d–f).

The routine test results indicated that scheelite exhibits a greasy luster, varying degrees of transparency, a range from translucency to transparency, with density falling within the normal range for scheelite (5.8–6.2) g/cm^3 [15]. Through an orthogonal polarizing microscope, non-homogeneous characteristics were suggested by four bright and four dark extinction phenomena observed in this gemstone. Grating the spectroscope analysis revealed absorption lines primarily in the yellow and green regions, with a weak Nd element absorption line detected at 584 nm. The nearly colorless scheelite sample showed no absorption line in the yellow region. However, the yellow tone scheelite sample exhibited absorption lines, regardless of whether the color tone was dark or light (see Table 1).

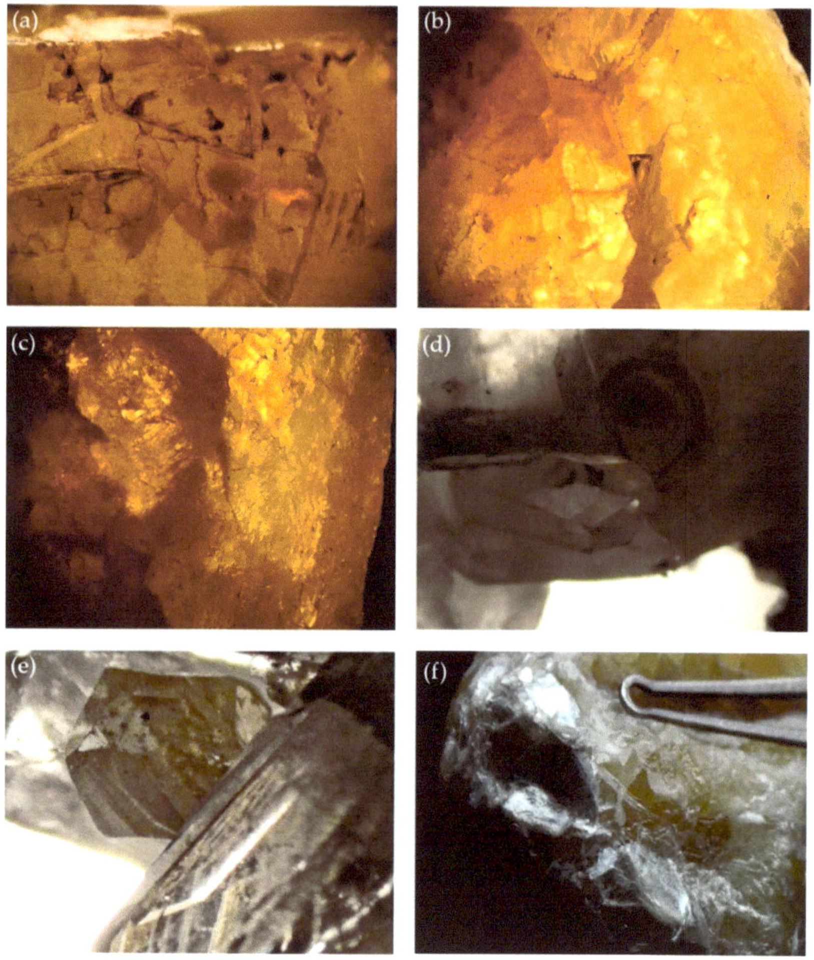

Figure 2. Observation of the internal characteristics by using a gemstone microscope. (**a**) Sample Sch−1 exhibited clear color bands. (**b**) Sample Sch−2 was relatively clean internally. (**c**) Sample Sch−3 was observed to be relatively clean internally. (**d**) Growth morphology was observed in sample Sch−4. (**e**) Sample Sch−4 revealed dark inclusions and a ring structure upon observation. (**f**) The naked eye observation of sample Sch−5 identified the presence of muscovite as an accessory mineral.

Table 1. Basic gemological mineralogical properties of scheelite from Xuebaoding.

Basic Properties	Sch−1	Sch−2	Sch−3	Sch−4	Sch−5
Color	Light yellow to yellow	Yellow	Yellow to orange	Nearly colorless	Light yellow
Relative density	6.07	6.14	6.11	5.81	6.13
Polarscope	Four bright and four dark	Four bright and four dark	Four bright and four dark	Four bright and four dark	Four bright and four dark
Zoomed in and observe	Color band	Clean	Clean	Dark inclusion and ring structure	Accessory mineral

All samples were found to exhibit significant fluorescence. The R−Sch samples with light yellow to orange tone showed no fluorescence at long wavelengths (L.W.), whereas medium to strong blue and white fluorescence was observed at short wavelengths (S.W.). In the A−Sch samples, white fluorescence was observed in Sch−4 (transparent and nearly colorless), while Sch−5 exhibited light blue fluorescence (translucent and light yellow tone) (refer to Table 2). According to previous research [1], it is known that as the color deepens, scheelite exhibits light yellow to white fluorescence. The observation of light blue fluorescence indicates the presence of 0.5 wt.% Mo in scheelite. In contrast, the observation of white fluorescence suggests that the Mo content exceeded 4.8 wt.%.

Table 2. Fluorescence characteristics and absorption spectra of scheelite from Xuebaoding.

Basic Properties	**Sch−1**	**Sch−2**	**Sch−3**	**Sch−4**	**Sch−5**
Fluorescence	L.W. no fluorescence S.W. medium blue to white fluorescence	L.W. no fluorescence S.W. strong blue to white fluorescence	L.W. no fluorescence S.W. strong blue to white fluorescence	L.W. no fluorescence S.W. white fluorescence	L.W. no fluorescence S.W. weak blue to white fluorescence
Grating spectroscope	A weak absorption line in the yellow zone	Two weak absorption lines in the yellow zone	Two weak absorption lines in the yellow zone A weak absorption line in the green zone	A weak absorption line in the green zone	A weak absorption line in the yellow zone

4.1.2. Mineralogical Characteristics

The R−Sch samples with light yellow to orange tone were observed using a polarized light microscope. The granular structure was found to be evenly, uniformly distributed using single polarized light, with mineral defects and cracks evident and all resulting from the grinding of thin slices (Figure 3a,c,e). Additionally, R−Sch samples with light yellow to orange tone displayed interference colors of light blue, blue, and yellow when examined using orthogonal polarized light (Figure 3b,d,f). Single polarized light microscopy revealed that the A−Sch samples exhibited a uniformly distributed granular structure without noticeable discontinuities and includes muscovite as an accessory mineral (Figure 4a,c). Orthogonal polarized light microscopy allowed for observation of interference colors of blue and light blue in the A−Sch samples (Figure 4b,d).

Figure 5 illustrated the back scattered electron images of two groups of scheelite samples. The R−Sch samples with light yellow to orange tone demonstrate a uniform internal mineral distribution without any apparent discontinuities (Figure 5a–c). The A−Sch samples have been observed to contain scheelite, quartz, and some black masses. The black masses are due to the absence of minerals, and the adhesive used to fix the thin film turns black after carbon spraying. Interestingly, the black inclusions that resemble bubbles or gas liquid inclusions are actually bubbles within the glue (Figure 5d). Furthermore, observations of fluorite and calcite confirm that scheelite from Xuebaoding predominantly occupies tensile fractures of the marble between the Pankou and Pukouling granites (Figure 5e,f).

4.1.3. Crystallization Degree

The X-ray powder diffraction (XRD) patterns of the A−Sch samples are presented in Figure 6. Black minerals can be seen on the surface of sample Sch−4. In order to confirm which minerals they were, we conducted XRD tests on these minerals, and the results showed that the black minerals were found to consist of muscovite (M_1) $KAl_2(Si_3Al)O_{10}(OH)_2$, muscovite ($M_2$) $KAl_2(Si_3Al)O_{10}(OH,F)_2$, muscovite ($M_3$) $(K_{0.82}Na_{0.18})(Fe_{0.03}Al_{1.97})(AlSi_3)O_{10}(OH)_2$, and illite (I) $(K,H_3O)Al_2Si_3AlO_{10}(OH)_2$ (Figure 6a).

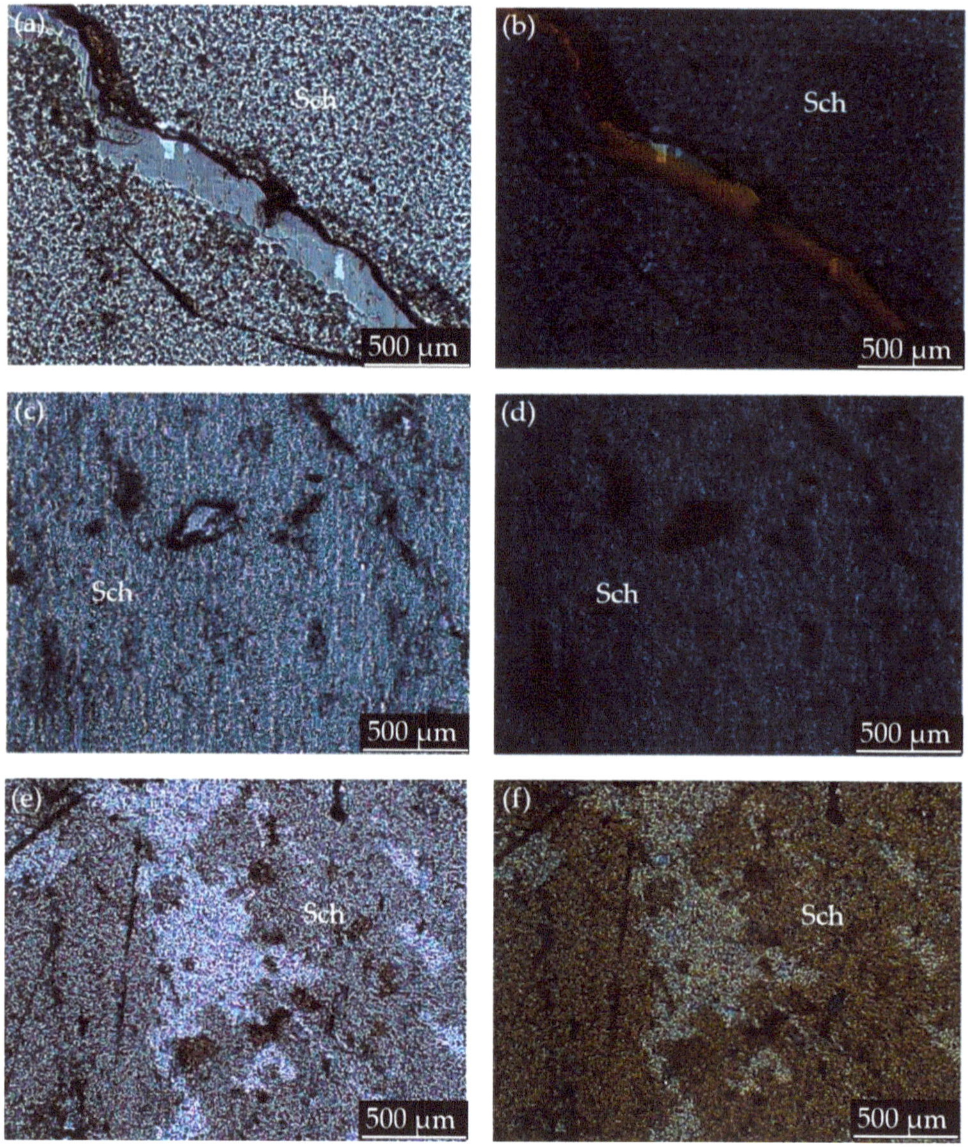

Figure 3. Observation of the R−Sch samples with light yellow to orange tones using a polarizing microscope. (**a**) Use of single polarized light and sample Sch−1 exhibited a uniformly distributed granular structure, with mineral deficiencies caused by grinding thin slices. (**b**) Employing orthogonal polarized light revealed an interference color of blue in sample Sch−1. (**c**) Sample Sch−2 displayed a uniformly distributed granular structure accompanied by cracks when observed using single polarized light. (**d**) Orthogonal polarized light showed an interference color of light blue in sample Sch−2. (**e**) Sample Sch−3 exhibited a uniformly distributed granular structure accompanied by cracks when observed using single polarized light. (**f**) Orthogonal polarized light revealed an interference color of yellow in sample Sch−3. Sch: scheelite.

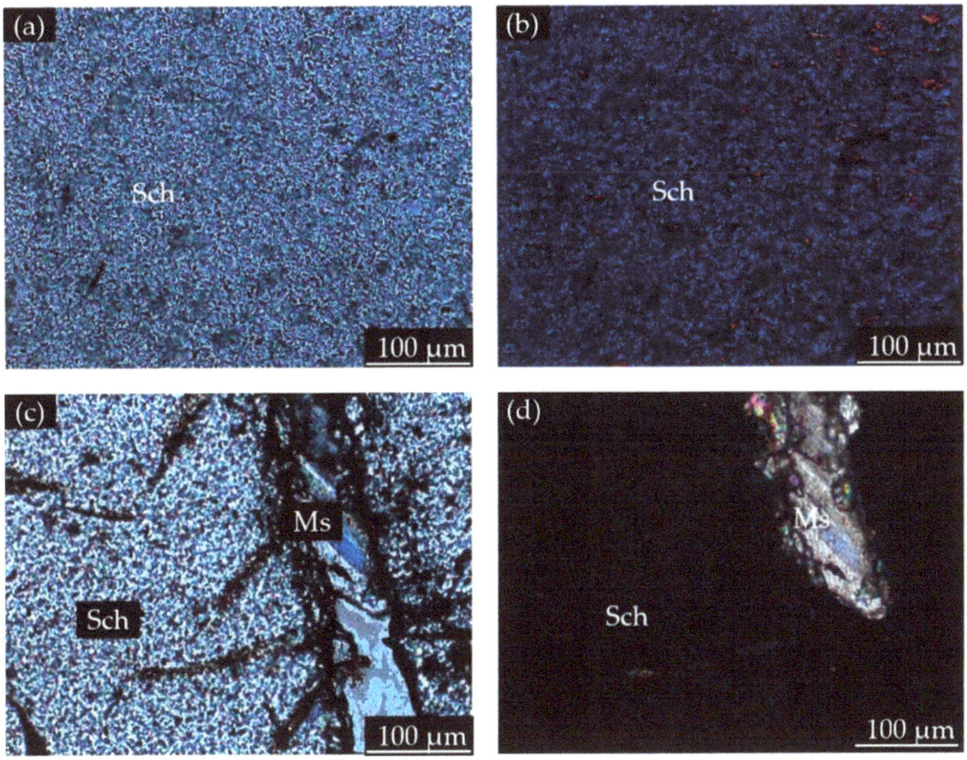

Figure 4. Observations of the A−Sch samples were made using the polarizing microscope. (**a**) Sample Sch−4 exhibited a uniform distribution of a granular structure without any noticeable discontinuities when observed with single polarized light. (**b**) Examining sample Sch−4 with orthogonal polarized light displayed interference colors of blue. (**c**) By utilizing single polarized light, sample Sch−5 showed a uniform distribution of granular structure without any apparent discontinuities, with muscovite identified as the accessory mineral. (**d**) Sample Sch−5 revealed interference colors of blue when viewed with orthogonal polarized light. Sch: scheelite Ms: muscovite.

The peaks indicated in Figure 6 are all characteristic peaks. However, the intensity of some peaks was too low and not considered to be analyzed as characteristic peaks in this study, therefore, they were not labeled. Characteristic strong spectral peaks of muscovite (M_1) with peak positions at 8.8°, 17.7°, 26.7°, and 45.2° can be seen in Figure 6a. Similarly, illite (I) exhibited characteristic strong spectral peaks at 8.8° and 45.2°, which can be detected in sample Sch−4. These peaks correspond to the crystal plane indices of (002), (004), (006), and (136). Additionally, muscovite (M_1) exhibited characteristic weak spectral peaks at 25.4°, 31.9°, 37.7°, 52.7°, 57.4°, 61.9°, 64.0°, 65.0°, and 69.3°, corresponded to crystal plane indices expressed as (−114), (−116), (133), (−139), (313), (156), (316), (333), and (−1313). Equally, muscovite (M_2) displayed characteristic weak spectral peaks at 19.8°, 23.8°, 35.8°, 37.7°, 42.3°, 52.8°, and 55.8° with crystal plane indices expressed as (−111), (023), (−117), (133), (135), (−139) and (139). Furthermore, the characteristic weak spectral peak of muscovite (M_3) were observed at 29.5°, 55.1° and 76.0°, corresponding to crystal plane indices of (025), (−314), and (264). Illite (I) exhibited characteristic weak spectral peaks at 22.8°, 27.8°, 31.1°, and 35.0°, with crystal plane indices expressed as (−113), (114), (115), and (131).

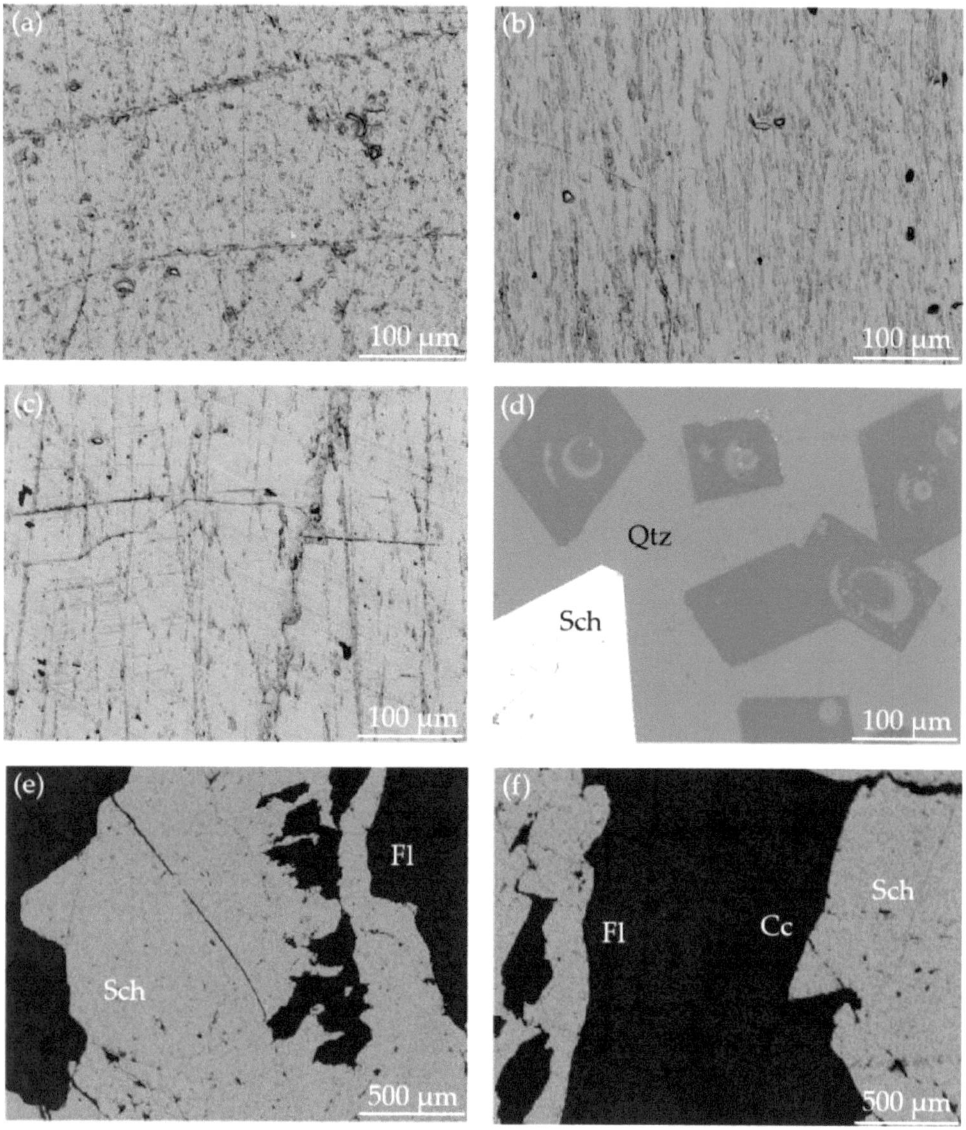

Figure 5. Back-scattered electron images of scheelite from Xuebaoding. (**a**) Sample Sch−1 exhibited a homogeneous internal mineral distribution without any noticeable interruptions. (**b**) Sample Sch−2 displayed a uniform internal mineral distribution devoid of any notable interruptions. (**c**) Sample Sch−3 demonstrated a homogeneous and uninterrupted internal mineral distribution. (**d**) In the case of sample Sch−4, accessory mineral quartz was observed, with the darker parts indicating potential mineral deficiencies. (**e**) Fluorite was detected within Sch−5. (**f**) Calcite was identified inside sample Sch−5. Sch: scheelite Qtz: quartz Fl: fluorite Cc: calcite.

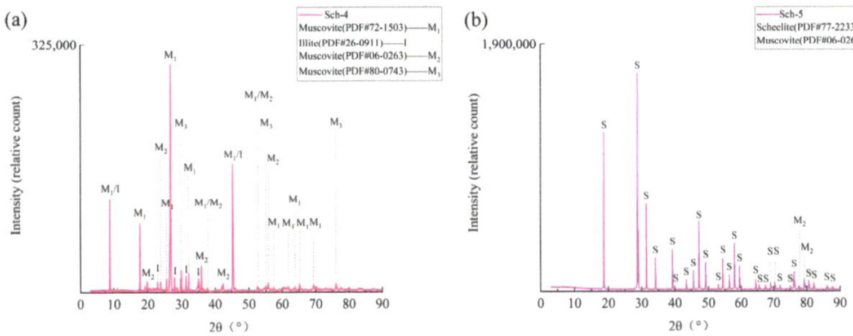

Figure 6. Detection of scheelite from Xuebaoding using X-ray powder diffraction grams. (**a**) Sch−4 exhibits muscovite and illite as accessory minerals. (**b**) Sch−5 shows muscovite as an accessory mineral.

The diffraction pattern of the sample Sch−5 (translucent and light yellow tone) (see Figure 6b) showed a remarkable similarity to the scheelite peak (PDF #77-2233) and contained trace amounts of muscovite (M_2) $KAl_2(Si_3Al)O_{10}(OH, F)_2$. Notably, the spectrum peak of scheelite (PDF #77-2233) utilized in study differed from previous investigations and aligned more closely with the sample employed herein. The characteristic strong spectral peaks of scheelite included 18.6°, 28.7°, 31.4°, 34.1°, 39.1°, 47.0°, 54.2°, 57.8°, and 59.4°, whereas the weaker spectral peaks encompassed 39.9°, 43.3°, 45.4°, 49.0°, 52.9°, 56.2°, 64.5°, 65.5°, 67.5°, 69.1°, 70.4°, 71.9°, 75.0°, 76.1°, 80.6°, 82.1°, 86.0°, and 87.5° all detected within sample Sch−5. Represented by the strong spectral peaks are the crystal plane indices (101), (112), (004), (200), (312), (211), (204), (116), (312), and (224). The crystal plane indices represented by weak spectral peaks are (114), (105), (123), (220), (301), (125), (321), (008), (305), (233), (127), (400), (141), (136), (404), (420), (228), and (129). Additionally, the characteristic spectral peak of muscovite (M_2) is observed at 77.8° and 79.2°, the crystal plane indices are expressed as (352) and (353). Since the main component of sample Sch−5 is scheelite, the crystallinity obtained by fitting the XRD pattern of sample Sch−5 using MDI Jade 6 software is considered to be the crystallinity of scheelite. Among the characteristic spectral peaks of scheelite, the crystallinity of each strong spectral peak were determined to be 85.48%, 77.93%, 91.12%, 96.71%, 95.73%, 92.58%, 96.64%, 95.03%, and 97.48%. Similarly, the crystallinity values of each spectral weak peak were found to be 99.67%, 99.01%, 98.06%, 97.09%, 99.48%, 98.48%, 98.96%, 99.46%, 99.52%, 99.27%, 99.25%, 99.52%, 99.61%, 98.06%, 99.02%, 99.24%, 99.52% and 99.75%. The higher crystallinity observed in each respective peak indicates a high degree of crystallinity within scheelite.

4.2. Geochemical Characteristics of Scheelite from Xuebaoding

4.2.1. Main Element Compositions

The R−Sch samples with light yellow to orange tone were subjected to analysis using an X-ray fluorescence spectrometer (XRF). The findings demonstrated the presence of WO_3, CaO, and P_2O_5, in the ranges of 79.16–80.86 wt.%, 17.17–17.60 wt.% and 0.75–1.64 wt.%, in all samples examined. Among them, the main constituents, WO_3 and CaO, constituted approximately 96 wt.% of the total mass, while also containing trace amounts of P_2O_5, SO_3, ReO_2, and SrO, along with rare earth elements such as Y, Tm, etc. (see Table 3).

The data of two groups of scheelite samples detected by the electron microprobe showed that the main elements WO_3 and CaO, in the range of 79.92–80.95 wt.% and 19.37–19.68 wt.%, accounted for about 99 wt.% of the total mass, accompanied by trace amounts of SiO_2, MnO, K_2O, Na_2O, MgO, and TiO_2. Ideally, the mass ratios of WO_3 and CaO in $CaWO_4$ should be 80.56 wt.% and 19.44 wt.%, respectively. However, it was observed that the samples Sch−1 and Sch−2 exhibited a complete attainment or surpassing of the ideal values for WO_3 and CaO content, while the sample Sch−3 fell below the ideal value for WO_3 content alone. Furthermore, the associated mineral with scheelite samples demonstrated WO_3 and CaO content either below or proximate to their respective ideal values (see Table 4).

Table 3. Chemical composition analysis of scheelite derived from Xuebaoding (wt.%).

Chemical Components	Sch−1	Sch−2	Sch−3
WO_3	80.86	79.16	79.47
CaO	17.43	17.60	17.17
P_2O_5	1.29	0.75	1.64
SO_3	0.37	0.40	-
Y_2O_3	0.05	-	0.04
ReO_2	-	1.32	1.57
Tm_2O_3	-	0.77	-
SrO	-	-	0.11
Total	100.0	100.0	100.0

Note: "-" indicates below detection limit or not detected.

Table 4. Main constituent composition of scheelite derived from Xuebaoding (wt.%).

Main Constituent	Sch−1	Sch−2	Sch−3	Sch−4	Sch−5
WO_3	80.62	80.95	79.92	80.73	80.28
CaO	19.68	19.61	19.54	19.37	19.68
SiO_2	0.23	0.25	-	-	0.25
MnO	0.01	0.02	0.01	-	0.07
K_2O	0.03	0.01	0.01	-	-
Na_2O	0.02	0.05	0.02	0.04	0.01
MgO	0.01	0.02	0.01	-	-
TiO_2	0.01	0.02	-	-	-
Total	100.61	100.93	99.50	100.14	100.29

Note: "-" indicates below detection limit or not detected.

4.2.2. In Situ Trace Element Characteristics

The two sample groups were analyzed at 30 points, and the trace element compositions of scheelite are shown in Table 5. In the R−Sch samples with light yellow to orange tone and the A−Sch sample Sch−5 (translucent and light yellow tone), the contents of trace elements Si, Sr, Fe, Y and Nb exceeded 10×10^{-6}, whereas the contents of trace elements Mg, Mn, Ge and Pb were below 10×10^{-6} but above 1×10^{-6}. However, the contents of trace elements Si, Sr, Fe, and Y in the A−Sch sample Sch−4 (transparency and nearly colorless) exceeded 10×10^{-6},while the contents of trace elements Li, P, Nb, and Mo were below 10×10^{-6} but above 1×10^{-6}. Among the trace elements mentioned above, Sr and Nb could exhibit isomorphism with Ca^{2+}, while Nb and Mo could exhibit isomorphism with W^{6+}. The A−Sch sample Sch−4 (transparent and nearly colorless) contained the most Mo, and a ring structure was visible (Figure 3e). Due to the ubiquitous presence of Mo elements in scheelite, approximately 25 wt.% of W within the mineral lattice was substituted by Mo in isomorphism. As a result, scheelite containing Mo was presented in two different particle forms, light and dark. It often manifests as light-colored scheelite surrounded by dark-colored scheelite, displaying an uneven ring distribution. Consequently, there exists a significant inverse relationship between the content of the W and Mo elements [16]. The scheelite from Xuebaoding region exhibits a relatively high abundance of the lithophile element Sr, with an average value of 166.93×10^{-6} and content ranging from 13.58 to 541.99×10^{-6}. The ratios of Rb/Sr, Nb/Ta, and Zr/Hf were within the ranges of 0–0.02×10^{-6}, 21.70–76.37×10^{-6}, and 8–63×10^{-6}, respectively, with mean values of approximately 0.01×10^{-6}, 48.36×10^{-6}, and 26.78×10^{-6}. Of particular note is the remarkably low Rb/Sr ratio which may be due to the isomorphism of Sr and Ca^{2+}. Furthermore, the lower Rb/Sr and Nb/Ta ratios may indicate that the ore-forming materials of scheelite exhibit characteristics indicative of a shell source.

Table 5. Trace element composition of scheelite from Xuebaoding ($\times 10^{-6}$).

Trace Element	Sch−1	Sch−2	Sch−3	Sch−4	Sch−5	Average Value
Si	922.30	569.87	701.26	619.13	588.76	680.26
P	3.54	0.10	1.14	1.38	0.42	1.32
As	0.51	0.60	0.47	0.31	0.26	0.43
Li	1.86	0.34	0.21	8.52	0.41	2.27
Be	-	-	0.28	0.00	0.17	0.15
Na	3.17	1.90	8.53	0.51	0.94	3.01
Mg	2.14	3.67	1.30	0.49	3.49	2.22
K	0.23	0.46	0.25	0.19	0.26	0.28
Ga	0.84	0.91	0.89	0.04	0.60	0.66
Ge	2.81	3.44	2.60	0.24	1.75	2.17
Rb	0.25	0.21	0.30	0.05	0.29	0.22
Sr	42.17	15.29	221.63	541.99	13.58	166.93
Sn	0.03	0.04	0.03	0.01	0.02	0.03
Pb	3.52	3.83	2.17	0.78	3.36	2.73
Sc	0.01	0.02	0.00	0.02	0.01	0.01
Ti	0.04	0.20	0.01	0.03	0.03	0.06
V	0.01	0.00	0.00	0.01	0.01	0.01
Cr	0.37	0.45	0.21	0.30	0.51	0.37
Mn	5.17	11.27	6.17	0.17	9.44	6.44
Fe	20.74	21.26	21.51	19.91	20.42	20.77
Co	0.03	0.02	0.02	0.02	0.02	0.02
Ni	0.05	0.04	0.05	0.07	0.06	0.05
Y	58.76	26.19	62.87	13.92	20.86	36.52
Zr	0.56	0.67	0.38	0.47	0.63	0.54
Nb	47.98	35.13	36.67	1.91	18.26	27.99
Mo	0.46	0.69	0.45	4.36	0.81	1.35
W	147,244.50	146,973.75	147,473.88	138,478.00	139,973.43	144,028.71
Hf	0.07	0.06	0.08	0.01	0.01	0.05
Ta	0.94	0.46	1.69	0.06	0.30	0.69
Th	0.10	0.21	0.02	0.93	0.87	0.43
U	0.09	0.25	0.03	0.42	0.91	0.34
Rb/Sr	0.01	0.01	0.00	0.00	0.02	0.01
Nb/Ta	51.04	76.37	21.70	31.83	60.87	48.36
Zr/Hf	8.00	11.17	4.75	47.00	63.00	26.78

Note: "-" indicates below detection limit or not detected.

Chondrite normalized REE patterns for the scheelite from Xuebaoding are depicted in Figure 7, with the chondrite-normalized values utilized were obtained by previous researchers [17]. As can be observed in Figure 7a, the curve of the R−Sch samples with light yellow to orange tone evidently has a right-dipping trend and the Eu element depletion. Similarly, the A−Sch samples exhibits a clear right-dipping curve, also indicating Eu depletion (see Figure 7b), as can be seen from Table 6, with the characteristics of REE content. The total rare earth elements (∑REE) of scheelite from Xuebaoding range from 30.32 to 188.74 $\times 10^{-6}$, with an average value of 124.66 $\times 10^{-6}$, demonstrating a low REE content and a wide range of variation. Moreover, the LREE/HREE content ranges from 2.61–3.22 $\times 10^{-6}$, suggesting that the REE composition is more enriched in LREE than in HREE. Notably, Eu exhibits significant negative anomalies, with the δ Eu range from 0.48 to 0.88 $\times 10^{-6}$. However, Ce exhibits insignificant anomalies, with the δ Ce range from 0.89 to 1.01 $\times 10^{-6}$. The results indicate higher elemental contents of Nd and Sm in the R−Sch samples with light yellow to orange tone, while relatively low content was found in the A−Sch sample Sch−4 (transparent and nearly colorless), higher content was present in sample Sch−5 (translucency and light yellow tone). High contents of rare earth elements such as La, Ce, Nd, Sm, Gd, Dy, and Y were detected. Scheelite crystallizes in the tetragonal system and elements such as Na, Sr, Y, Pb and REE can substitute for Ca^{2+} in the scheelite structure [18]. In addition, Mo^{6+} and Nb^{5+} can also substitute for W^{6+} in the tetrahedral $WO_4{}^{2-}$ sites of scheelite [19].

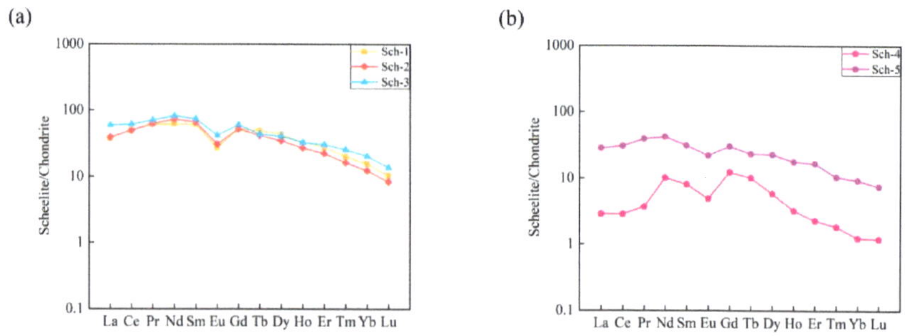

Figure 7. Chondrite-normalized REE patterns for the scheelite from Xuebaoding. (**a**) The R−Sch samples with light yellow to orange tones exhibit consistent curves and significant Eu elemental depletion. (**b**) In the A−Sch samples, a significant difference in content between the two samples can be seen, with the sample Sch−4 (transparent and nearly colorless) having a lower content.

Table 6. Rare earth element composition of scheelite from Xuebaoding ($\times 10^{-6}$).

Rare Earth Element	Sch−1	Sch−2	Sch−3	Sch−4	Sch−5	Average Value
La	11.89	12.13	18.53	1.98	8.83	10.67
Ce	40.58	39.85	49.33	10.05	24.75	32.91
Pr	7.48	7.68	8.64	2.64	4.78	6.24
Nd	37.11	43.90	49.36	6.09	25.23	32.34
Sm	12.08	12.94	14.50	1.58	6.05	9.43
Eu	2.01	2.27	3.08	0.36	1.61	1.87
Gd	13.56	13.70	15.69	3.21	7.81	10.79
Tb	2.36	1.97	2.06	0.48	1.10	1.59
Dy	14.09	11.20	13.17	1.89	7.32	9.53
Ho	2.39	1.95	2.37	0.42	1.27	1.68
Er	5.87	4.71	6.43	0.93	3.51	4.29
Tm	0.66	0.53	0.83	0.11	0.34	0.49
Yb	3.31	2.59	4.29	0.54	1.95	2.54
Lu	0.35	0.28	0.46	0.04	0.25	0.28
Y	58.76	26.19	62.87	13.92	20.86	36.52
∑RFF + Y	212.50	181.89	251.61	44.24	115.66	161.18
∑REE	153.74	155.7	188.74	30.32	94.80	124.66
LREE	111.15	118.77	143.44	22.70	71.25	93.46
HREE	42.59	36.93	45.30	7.62	23.55	31.20
LREE/HREE	2.61	3.22	3.17	2.98	3.03	3.00
$(La/Yb)_N$	2.42	3.16	2.91	2.47	3.05	2.80
$(La/Sm)_N$	0.62	0.59	0.80	0.79	0.92	0.74
$(Gd/Yb)_N$	3.31	4.22	2.95	4.80	2.04	3.46
δEu	0.48	0.52	0.62	0.48	0.88	0.59
δCe	1.01	0.97	0.93	0.89	0.91	0.94

4.2.3. In Situ Sr-Nd Isotope Characteristics

Sr, Sm, and Nd elemental and isotopic characteristics of scheelite from Xuebaoding were analyzed, and the results are presented in Table 7. At the same times, multiple samples were tested at 15 different points. Finally, the findings indicated that the Sr elemental content of scheelite ranges from 13.58 to 541.99 $\times 10^{-6}$, while the variation range of $^{87}Rb/^{86}Sr$ and $^{87}Sr/^{86}Sr$ values was observed to be between 0.00003 to 0.00149 and 0.71103 to 0.75407, respectively. Furthermore, the Sm elemental content ranged from 1.58 to 14.5 $\times$ 10^{-6} and Nd element content ranged from 6.09 to 49.36 $\times$ 10^{-6} in scheelite samples obtained from the Xuebaoding region. Similarly, the range of $^{147}Sm/^{144}Nd$ and $^{143}Nd/^{144}Nd$ values varied from 0.1690 to 0.4042 and from 0.51147 to 0.54555, respectively. However, the Sm/Nd ratio of scheelite from Xuebaoding ranged between 0.24 and 0.33, The application of the Sm-Nd isotopic system for scheelite dating has been successfully demonstrated by previous research [20,21].

Table 7. Sr-Nd Isotopes of scheelite from Xuebaoding.

Isotope	Sch−1	Sch−2	Sch−3	Sch−4	Sch−5
$Sr/10^{-6}$	42.17	15.29	221.63	541.99	13.58
$Sm/10^{-6}$	12.08	12.94	14.50	1.58	6.05
$Nd/10^{-6}$	37.11	43.9	49.36	6.09	25.23
Sm/Nd	0.33	0.29	0.29	0.26	0.24
$^{87}Rb/^{86}Sr$	0.00097	0.00149	0.00019	0.00003	0.00148
$^{87}Sr/^{86}Sr$	0.71789	0.71920	0.71103	0.75407	0.71477
$\pm 2\sigma$	0.0006	0.0010	0.0001	0.0001	0.0001
$^{147}Sm/^{144}Nd$	0.2092 0.2189 0.2221	0.2307 0.2059	0.1690 0.1778	0.4042	0.1614
$\pm 2\sigma$	0.0005	0.0011	0.0005	0.0239	0.0004
$^{143}Nd/^{144}Nd$	0.51199 0.51147 0.51190	0.51196 0.51181	0.51168 0.51199	0.54555	0.51177
$\pm 2\sigma$	0.0003	0.0004	0.0004	0.0214	0.0005
ε_{Nd} (0)	−12.64 −22.78 −14.40	−13.23 −16.15	−18.69 −12.64	641.95	−16.93
ε_{Nd} (*t*)	−10.77 −10.54 −10.47	−10.27 −10.85	−11.71 −11.51	−6.21	−11.89

Both groups of tested samples exhibit a linear distribution trend, however, only the R−Sch samples with light yellow to orange tones and the A−Sch sample Sch−5 could demonstrate a good straight line of fit. The Isoplot program [22] was used for analyzing Sm-Nd isotopic ages, with a decay constant value of λ = 6.54 $\times$ 10^{-12} a^{-1} for ^{147}Sm utilized in the calculation. Chondrite Uniform Reservoir (CHUR) values of 0.1967 for $^{147}Sm/^{144}Nd$ and 0.512637 for $^{143}Nd/^{144}Nd$ were adopted [23]. Thus, the results indicate an isochronal age of (183 $\pm$ 8.6) Ma (n = 8, MSWD = 1.1) with an initial ratio of $^{143}Nd/^{144}Nd$ as (0.5116 $\pm$ 0.0012) for scheelite from Xuebaoding, all within the error range (see Figure 8). According to the ICS international chronostratigraphic chart [24], the age of scheelite can be inferred as dating from the Toarcian stage in the Lower Jurassic Epoch. During the process, the ε_{Nd} (*t*) and the ε_{Nd} (0) values were calculated for scheelite from Xuebaoding at t = 183 Ma, yielding ranges of −10.27 to −11.89 (average: −11.00), and −12.64 to −22.78 (average: −15.93), respectively. Because of the extremely low measured Rb element content amounts in scheelite, all below 1 $\times$ 10^{-6}, the contribution of ^{87}Sr from the decay of ^{87}Rb can be considered negligible. Based on this, the measured values of $(^{87}Sr/^{86}Sr)_0$ can be replaced with $^{87}Sr/^{86}Sr$ to elucidate the provenance of ore-forming materials.

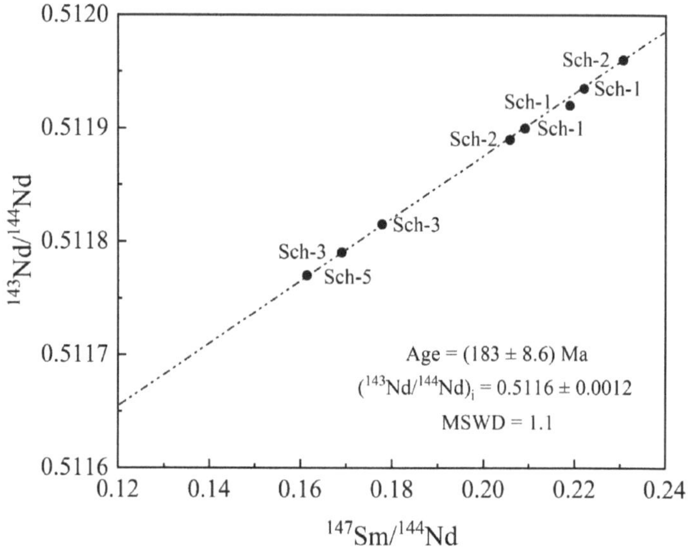

Figure 8. An Sm-Nd isotope isochronal chart of scheelite from Xuebaoding, indicating a metallogenic age of 183 Ma, formed in the Lower Jurassic Epoch.

5. Discussion

5.1. Rare-Earth Patterns of Scheelite from Xuebaoding

The rare-earth patterns of scheelite from the Xuebaoding W-Sn-Be deposit, as studied in this paper, were basically consistent with previous researchers [25]. A relatively smooth transition pattern was exhibited, with a slight enrichment of LREEs and a minor loss of HREEs. There were no significant anomalies observed for Ce, while an evident moderately negative anomaly was observed for Eu. However, it should be noted that the total content of rare earths was comparatively low compared to previous studies. The moderately negative anomalies exhibited by Eu elements (δ Eu = 0.48–0.88 $\times 10^{-6}$, average: 0.59 $\times 10^{-6}$) were similar to those found in the oxide stage scheelite from the Donggushan tungsten deposit in Anhui Province (δ Eu = 0.30–0.68 $\times 10^{-6}$, average: 0.51 $\times 10^{-6}$). Additionally, the rare-earth patterns displayed clear enrichment in LREEs and deficiency in HREEs [26]. They were clearly different from scheelite in other regions, such as the three generations of scheelite from Guilin Zhengfu Molybdenum in the Jiangnan tungsten belt (δEu = 0.08–0.99 $\times 10^{-6}$, 0.09–0.26 $\times 10^{-6}$, and 0.11–0.36 $\times 10^{-6}$), scheelite from Zhaishang gold mine in the West Qinling Mountains (δ Eu = 0.75–1 $\times 10^{-6}$), scheelite from the Xianggou gold and tungsten mines in the South Qinling Mountains (δ Eu = 0.61–0.83 $\times 10^{-6}$, average: 0.72 $\times 10^{-6}$), and scheelite from the Baojinshan mine in Hunan Province (δ Eu = 0.87–1.44 $\times 10^{-6}$) [27–30]. Thus, the rare-earth patterns of scheelite from Xuebaoding exhibit a difference from the conventional MREE enriched type and instead display a flat type without MREE enrichment, accompanied by negative Eu anomalies. Based on this, the tetrad effect in the rare-earth patterns is characterized by a gradual decrease in LREEs and HREEs [25]. Of course, the salinity of the ore-forming fluid in the Xuebaoding beryl scheelite deposit is classified as low salinity within the NaCl-CO_2-H_2O system, with scheelite exhibiting a predominant salinity range between 3% and 6% (NaCl) [12]. However, both sample groups utilized in this study lacked sodium minerals, displaying a lower chemical composition content of Na_2O ranging from 0.01 to 0.05%, along with the trace element Na at levels of 0.51–8.53 $\times 10^{-6}$. Therefore, the above findings suggest that the scheelite from Xuebaoding was formed in a low-sodium environment. Moreover, the influencing factor of rare-earth patterns is the partition coefficient of rare earths in the hydrothermal fluid rather than crystal chemistry.

The tetrad effect of rare earths refers to the demarcation point represented by Nd/Pm, Gd, Ho/Er, where every four rare earths constitute a distinct group; that is La-Ce-Pr-Nd, Pm-Sm-Eu-Gd, Gd-Tb-Dy-Ho, and Er-Tm-Yb-Lu. According to the arrangement of atomic numbers, specific physical and chemical properties or their natural composition gives rise to four groups of convex or concave curves. The convex curve represents the M type rare earth tetrad effects, while the concave curve corresponds to the W-type rare earth tetrad effects [31]. By quantitatively identifying the rare earth tetrad patterns of scheelite from Xuebaoding as either M type or W type, based on the parameters proposed by Irber [32] and Monecke et al., [33] we utilized the central elements (Ce, Pr or Tb, Dy) of each quaternary subgroup from both sample groups. Using the prescribed methodology, the normalized values of the chondrites (Ce_N, Pr_N or Tb_N, Dy_N)can be computed, along with the interpolation ratio (Ce_N/Ce^*, Pr_N/Pr^*) for the initial elements (La, Gd) and final elements' (Nd, Ho) lines within each of the four subgroups. As a result, the M-type rare earth tetrad types exhibit a ratio greater than one and a convex curve, whereas the W type rare earth tetrad types demonstrate the opposite [34]. The intensities were measured using Irber and Monecke's method to calculate T_1, T_3, t_1, t_3, and $TE_{1,3}$, and the corresponding calculation results are shown in Table 8. From the obvious tetrad effect of $TE_{1,3} > 1.10$, the samples Sch−1 and Sch−4 exhibit the rare earth tetrad effects is clearly within the two groups of scheelite samples, while samples Sch−2, Sch−3, and Sch−5 do not display a pronounced effect. Although the method can be applied to granite, the scheelite form Xuebaoding mainly occurs in the tensile fractures of marble between the Pankou and Pukouling granite, thus the method is applicable.

Table 8. The M and W type four group effect parameters of scheelite from Xuebaoding.

Four Group Effect	Sch−1	Sch−2	Sch−3	Sch−4	Sch−5
Ce_N/Ce^*	1.12	1.02	0.92	1.67	0.94
Pr_N/Pr^*	1.16	1.06	0.96	2.49	1.06
Tb_N/Tb^*	1.11	0.99	0.88	1.05	0.92
Dy_N/Dy^*	1.13	1.03	1.01	0.78	1.08
Tm_N/Tm^*	1.01	1.01	1.09	1.18	0.82
Yb_N/Yb^*	1.09	1.06	1.14	1.39	0.95
T_1	0.18	0.06	0.07	1.56	0.07
T_3	0.13	0.03	0.06	0.25	0.15
t_1	1.14	1.04	0.94	2.04	1.00
t_3	1.12	1.01	0.94	0.91	0.99
$TE_{1,3}$	1.13	1.02	0.94	1.36	1.00

Note: $Ce^* = La_N^{2/3} \times Nd_N^{1/3}$; $Pr^* = La_N^{1/3} \times Nd_N^{2/3}$; $Tb^* = Gb_N^{2/3} \times Ho_N^{1/3}$; $Dy^* = Gb_N^{1/3} \times Ho_N^{2/3}$; $Tm^* = Er_N^{2/3} \times Lu_N^{1/3}$; $Yb^* = Er_N^{1/3} \times Lu_N^{2/3}$ [34]. $TE_{1,3}$ is defined as the degree of tetrad effect.

5.2. *The Metallogenic Age of Scheelite from Xuebaoding*

The interior of the scheelite crystals in both sample groups exhibited uniform mineral composition and no growth discontinuities. Meanwhile, Sm-Nd isotope dating showed represent a metallogenic age of 183 Ma for the formation of scheelite, which was occurring during the Lower Jurassic Epoch, which aligns well with previous studies [25]. In addition, the $^{40}Ar/^{39}Ar$ fast neutron activation dating of quartz was employed to determine a metallogenic age of 186 Ma for the Xuebaoding deposit, which belongs to the early Yanshan period. Moreover, the mineral crystallization solution was formed during the magmatic period (224 Ma), while the hydrothermal activity associated with mineralization primarily took place after this period (186–192 Ma) [12]. However, due to limitations in using only quartz for determining metallogenic age and considering that the P-T conditions for quartz formation span a wide range, the direct utilization of quartz for the age of the Xuebaoding deposit is not feasible. Although both sample groups used in the article are from Xuebaoding, but do not belong to the same vein and lack accessory minerals such as cassiterite, beryl, and other minerals, so they only represent the metallogenic age of scheelite. Nevertheless, the main metallogenic age of the Xuebaoding W-Sn-Be rare metal deposit is represented by a metallogenic age of 182 Ma observed in scheelite deposits within the second belt of Xuebaoding vein due to its close symbiotic relationship with minerals like cassiterite and beryl [25].

5.3. Genesis Types of Scheelite from Xuebaoding

According to previous research, the Xuebaoding W-Sn-Be deposit may represent a novel type of deposit related to A type granite [25]. Of course, four important indicators of A type granite have been discovered. Firstly, the presence of enriched Si, Na, K elements and the absence of Ca, Mg and Al elements are present in the rocks. Secondly, there are high ratios of (K_2O + Na_2O)/Al_2O_3, FeO/MgO and Ga/Al values. Thirdly, enrichment in high-field-strength elements such as Rb, Th, Nb, Ta, Zr, Hf, Ga, Y is observed while Sr, Ba, Cr, Co, Ni, V, Eu, Ti and P elements are barren. Fourthly, the right dipping seagull type REE distribution pattern and significant negative Eu anomalies are evident [35]. Based on the study, the primitive mantle standardized spider graph of trace elements (see Figure 9) in two groups of sample scheelite from the Xuebaoding demonstrates that the scheelite is enriched in elements such as La, Nd, and Y, while exhibiting a significant lack of high-field-strength elements including P, Zr, and Ti, as well as large ion lithophile elements like K and Ba. In addition, the range of values for $10^4 \times$ Ga/Al values between 3.50 and 5.67, which is notably higher than that of the I type (average: 2.1) and S type (average: 2.28) granite, indicate a consistency with A-type granite (>2.6) [36]. However, on the graphs of Nb, Ce, Zr, Y$-10^4 \times$ Ga/Al (see Figure 10), both sample groups are classified within the A type region. In summary, the genesis type of scheelite from Xuebaoding is A-type granite, which serves as the ore-bearing host rock of scheelite, which can be mutually confirmed with the tensile fractures of marble mainly occurring between the Pankou and Pukouling granites.

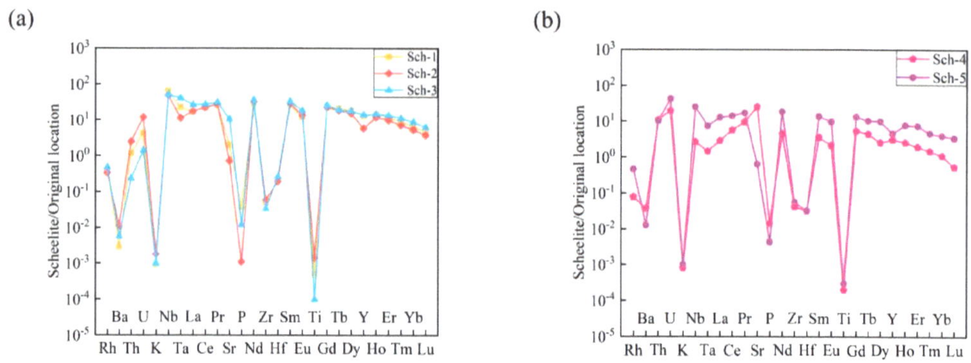

Figure 9. Primitive mantles standardized the spider graph of trace elements in scheelite from Xuebaoding. (**a**) In R−Sch samples with light yellow to orange tone, the curve fluctuations exhibit relatively consistent, characterized by barren high-field-strength elements such as P, Zr, Ti, and large ion lithophile elements like K and Ba. (**b**) In the A−Sch samples, both samples demonstrate similar contents of barren high-field-strength elements including P, Zr, and Ti.

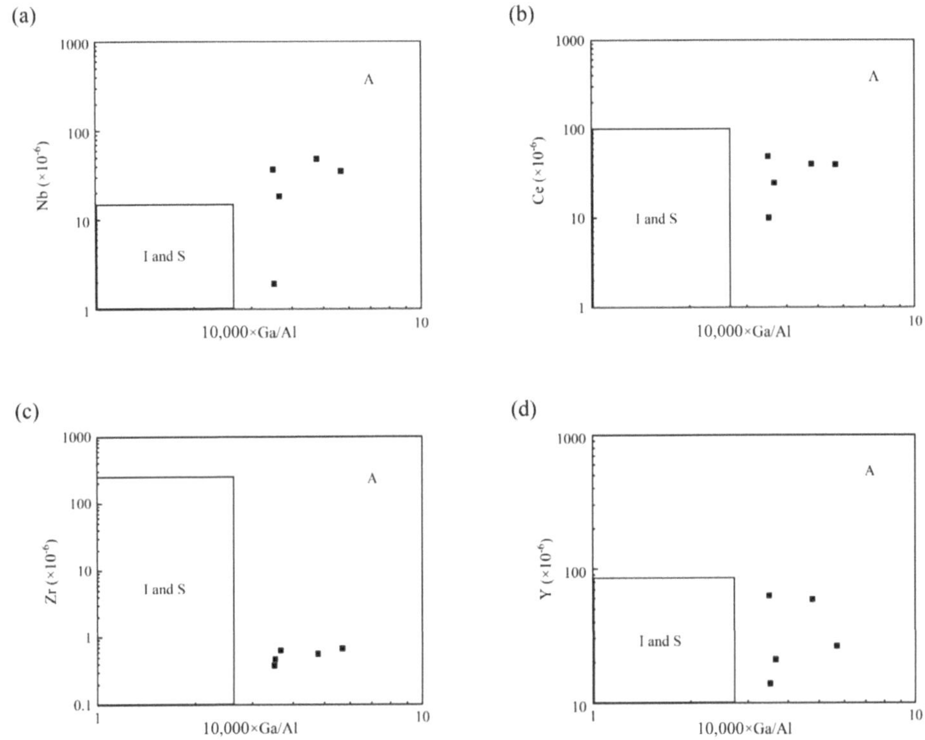

Figure 10. Diagram of Nb, Ce, Zr, Y-$10^4 \times$ Ga/Al in scheelite from Xuebaoding according to previous studies [36]. (**a**) The diagram of Nb$-10^4 \times$ Ga/Al in two groups of scheelite samples, it can be seen that both fall into the A region but are dispersed in distribution. (**b**) Diagram of Ce$-10^4 \times$ Ga/Al in two groups of scheelite samples shows that the samples all fall into the A region and are more dispersed. (**c**) As shown in the diagram of Zr$-10^4 \times$ Ga/Al in two groups of scheelite samples, both fall into the A region and the distribution is concentrated. (**d**) Y$-10^4 \times$ Ga/Al diagram of two groups of scheelite samples both fall into the A region, with a relatively dispersed distribution.

5.4. The Source of Ore-Forming Materials of Scheelite from Xuebaoding

The Sr-Nd isotopes in scheelite is very effective indexes to confirm the source of the ore-forming minerals [37]. The $\varepsilon_{Nd}(t)$ values of scheelite from Xuebaoding range from -10.27 to -11.89, while the $(^{87}Sr/^{86}Sr)_0$ values span between 0.71103 and 0.71920, with an average value of 0.71512. However, the average $^{87}Sr/^{86}Sr$ ratio of crustal derived magma is typically 0.719, while magma with ratio ranging from 0.706 to 0.719 originates from the source region where crust and mantle mix. At the same time, the value of $\varepsilon_{Nd}(t)$ is bounded by 0, with positive values indicating magma derived from the depleted mantle and negative values representing the magma sourced crust or enriched mantle [38]. From the range of endpoint values in the $(^{87}Sr/^{86}Sr)_0$-$\varepsilon_{Nd}(t)$ relationship diagram, it can be found that the scheelite from Xuebaoding mainly originates from the upper crust and a subset of samples fall within the younger crust region (Figure 11). Samples Sch$-$1, Sch$-$2, and Sch$-$3 are located in the upper crust, sample Sch$-$5 is located in the younger crust, and sample Sch$-$4 is not within the range shown in the figure.

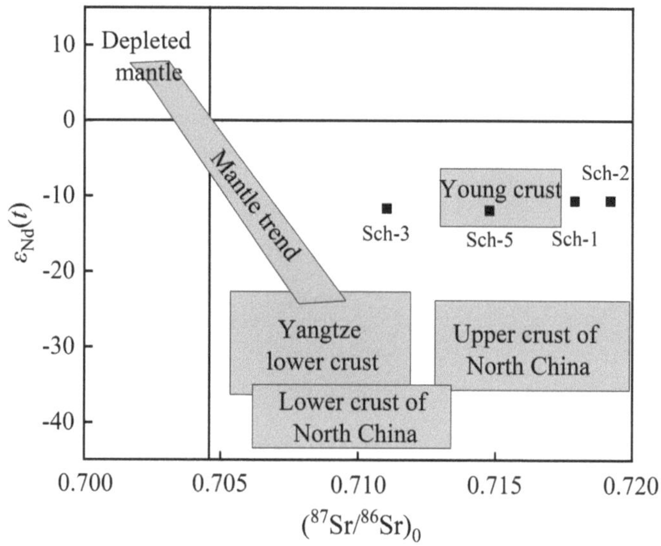

Figure 11. A diagram illustrating the $(^{87}Sr/^{86}Sr)_0$-$\varepsilon_{Nd}(t)$ relationship with scheelite from Xuebaoding, indicating the predominant occurrence in the upper crust and partial presence in the younger crust section according to predecessors [39].

6. Conclusions

(1) Scheelite from Xuebaoding exhibits a weak dichroism in a light yellow tone, and the density decreases with increasing Mo content. In addition, L.W. displays inert fluorescence, while S.W. demonstrates moderate to strong blue white fluorescence. Clear color bands and ring structures are observed inside. The scheelite crystals possess crystallinity with a distinct granular structure, and the interference color falls into three categories: light blue, blue, and yellow. The mineral components are relatively concentrated, primarily composed of muscovite and illite as associated mineral along with minor amounts of fluorite and calcite.

(2) The mass ratio of the main chemical components WO_3 and CaO in scheelite closely approximates or exceeds the ideal value. The content of trace elements Si, Sr, and Fe is relatively high. The REE composition indicates a higher enrichment of LREE compared to HREE, with significant Eu negative anomalies and insignificant Ce anomalies. The formation age of scheelite can be traced back to the Early Jurassic period, with an age of 183 Ma.

(3) The scheelite from Xuebaoding was formed in a low sodium environment, and the rare earth patterns exhibit the tetrad effect, which is influenced by the distribution coefficient of rare earths in the hydrothermal solution rather than crystal chemistry. The genesis type corresponds to A type granite, which serves as the host rock for scheelite ore. Sr-Nd isotopes indicate that the material source primarily originates in the upper crust, with some coming from the young crust.

Author Contributions: Q.C. and M.S. designed the experiments; Q.C. and Y.Y. performed the experiments; S.M. resources; H.L. investigation; Q.C. analyzed the data and wrote the paper. All authors have read and agreed to the published version of the manuscript.

Funding: The research was funded by National Natural Science Foundation of China (Grant No. 42002156), Natural Science Foundation of Hebei Province of China (Grant No. D2021403015), Science and Technology Project of Hebei Education Department (Grant No. QN2021027), Hebei GEO University National Pre-research Project Funding (Grant No. KY2024YB01), Graduate Innovation Project of Hebei Province of China (Grant No. CXZZSS2023134), the Opening Foundation of Hebei Key Laboratory of Strategic Critical Mineral Resources (Grant No. HGU-SCMR2310 and HGU-SCMR2309). The APC was funded by [National Natural Science Foundation of China], grant number [42002156], [Natural Science Foundation of Hebei Province of China], grant number [D2021403015], and [Science and Technology Project of Hebei Education Department], grant number [QN2021027].

Data Availability Statement: All data are contained within the work.

Acknowledgments: We are grateful to the Editor-in-Chief and the Academic Editors, as well as two anonymous reviewers for their constructive comments and suggestions that greatly improve this manuscript.

Conflicts of Interest: The authors declare no conflict of interest.

References

1. Li, S.Y. *Crystallography and Mineralogy*; Geological Publishing House: Beijing, China, 2008; p. 284. (In Chinese)
2. Palmer, M.C.; Scott, J.M.; Luo, Y.; Sakar, C.; Pearson, D.G. In-situ scheelite LASS-ICPMS reconnaissance Sm-Nd isotope characterisation and prospects for dating. *J. Geochem. Explor.* **2021**, *224*, 106760. [CrossRef]
3. Poulin, R.S.; McDonald, A.M.; Kontak, D.J.; McClenaghan, M.B. On the relationship between cathodoluminescence and the chemical composition of scheelite from geologically diverse ore-deposit environments. *Can. Mineral.* **2016**, *54*, 1147–1173. [CrossRef]
4. Poulin, R.S.; Kontak, D.J.; McDonald, A.; McClenaghan, M.B. Assessing scheelite as an ore-deposit discriminator using its trace-element and REE chemistry. *Can. Mineral.* **2018**, *56*, 265–302. [CrossRef]
5. Scanlan, E.J.; Scott, J.M.; Wilson, V.J.; Stirling, C.H.; Reid, M.R.; Roux, P.J. In situ ^{87}Sr /^{86}Sr of scheelite and calcite reveals proximal and distal fluid-rock interaction during orogenic W-Au mineralization, Otago Schist, New Zealand. *Econ. Geol.* **2018**, *113*, 1571–1586. [CrossRef]
6. Song, G.X.; Xiong, Y.X. Advances and Prospects of Scheelite Genetic Mineralogy. *Shandong Land Resour.* **2021**, *37*, 1–9.
7. Liu, Y. Mineralogical characteristics and genetic mechanism of the Xuebaoding deposit in northwestern Sichuan Province. *Acta Petrol. Mineral.* **2017**, *36*, 549–563.
8. Wu, D.W.; Li, B.H.; Du, X.F.; Dong, X.Y.; Fu, T.Y.; Xu, L. Fluid inclusion study and its geological significance on Xuebaoding W-Sn-Be deposit, Sichuan. *Miner. Depos.* **2015**, *34*, 745–756.
9. An, N.; Guo, Q.F.; Liao, L.B. Gemological and Spectroscopic Characteristics of Scheelite from Xuebaoding, Sichuan. In Proceedings of the China International Jewelry and Jewelry Academic Exchange Conference, Beijing, China, 13 November 2019; pp. 173–176.
10. Liu, Y.; Deng, J.; Xing, Y.Y.; Jiang, S.Q. Vibrational Spectra of Scheelite and Its Color Genesis. *Spectrosc. Spectr. Anal.* **2008**, *28*, 121–124.
11. Li, J.K.; Liu, S.B.; Wang, D.H.; Fu, X.F. Metallogenic epoch of Xuebaoding W-Sn-Be deposit in northwest Sichuan and its tectonic tracing significance. *Miner. Depos.* **2007**, *26*, 557–562.
12. Cao, Z.M.; Li, Y.G.; Ren, J.G.; Li, B.H.; Xu, S.J.; Wang, R.C.; Zheng, L.C.Y.; Jin, T.B.Z.; Xiao, L.X.Y. Characteristics and tracing and dating of volatile ore-forming fluids in the Xuebaoding beryl scheelite vein deposit. *Sci. China (Ser. D)* **2002**, *32*, 64–72.
13. Li, B.H.; Cao, Z.M.; Li, Y.G.; Yi, H.S. CO_2 from Fluid Inclusions in Xuebaoding W-Sn-Be Deposit and Its Mineralization Significance. *Miner. Depos.* **2002**, *21*, 397–400.
14. Zhu, X.X.; Liu, Y. Chemical Composition of Minerals in Xuebaoding W-Sn-Be Deposit, Sichuan Province: Constraints on One Genesis. *Rock Miner. Anal.* **2021**, *40*, 296–305.
15. Zhang, B.L. *System Gemology*; Geological Publishing House: Beijing, China, 2006; pp. 523–524. (In Chinese)
16. Li, M.R.; Wang, H.L.; Li, B.; Che, W.F.; Liang, D.Y.; Meng, Q.B. Mineralogical features, crystal structure, and floatability of the Mo-bearing scheelite. *Acta Metall. Sin.* **2022**, *42*, 213–221.
17. Sun, S.S.; McDonough, W.F. Chemical and isotopic systematics of oceanic basalts: Implications for composition and processes. In *Magmatism in the Ocean Basin*; Geological Society Special Publication: London, UK, 1989; Volume 42, pp. 313–345.
18. Brugger, J.; Bettiol, A.A.; Costa, S.; Lahaye, Y.; Bateman, R.; Lambert, D.D.; Jamieson, D.N. Mapping REE distribution in scheelite using luminescence. *Mineral. Mag.* **2000**, *64*, 891–903. [CrossRef]
19. Nevin, C.G.; Pandalai, H.S. Rare earth element geochemistry and fluid characteristics of scheelite in the Hutti gold deposit, Hutti-Maski schist belt, Raichur district, Karnataka, India. *J. Asian Earth Sci.* **2020**, *189*, 104–161. [CrossRef]
20. Bell, K.; Anglin, C.D.; Franklin, J.M. Sm-Nd and Rb-Sr isotope systematics of scheelites: Possible implications for the age and genesis of vein-hosted gold deposits. *Geology* **1989**, *17*, 500–504. [CrossRef]
21. Eichhorn, R.; Höll, R.; Jagoutz, E.; Schärer, U. Dating scheelite stages: A strontium, neodymium, lead approach from the Felhertal tungsten deposit, Central Alps, Austria. *Geochim. Et Cosmochim. Acta* **1997**, *61*, 5005–5022. [CrossRef]
22. Vermeesch, P. IsoplotR: A free and open toolbox for geochronology. *Geosci. Front.* **2018**, *9*, 1479–1493. [CrossRef]
23. Steiger, R.H.; Jager, E. Subcommission on geochronology: Convention on the use of decay constants in geo-and cosmochronology. *Earth Planet. Sci. Lett.* **1977**, *36*, 359–362. [CrossRef]
24. Cohen, K.M.; Finney, S.C.; Gibbard, P.L.; Fan, J.-X. The ICS International Chronostratigraphic Chart. *Epis. J. Int. Geosci.* **2023**, *36*, 199–204. [CrossRef]

25. Liu, Y.; Deng, J.; Li, C.F.; Shi, G.H.; Zheng, A.L. Rare earth geochemistry and Sm-Nd isotope dating of the Xuebaoding scheelite deposit in Sichuan. *Chin. Sci. Bull.* **2007**, *52*, 1923–1929.
26. Nie, L.Q.; Zhou, T.F.; Zhang, Q.M.; Zhang, M.; Wang, L.H. Trace elements and Sr-Nd isotopes of scheelites: Implications for the skarn tungsten mineralization of the Donggushan deposit, Anhui Province. *Acta Petrol. Sin.* **2017**, *33*, 3518–3530.
27. Ren, K.D.; Zhang, D.Y.; Meng, X.; Zhang, F.; Wang, J.; Xi, X.C.; Zhang, Y.W.; Su, H.B. Enrichment mechanism of rare earth (REE) of high-Mo scheelite in Guilinzheng deposit from Jiangnan tung mineralization belt and its geological significance. *Miner. Depos.* **2022**, *41*, 859–877.
28. Wei, Y. The Mineral Characteristics and Metallogenic Mechanism of Scheelite in the Zhaishang Gold Deposit in the Western Qinling Orgenic Belt. Master's Thesis, China University of Geosciences, Beijing, China, 2014.
29. Xue, Y.S.; Liu, X.W.; Hu, X.S.; Cun, X.N.; Yang, H.T. Scheelite geochemical signature and calcite Sm-Nd dating of the Xianggou Au-W deposit in south Qinling orogen, central China: Constraints, on the ore-forming process. *Geol. Chin.* **2021**, *48*, 1818–1837.
30. Peng, J.T.; Wang, C.; Li, Y.L.; Hu, A.X.; Lu, Y.L.; Chen, X.J. Geochemical characteristics and Sm-Nd Geochronology of scheelite in the Baojinshan ore district, central Human. *Acta Petrol. Sin.* **2021**, *37*, 665–682.
31. Masuda, A.; Kawakami, O.; Dohmoto, Y.; Takenaka, T. Lanthanide tetrad effects in nature: Two mutually opposite types, W and M. *Geochem. J.* **1987**, *21*, 119–124. [CrossRef]
32. Irber, W. The lanthanide tetrad effect and its correlation with K/Rb, Eu/Eu*, Sr/Eu, Y/Ho, and Zr/Hf of evolving peraluminous granite suites. *Geochim. Et Cosmochim. Acta* **1999**, *63*, 489–508. [CrossRef]
33. Monecke, T.; Kempe, U.; Monecke, J.; Sala, M.; Wolf, D. Tetrad effect in rare earth element distribution patterns: A method of quantification with application to rock and mineral samples from granite-related rare metal deposits. *Geochim. Et Cosmochim. Acta* **2002**, *66*, 1185–1196. [CrossRef]
34. Zhao, Z.H.; Bao, Z.W.; Qiao, Y.L. A peculiar composite M- and W-type REE tetrad effect: Evidence from the Shuiquangou alkaline syenite complex, Hebei Province, China. *Chin. Sci. Bull.* **2010**, *55*, 1474–1488. [CrossRef]
35. Zhou, J.Y.; Tan, H.Q.; Gong, D.X.; Zhu, Z.M.; Luo, L.P. Wulaxi Aluminous A-type Granite in Western Sichuan, China: Recording Early Yanshanian Lithospheric Thermo-upwelling Extension of Songpan-Garze Orogenic Belt. *Geol. Rev.* **2014**, *60*, 348–362.
36. Whalen, J.B.; Currie, K.L.; Chappell, B.W. A-type granites: Geochemical characteristics, discrimination and petrogenesis. *Contrib. Mineral. Petrol.* **1987**, *95*, 407–419. [CrossRef]
37. Song, G.X.; Qin, K.Z.; Li, G.M.; Evans, N.J.; Chen, L. Scheelite elemental and isotopic signatures: Implications for the genesis of skarn-type W-Mo deposits in the Chizhou Area, Anhui Province, Eastern China. *Am. Mineral.* **2014**, *99*, 303–317. [CrossRef]
38. Shao, J.A.; Mou, B.L.; Zhu, H.Z.; Zhang, L.Q. Material source and tectonic settings of the Mesozoic mineralization of the Da Higgan Mts. *Acta Petrol. Sin.* **2010**, *26*, 649–656.
39. Guo, Z.J.; Li, J.W.; Huang, G.J.; Guan, J.D.; Dong, X.Z.; Tian, J.; Yang, X.C.; She, H.Q.; Xiang, A.P.; Kang, Y.J. Sr-Nd-Pb-Hf isotope characteristics of ore-bearing granites in the Honghuaerji scheelite deposit, Inner Mongolia. *Geol. Chin.* **2014**, *41*, 1226–1241.

minerals

MDPI

Article

Genesis of the Daliuhang Gold Deposit, Jiaodong Peninsula, Eastern China: Constraints from H-O-S-Pb-He-Ar Isotopes, and Geochronology

Jiepeng Tian [1,2,3,4,*], Junjian Li [4], Xuan Wu [5], Chao Fu [4], Zhicai Dang [4], Pengpeng Zhang [6], Jiangtao He [7], Wenlong Tang [4] and Ruicong Tian [8]

1 Key Laboratory of Building Structural Retrofitting and Underground Space Engineering of Ministry of Education, Jinan 250101, China
2 School of Civil Engineering, Shandong Jianzhu University, Jinan 250101, China
3 Key Laboratory of Gold Mineralization Processes and Resource Utilization Subordinated to the Ministry of Land and Resources, Jinan 250013, China
4 Tianjin Center, China Geological Survey, Tianjin 300170, China; fuchaocugb@163.com (C.F.); dangzhicai@163.com (Z.D.); twl011562@126.com (W.T.)
5 801 Institute of Hydrogeology and Engineering Geology, Shandong Provincial Bureau of Geology & Mineral Resources, Jinan 250014, China; xuanwu19921004@163.com
6 Institute of Geophysical and Geochemical Exploration, Chinese Academy of Geological Science, Langfang 065000, China; zhangpengpeng@mail.cgs.gov.cn
7 Institute of Geological Surveying, Hebei GEO University, Shijiazhuang 050031, China; jiangtaohe@hgu.edu.cn
8 School of Geography and Tourism, Qilu Normal University, Jinan 250200, China
* Correspondence: tianjiepeng20@sdjzu.edu.cn

Abstract: The Daliuhang gold deposit in the Qipengfu (Qixia–Penglai–Fushan) ore concentration area is a typical gold deposit of medium-low temperature hydrothermal veins. Uncertainties regarding the primary sources of ore-forming fluids, as well as whether host rocks contribute materials to the mineralization of the gold deposits in the Jiaodong Peninsula, are still subject to intense debate. Hydrogen–oxygen isotope results show that atmospheric water is involved in ore-forming fluids. According to the results of the helium–argon isotopes of pyrite, it is hypothesized that the initial fluid source was located in the oceanic crust or upper mantle lithosphere above the Early Cretaceous Paleo-Pacific Plate, as it was subducted into the eastern part of the eastern North China Craton. In situ sulfur isotope results show that high $\delta^{34}S$ values characterize the pyrite in the main mineralization period. It is inferred that during the thinning and melting process of the lithospheric mantle, the volatile components enriched in pyrite contributed to the release of $\delta^{34}S$. At the same time, when the fluids ascended to the weak zones, such as fissures of ore-endowed peripheral rocks, the $\delta^{34}S$ in the peripheral rocks were extracted, and the two processes acted together to cause high $\delta^{34}S$ values to occur. Similarly, the lead and strontium isotopic compositions indicate a crust–mantle mixing attribute of the mineralized material source. The zircon U–Pb age of the ore-hosting granodiorite was 130.35 ± 0.55 Ma, and the Rb–Sr isochron age of the pyrite from the main mineralization period was 117.60 ± 0.10 Ma, which represents the timing of felsic magmatism and gold mineralization, respectively, with at least 10 Ma between the magmatism and mineralization. The magma gradually cooled over time after its formation, and when the granodiorite cooled down to 300 ± 50 °C, the temperature and pressure conditions were most conducive to the precipitation of gold. It is inferred that gold-rich initial mantle fluids with volatile components, rising along tectonically weak zones, such as fractures, underwent fluid phase separation in the fractured position of the granite and extracted the gold from the granodiorite, forming gold deposits.

Keywords: Daliuhang gold deposit; in situ sulfur isotopes; H–O–Pb–He–Ar isotopes; pyrite Rb–Sr isochron age; zircon U–Pb age

check for updates

Citation: Tian, J.; Li, J.; Wu, X.; Fu, C.; Dang, Z.; Zhang, P.; He, J.; Tang, W.; Tian, R. Genesis of the Daliuhang Gold Deposit, Jiaodong Peninsula, Eastern China: Constraints from H-O-S-Pb-He-Ar Isotopes, and Geochronology. *Minerals* **2023**, *13*, 1339. https://doi.org/10.3390/min13101339

Academic Editor: Ryan Mathur

Received: 8 September 2023
Revised: 4 October 2023
Accepted: 16 October 2023
Published: 19 October 2023

1. Introduction

The Jiaodong Peninsula is located on the southeastern edge of the North China Craton (NCC), and it is China's most productive gold province, with more than 5400 t of gold resources [1]. The preserved resource reserves exceed 4000 t, which accounts for more than one-quarter of China's total. The Jiaodong gold province is the third-largest gold mining area in the world [2–9]. There are three ore concentration areas in the region: the Jiaoxibei (Laizhou–Zhaoyuan) ore concentration area, the Qipengfu (Qixia–Penglai–Fushan) ore concentration area, and the Muru (Muping–Rushan) ore concentration area [8,10]. The Daliuhang gold field is located in the northern part of the Qipengfu ore concentration area. It consists of several gold deposits, including Daliuhang, Menlou, Ankou, Hexi, and other gold deposits. The ore bodies exist in Early Cretaceous granodiorite. They belong to middle-low-temperature hydrothermal vein-type gold deposits, and they are one of the most important types of gold deposits in the Jiaodong Peninsula.

Early studies have suggested several genetic types of gold deposits in the Jiaodong Peninsula, but its genesis remains enigmatic. Studies have indicated that the gold mineralization in Jiaodong Peninsula, and even eastern North China, is closely related to the destruction of the eastern North China Craton, and it is proposed that the eastern North China gold deposits belong to the "Decratonic gold deposits" [11]. During the Early Cretaceous, against a strong extensional tectonic background, the Paleo-Pacific plate was subducted beneath the eastern North China Craton. The lithospheric mantle, which was strongly metasomatized, underwent large-scale partial melting [11,12]. The produced basaltic magma reacted with the strongly enriched mantle and the lower crust, and the interaction between the fluids and crustal-source felsic magma formed gold-rich magma [11]. The auriferous fluids were exsolved from the magma and uplifted along the deep faults, which were mineralized by filling in the subtectonic levels [11]. Other works believe that the gold deposits formed during the dehydration and metamorphism of the subducting slab of the Paleo-Pacific Plate. Gold may have come from the devolatilization of stalled subducting slabs and oceanic crustal sediments [7,13,14]. Tectonic stress conversion resulted in ultra-high pressure gold-rich fluids, which migrated along the abyssal fault and other crustal hierarchical rupture systems and were further deposited to form the minerals, thus being considered "orogenic gold deposits" [7,15]. Additionally, some other studies also suggest that these gold deposits belong to the greenstone belt gold deposit, and its mineralization is likely related to the Mesozoic granitic magmatism. This indicates that the Jiaodong gold deposit may represent post-magmatic hydrothermal deposits [14,16]. However, some recent studies suggest that the gold deposits in this area are "Jiaodong-type gold deposits", in contrast to all the above types [6,17–20]. One of the focal points of the various genetic models mentioned above is the origin of gold-bearing fluids. Therefore, the study of the geological characteristics, the origin of the ore-forming fluids and materials, and the mineralization chronology of the Daliuhang gold deposit can enrich the theoretical study of medium-low temperature hydrothermal vein-type gold deposits, enhance our understanding of the origin of ore-forming fluids in the Qipengfu (Qixia–Penglai–Fushan) ore concentration area, and provide a theoretical basis for prospecting work on the same types of gold deposits in this area.

In this study, the Rb–Sr ages of pyrite in ores were obtained to determine the precise timing of gold mineralization in the Daliuhang gold deposit. In addition, the U–Pb ages of zircon from ore-hosting granodiorite were obtained to determine the timing of felsic magmatic events. In situ sulfur isotopic analyses of gold-hosted pyrite, hydrogen–oxygen isotopic analyses of quartz, lead isotopic analyses of pyrite, and helium–argon isotopic analyses of primary ore-forming fluids entrapped in pyrite were conducted in order to place constraints on the sources and temporal evolution of the ore-forming fluids, as well as on the probable origins of the sulfur and ore metals. The integrated results contribute to a better understanding of the genesis of the Daliuhang deposit.

2. Regional Geology

The Jiaodong Peninsula consists of three parts: the Jiaobei Terrane in the north, the Weihai Terrane in the east, and the JiaoLai Basin in the south. The Qipengfu ore concentration area is located in the north-central part of the Jiaobei Terrane, which mainly consists of Precambrian basement rocks and Mesozoic igneous intrusions (Figure 1). The Precambrian basement rocks mainly include the Archean granite–greenstone belt (Tonalite–Trondhjemite–Grandiorite (TTG) gneisses, known as the Jiaodong Group) and Paleoproterozoic metamorphic strata. The Mesozoic igneous intrusions are divided into the following groups: Late Jurassic (165–150 Ma) granite (Linglong granite), Early Cretaceous (132–123 Ma) porphyritic granodiorite (Guojialing granodiorite), Early Cretaceous (118–110 Ma) granite (Weideshan granite) [21–24], and various kinds of dikes dominated by lamprophyres and diorite–porphyrites. The structures in the area are mainly fault structures, divided into three groups: NNE-, NE-, and NNW-trending crustal-scale fault zones. The NNE- and NE-trending fault zones are the most developed and the main ore-controlling structures in the area, including the Sanshandao, Jiaojia, Zhaoyuan–Pingdu, Qixia, and Muping–Rushan fault zones.

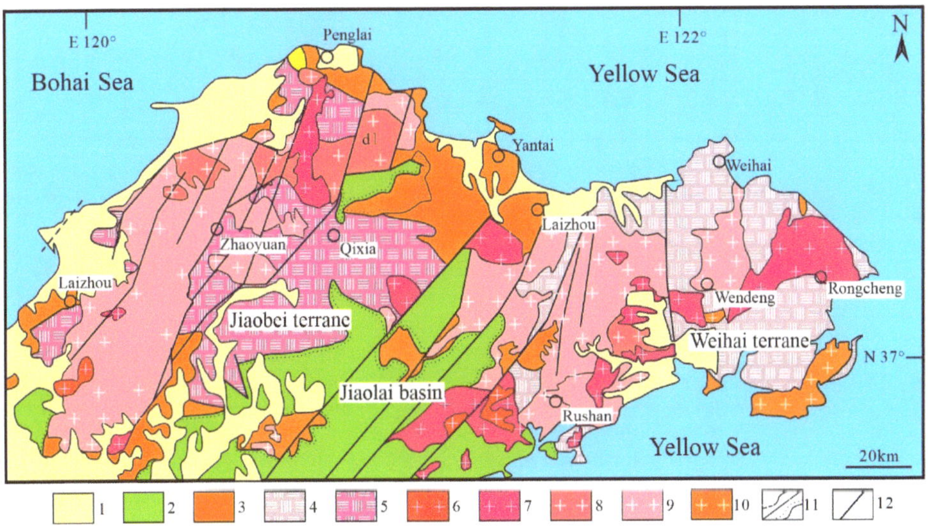

Figure 1. Map showing the regional geology in the Jiaodong Peninsula (modified after [9,10]). (1) Quaternary sediments; (2) Cretaceous sedimentary basin; (3) Paleoproterozoic and Neoproterozoic; (4) Neoproterozoic bearing eclogite granitic gneiss; (5) Archean granite–greenstone belt; (6) Late Cretaceous granite (Laoshan granite); (7) Early Cretaceous granite (Late, Weideshan granite); (8) Early Cretaceous granite (Early, Guojialing granite); (9) Late Jurassic granite (Linglong granite); (10) Triassic granitoid; (11) geological boundary of conformity/unconformity; (12) fault.

3. Ore Deposit Geology

The Daliuhang deposit mainly developed from the Late Jurassic biotite granite and Early Cretaceous porphyritic granodiorite (Figure 2). Ultramafic to intermediate dikes are also developed, including granite pegmatite, lamprophyre, and quartz–diorite porphyrite.

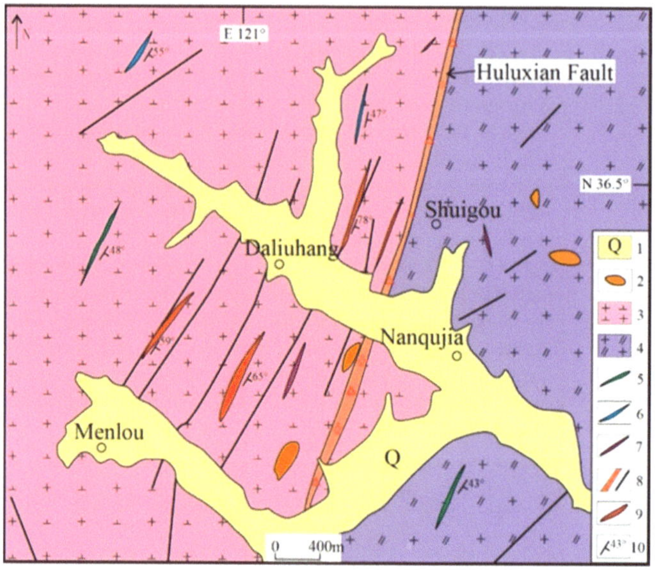

Figure 2. Geological map of the Daliuhang deposit. (1) Quaternary sediments; (2) Neoproterozoic; (3) Early Cretaceous porphyritic granodiorite (Guojialing granodiorite); (4) Late Jurassic granite (Linglong biotite granite); (5) quartz diorite porphyrite; (6) lamprophyre; (7) granite pegmatite; (8) fault; (9) ore vein; (10) occurrence.

The Huluxian Fault exerts full control over the ore veins in the area, and the occurrence of ore veins is in complete alignment with it. The ore veins consistently strike 20–30° and dip 59–78° SE, with a gentle upward slope and a steep downward slope. The ore-hosting enclosing rocks mainly consist of pyrite–seritic cataclastic rock and porphyritic granodiorite. The width of quartz–pyrite veins is generally more than ten to dozens of centimeters (Figure 3a), with uneven development. The gold grade is generally higher in the areas where quartz–polymetallic sulfide veins are well-developed (Figure 3b). The mineral composition of the ore comprises two main parts: metallic minerals and non-metallic minerals. Sulfides, including pyrite, and a small amount of galena, dominate the metallic minerals. The non-metallic minerals include quartz, feldspar, calcite, sericite, chlorite, etc. (Figure 3c–f). The main ore textures are automorphic granular, pressure, and hypidiomorphic–itomorphic granular, etc. (Figure 3e,f), and they exhibit veinlet-disseminated, mottled, massive, and other structures (Figure 3e,f). The main alterations of the surrounding rocks include potashization, pyriteization, silicification, and sericitization, followed by carbonatization, chloritization, and kaolinization. Sericitization of the surrounding rocks is the most developed alteration phenomenon in the deposit. The width of its alteration zone ranges from tens of centimeters to several meters. The alteration is more substantial closer to the main fault surface and gradually weakens on both sides.

Based on the interpenetration relationship of ore veins, mineral assemblage characteristics, and observations using backscattered electron (BSE) images, the mineralization can be roughly divided into four stages: Stage I consists of pyrite and quartz, with a small amount of medium and fine-grained pyrite. Stage II contains quartz and pyrite, where white-grey quartz fills cavities, and pyrite occurs as coarse euhedral cubes (Py2), displaying a smooth surface and a single component in the backscattered electron (BSE) images (Figure 4a,b). Stage III is characterized by gold, quartz, and polymetallic sulfide phases. The quartz is grey to pale with relatively coarse particles, and the pyrite (Py3) and galena are medium-coarse-grained and more automorphic. Locally, Stage III minerals are superimposed on Py2, forming an irregular, dark inner core (Py3a) with a bright outer rim

(Py3b) in the BSE images (Figure 4c,d). Milky carbonate–quartz veins represent Stage IV. Stages II and III represent the main gold mineralization episodes.

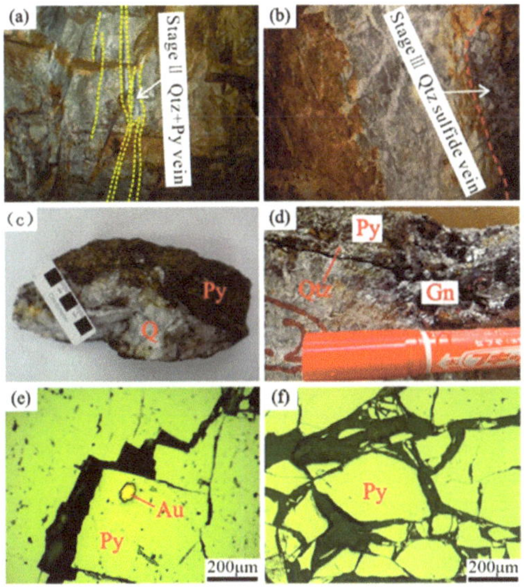

Figure 3. Photographs of major wall rocks and various mineralized veins in the Daliuhang gold deposit. (**a**) Pyrite sericite cataclastic rock and quartz–pyrite veins (Stage II); (**b**) quartz vein (Stage II) and its adjacent quartz–galena–pyrite vein (Stage III); (**c**) representative specimens of quartz–pyrite veins (Stage II); (**d**) representative specimens of quartz–galena–pyrite veins (Stage III); (**e**) euhedral pyrite and gold grains occur inside pyrite crystals; (**f**) pyrite with an automorphic-semi-automorphic structure. Au, gold; Gn, galena; Py, pyrite; Qtz, quartz.

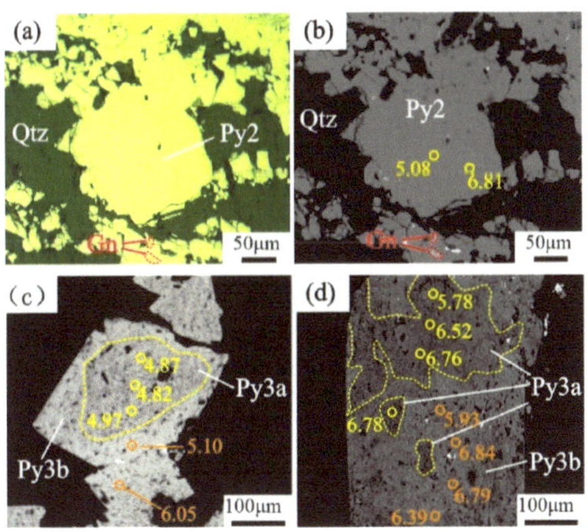

Figure 4. Reflected light photomicrographs (**a**) and backscattered electron (BSE) images (**a**–**d**) showing textural features of pyrite from main mineralizing stages II and III. (**a**) Cataclastic texture of pyrite, pyrite intergrowth with galena; (**b**) Py2 from quartz–pyrite veins (Stage II) is homogeneous in composition; (**c**) regular Py3a is overgrown by zoned and brighter Py3b from quartz polymetallic sulfide veins (Stage III); (**d**) irregular Py3a is overgrown by zoned and brighter Py3b from quartz polymetallic sulfide veins (Stage III). Gn, galena; Py, pyrite; Qtz, quartz.

4. Sampling and Analytical Methods

A total of 23 samples were collected from the pit at a depth of +200 m of the Daliuhang gold deposit, including pyrite, galena, and other sulfides, which were intergrown with quartz veins.

4.1. Isotopes Analysis

Quartz grains were crushed into 40–60 mesh and handpicked from under a binocular (purity > 99%). Hydrogen and oxygen isotopic composition analyses were performed at the MLR Key Laboratory of Metallogeny and Mineral Assessment, Institute of Mineral Resources, CAGS. The samples were heated to approximately 500 °C in an induction furnace to release moisture. The samples were heated to 180–200 °C until the vacuum degree was less than 10^{-1} Pa, and then the volatile substances in the sample were degassed. At a temperature of 400 °C, water was converted into hydrogen gas by a heated zinc powder, which was then analyzed with a MAT–253 mass spectrometer. The analysis results of the standard water samples show that the accuracy of the δD was ±2‰. The pure mineral was crushed into 200 mesh, and then the crushed material was reacted with BrF5 at 500–600 °C for 14 h to produce O_2, and then reacted with graphite at 700 °C to produce CO_2 under the action of a platinum catalyst. The carbon dioxide's oxygen isotopes were then measured using a MAT–253 mass spectrometer. The reproducibility of the isotopically homogeneous pure quartz was approximately ±0.2‰. More detailed analytical conditions and procedures are described by previous researchers [25].

In situ sulfur isotopic compositions of four pyrite samples from Stage II in thin sections were analyzed using a 193 nm ArF excimer laser ablation system (RESOlution S–155) connected to an MC–ICP–MS (Nu Plasma II) at the State Key Laboratory for Geological Processes and Mineral Resources at the China University of Geosciences (Wuhan), China. Helium delivered the ablative material into the plasma at a gas flow rate of 0.4 L/min. Argon (0.8 L/min) and a small amount of nitrogen (~4 mL/min) were mixed with the carrier gas before entering the MC–ICPMS. The addition of a small amount of N_2 gas is effective in removing polyatomic interferences (e.g., $^{16}O_2{}^{+}$) from ^{32}S [26,27]. The laser diameter was 23 μm, and the laser repetition frequency was 8 Hz for single-site analysis. The energy flux of the laser was about 3–4 J/cm^2. The internal pyrite standard (WS–1) consisted of natural pyrite crystals from the Wenshan polymetallic deposit in Yunnan Province, southern China, and was used to calibrate the mass bias of the S isotopes. The $\delta^{34}S_{VCDT}$ value of WS–1 (1.1 ± 0.2‰) was determined by SIMS at the China Institute of Geochemistry, Guangzhou [28]. The analytical precision was evaluated to be better than 0.5‰.

Lead isotope analysis was performed at the Beijing Research Institute of Uranium Geology, Beijing. Pb isotope samples were taken from pyrite and galena in Stage III mineralization. After dissolution and separation of the samples, the lead isotope analyses were conducted using a thermal ionization mass spectrometer (ISOPROBE–T) produced by GV, UK, at a relative humidity of 36% and a room temperature of 20 °C. The measured Pb isotopic ratios of the international standard NBS981 are approximately 0.2‰. The analytical results of the standard NBS981 are as follows: $^{206}Pb/^{204}Pb$ = 16.937 ± 0.002, $^{207}Pb/^{204}Pb$ = 15.457 ± 0.002, and $^{208}Pb/^{204}Pb$ = 272 36.611 ± 0.004.

Helium–argon isotope analyses were performed at the MLR Key Laboratory of Metallogeny and Mineral Assessment, Institute of Mineral Resources, CAGS. The samples were taken from pyrite in Stage III mineralization. The Helix SFT™ Split Flight Tube noble gas mass spectrometer was used. The system consists of crushing, purification, and mass spectrometry systems. Tests were performed under high-vacuum conditions, with the vacuum of the crushing and purification systems at 5×10^{-9}–10×10^{-9} mbar and the vacuum of the mass spectrometry system at 7×10^{-9}–9×10^{-10} mbar. The mass spectrometry ion source was Nier, with a sensitivity better than 2×10^{-4} amps/Torr for helium at a trap current of 800 μA, and 1×10^{-3} amps/Torr for argon at a trap current of 200 μA. The static rate of rise was less than 1×10^{-12} cm^3 STP/min for ^{40}Ar, and the background was less than

5×10^{-14} cm^3 STP for ^{36}Ar. The Faraday background was less than 5×10^{-14} STP/min for ^{36}Ar, the static rate of rise was less than 1×10^{-12} cm^3 STP/min, and the ^{36}Ar background was less than 5×10^{-14} STP/min. The Faraday cup resolution was >400 and the ion counter resolution was >700, which allowed for complete separation of the ^3He, ^4He, HD + H_3, and ^3He peaks. A program blank was performed before each measurement. The blanks for both helium and argon were negligible ($^3He < 3 \times 10^{-17}$ cm^3 STP, $^{40}Ar < 4 \times 10^{-17}$ cm^3 STP), approximately 0.1% of the sample signal. The sample test results were corrected using the test results of the day's air standards and the air standard values. The standard value of $^3He/^4He$ for air was 1.4×10^{-6} (Ra). The $^{40}Ar/^{36}Ar$ and $^{36}Ar/^{38}Ar$ standard values were adopted as 295.5 and 5.35, respectively.

4.2. Zircon U-Pb Dating

Zircon crystals for U–Pb dating were separated from the ore-hosting porphyritic granodiorite (Early Cretaceous granodiorite). Zircon LA–ICP–MS U–Pb dating was conducted using a Resonetics–M50 laser ablation system equipped with a 193 nm ArF excimer laser connected to a multi-collector inductively coupled plasma mass spectrometer (Neptune), at the Tianjin Center, China Geological Survey, Tianjin. Helium was used as the carrier gas to enhance the transport efficiency of the ablated material. The spot diameter was 35 μm, and the ablation frequency was 9 Hz. The silicate glass standard NIST SRM 612 was used to calibrate U, Th, and Pb concentrations. All measurements were performed using zircon GJ–1 as the external standard, with a recommended $^{206}Pb/^{238}U$ age of 609 Ma [29]. In contrast, the Plesŏvice zircon standard = 337.13 ± 0.37 Ma [30] was used as a blind sample to monitor the data quality. The raw data were processed using ICPMS DataCal software [31], and the ^{208}Pb method was used for the common Pb correction [32]. Concordia diagrams and weighted mean ages were generated using the ISOPLOT program version 3.75 [33]. Errors for individual analyses were at the 2σ level, and deviations of the pooled analyses are given at the 95% confidence level.

4.3. Rb-Sr Dating

The Rb–Sr isotope system was used to date five samples of pyrite from Stage III mineralization. The Rb–Sr isotope analysis was performed in the Laboratory of Isotope Chronology at the Tianjin Center, China Geological Survey, Tianjin. The pyrite samples were accurately weighed and dissolved with HNO_3 + $HClO_4$. The dissolved sample solution was evenly divided into two parts. One part was mixed with ^{87}Rb + ^{84}Sr diluent. After separation and purification, the rubidium and strontium contents were determined. The other portion was purified and used for strontium isotope ratio determination. To determine the $^{87}Sr/^{86}Sr$ isotope ratio, a mass fractionation correction was made using $^{88}Sr/^{86}Sr$ = 8.375209, and the instrumental state was monitored using the international SRM987 standard solution: the ratio obtained was $^{87}Sr/^{86}Sr$ = 0.710242 ± 0.000008 (2σ). The reliability of the analytical method was monitored using the international rock standard BCR–2, which gave a ratio of $^{87}Sr/^{86}Sr$ = 0.705015 ± 0.000009 (2σ). For the full-flow blank, the rubidium content was less than 103 pg, and the strontium content was less than 276 pg. The detailed experimental separation process and instrument test parameters followed those of previous researchers [24,34], and the ISOPLOT program was used to calculate the isochron age.

5. Results

5.1. Hydrogen and Oxygen Isotopes

This paper conducted H–O isotope analysis on quartz samples from the Daliuhang gold deposit during its main gold mineralization episodes and collected previous research results on the Jiaojia and Jinqingding gold deposits (Table 1). The $\delta^{18}O$ values of the hydrothermal quartz ($\delta^{18}O_{SMOW}$ ‰) ranged from 15.8 to 18.3‰, with a median value of 16.5‰. The hydrogen isotope composition δD_{smow} value varied from −101.0 to −84.0‰, with an average value of −89.0‰ (Table 1). Assuming that quartz is isotopically equilibrated with the ore-forming hydrothermal fluids, the oxygen isotope composition of the ore-forming hydrothermal fluids can be estimated following the temperature-dependent quartz–water oxygen isotope fractionation equation: $1000\ln_{\alpha}Q\text{–}W = 3.38 \times 10^6/T^2 - 2.90$ [35], that is, $\delta^{18}O_{H2O} = \delta^{18}O_{smow} - 1000\ln_{\alpha}Q\text{–}W$. The $\delta^{18}O_{H2O}$ value ranged from 7.63 to 10.90‰, with an average of 8.52‰. The hydrogen and oxygen isotope values of the Daliuhang gold deposit are relatively similar to those of the Jiaojia and Jinqingding gold deposits (Figure 5).

Table 1. Hydrogen and oxygen isotope compositions in the Daliuhang gold deposit.

Stage	Mineral	δD_{SMOW} (‰)	$\delta^{18}O_{SMOW}$ (‰)	$\delta^{18}O_{H2O}$ (‰)	Th-TOT (°C)	Source
Stage III	Quartz	−87	15.90	7.63	265	This study
	Quartz	−87	16.20	7.93	265	
	Quartz	−84	16.10	7.83	265	
	Quartz	−101	16.60	8.40	280	[32]
	Quartz	−88	18.30	10.90	300	
	Quartz	−87	15.80	8.40	300	

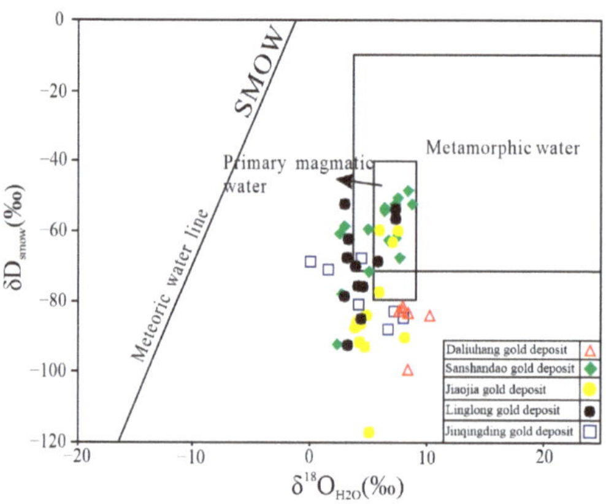

Figure 5. Isotopic compositions of oxygen and hydrogen in the Daliuhang gold deposit.

5.2. In Situ Sulfur Isotopic Compositions

The in situ sulfur isotopic compositions of pyrite from the ores are listed in Table 2. The $\delta^{34}S$ values of the pyrite ranged from 4.82‰ to 7.04‰ (average of 6.12‰), which were all positively biased away from meteoric sulfur and had a narrow variation, with a high degree of sulfur homogenization characterized by a mild enrichment of $\delta^{34}S$ (Figure 6). Based on the backscattered electron (BSE) images (Figure 4c,d), the pyrite can be divided into a dark inner core (Py3a) and a bright outer rim (Py3b). As shown in Figure 4c,d, the core (Py3a) had $\delta^{34}S$ values of 4.82–6.78‰ (average of 5.37‰), with an increase to 5.10–7.04‰ (average of 6.33‰) on the rim (Py3b).

5.3. Lead Isotope Compositions

The analytical results of the lead isotopes of ten sulfides separated from the Daliuhang gold deposit during Stage III mineralization are presented in Table 3 (attached at the end of the article). For comparison, five published Pb isotopic data of sulfides [36,37] are also listed in Table 3. The pyrite had $^{206}Pb/^{204}Pb$ ratios ranging from 17.365 to 17.566 (mean of 17.493), $^{207}Pb/^{204}Pb$ from 15.514 to 15.576 (mean of 15.536), and $^{208}Pb/^{204}Pb$ from 38.118 to 38.462 (mean of 38.258). The galena had $^{206}Pb/^{204}Pb$ ratios ranging from 17.503 to 17.577 (mean of 17.532), $^{207}Pb/^{204}Pb$ from 15.515 to 15.60 (mean of 15.548), and $^{208}Pb/^{204}Pb$ from 38.272 to 38.548 (mean of 38.374).

Table 2. In situ sulfur isotopic analyses in the Daliuhang gold deposit.

Stage	Sample Serial Number	Mineral	$\delta^{34}S_{CDT}$ (‰)
Stage II	SD45B1-1	Pyrite	6.81
	SD45B1-2	Pyrite	5.08
Stage III (Py3a)	SD44B1-5	Pyrite	6.78
	SD44B1-6	Pyrite	6.76
	SD44B1-7	Pyrite	6.52
	SD44B1-8	Pyrite	5.78
	SD46B1-3	Pyrite	6.64
	SD46B1-4	Pyrite	6.59
	SD47B1-1	Pyrite	4.87
	SD47B1-2	Pyrite	4.82
	SD47B1-3	Pyrite	4.97
Stage III (Py3b)	SD44B1-1	Pyrite	6.39
	SD44B1-2	Pyrite	6.79
	SD44B1-3	Pyrite	6.81
	SD44B1-4	Pyrite	5.93
	SD46B1-1	Pyrite	7.04
	SD46B1-2	Pyrite	6.49
	SD47B1-4	Pyrite	5.10
	SD47B1-5	Pyrite	6.05

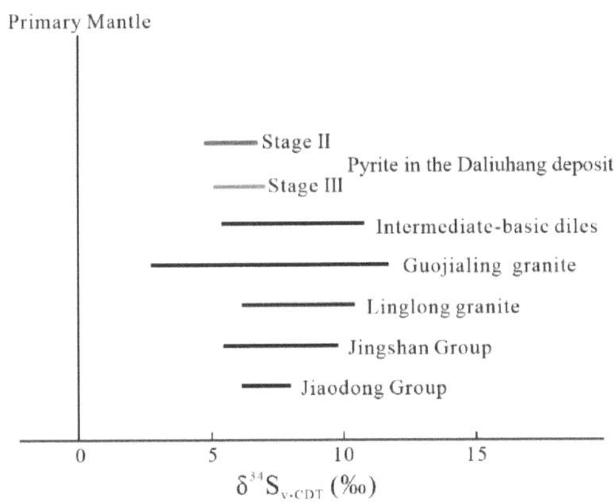

Figure 6. Sulfur isotopic composition of pyrite from the ore deposits in the Daliuhang gold deposit and regional rocks.

5.4. Helium-Ar Isotopes of Pyrite

The results of the He and Ar isotope analyses of fluid inclusions in pyrite are listed in Table 4 (attached at the end of the article). The contents of ^{4}He varied from 0.28 $\times$ 10^{-8} cm^3/gSTP to 8.63 $\times$ 10^{-8} cm^3/gSTP, and the ^{3}He/^{4}He ratios ranged mainly from 1.43 to 1.92 Ra, where Ra represents the standard value of helium in air (^{3}He/^{4}He = 1.4 $\times$ 10^{-6}). The ^{40}Ar/^{36}Ar ratios were in the range of 446.30–753.80.

Table 3. Pb isotope compositions of sulfides from the Daliuhang gold deposit.

Sample Serial Number	Mineral	$^{206}Pb/^{204}Pb$	$^{207}Pb/^{204}Pb$	$^{208}Pb/^{204}Pb$	$^{206}Pb/^{207}Pb$	t (Ma)	μ	ω	Th/U	V1	V2	Δβ	Δγ	Source
SD44B1-1	Pyrite	17.560	15.573	38.462	1.1276	741	9.53	41.53	4.22	84.65	41.4	15.85127	28.39572	This study
SD44B1-1a	Galena	17.516	15.515	38.272	1.1290	708	9.41	40.34	4.15	76.1	38.1	12.06784	23.31551	This study
SD44B1-2	Galena	17.577	15.600	38.548	1.1267	758	9.58	42.10	4.25	88.62	42.89	17.61252	30.69519	This study
SD44B1-2a	Galena	17.532	15.542	38.361	1.1280	726	9.47	40.92	4.18	80.18	39.55	13.82909	25.69519	This study
SD46B1a	Pyrite	17.533	15.576	38.340	1.1256	763	9.54	41.18	4.18	82.70	42.75	16.04697	25.13369	This study
SD46B1b	Pyrite	17.510	15.545	38.248	1.1264	745	9.48	40.58	4.14	78.34	40.87	14.02479	22.67380	This study
SD46B1-1	Pyrite	17.465	15.527	38.118	1.1248	757	9.45	40.09	4.11	74.89	40.32	12.85062	19.19786	This study
SD46B1-2	Pyrite	17.566	15.535	38.307	1.1307	695	9.45	40.37	4.13	77.26	39.98	13.37247	24.25134	This study
SD46B1-3	Pyrite	17.513	15.522	38.245	1.1283	718	9.43	40.31	4.14	76.16	38.98	12.52446	22.59358	This study
SD46B1-4	Pyrite	17.532	15.520	38.258	1.1296	702	9.42	40.22	4.13	75.70	38.85	12.39400	22.94118	This study
DL-9-1	Pyrite	17.477	15.527	38.267	1.1256	749	9.45	40.70	4.17	78.24	38.79	12.85714	23.17647	[36]
DL-9-1	Galena	17.503	15.536	38.314	1.1266	740	9.46	40.84	4.18	79.36	39.24	13.40509	24.42781	[36]
DL-9-3	Pyrite	17.398	15.522	38.172	1.1209	798	9.45	40.75	4.17	77.76	38.51	12.51142	20.65241	[36]
DL-3-5	Pyrite	17.504	15.532	38.296	1.1269	736	9.45	40.72	4.17	78.63	39.19	13.20287	23.95989	[36]
DL-8-6	Pyrite	17.365	15.514	38.125	1.1193	813	9.44	40.67	4.17	76.93	38.08	12.01566	19.38235	[36]

Note: t (Ma): model age; μ: initial $^{238}U/^{204}Pb$; ω: initial $^{232}Th/^{204}Pb$; t, μ, ω, Th/U, Δβ, Δγ parameters were all calculated by Geokit software [37].

Table 4. The He-Ar samples analysis results in the Dliuhang gold deposit. Note: $^{40}Ar^*(\%) = [(^{40}Ar/^{36}Ar)_{sample} - 295.5]/(^{40}Ar/^{36}Ar)_{sample} \times 100$.

Deposit	Mineral	$^{36}Ar/^{38}Ar$	^{40}Ar (10^{-8} cm^3STP/g)	^{4}He (10^{-8} cm^3STP/g)	$^{40}Ar/^{4}He$	$^{40}Ar/^{36}Ar$	$^{3}He/^{4}He$ (10^{-6})	R/Ra	$^{40}Ar^*$ (%)	$^{40}Ar^*/4He$	Mantle ^{4}He (Rm: 6)/%	Mantle ^{4}He (Rm: 9)/%
Daliuhang	Pyrite	5.33	0.05	0.28	19.57	753.80	1.92	1.37	60.80	11.90	21.55	14.29
	Pyrite	5.35	0.29	7.51	3.89	473.10	1.53	1.09	37.54	1.46	16.79	11.13
	Pyrite	5.36	0.28	7.87	3.53	461.20	1.58	1.13	35.93	1.27	17.48	11.59
	Pyrite	5.39	0.38	8.63	4.40	446.30	1.43	1.02	33.79	1.49	15.67	10.39
Magnanimity		(Rare gases in saturated air rainwater [38])				295.5	1					
Mantle		(Rare gases in the mantle [38])				>40,000	6~9					
Crust		(Radiogenic rare gases in the Earth's crust [38])				>45,000	≤0.1					

5.5. Zircon U-Pb Age of Granodiorite

Zircon crystals from the Early Cretaceous porphyritic granodiorite are generally euhedral, pale yellow, and translucent. Most of them are equiprismatic to prismatic, with a few short columnar grains. Most of the grains are 100 to 200 μm in diameter, but a few grains are 300 to 400 μm in diameter. Based on cathodoluminescence (CL) images (Figure 7), magmatic oscillatory zoning is present in most grains. However, some zircon cores are residual. Many of the residual cores show signs of partial melting and are clearly bounded by oscillatory zones. A total of 26 spots were analyzed in the regions with oscillatory zoning. All analyses agree on a weighted average $^{206}Pb/^{238}U$ age of 130.35 ± 0.55 Ma (Figure 7).

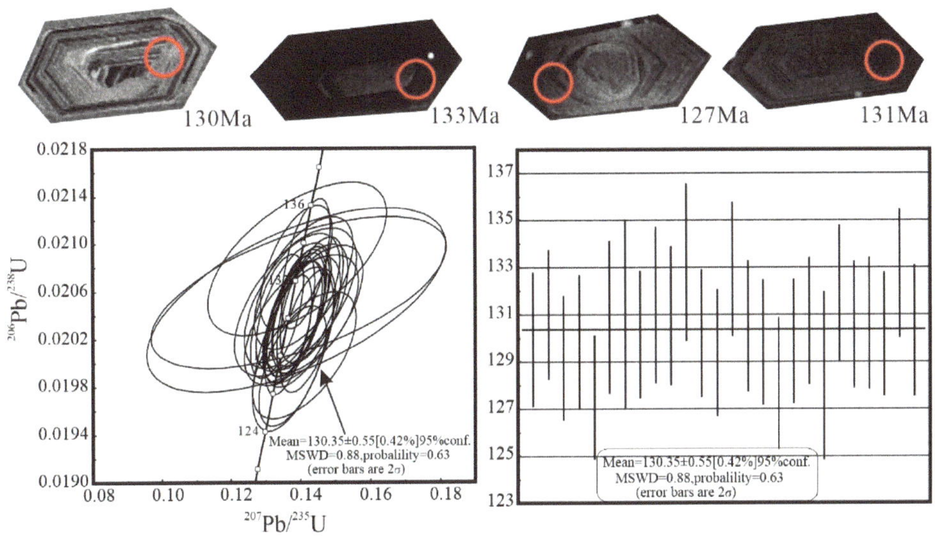

Figure 7. Representative zircon crystals and U–Pb Concordia diagram for granodiorite.

5.6. Pyrite Rb-Sr Age

The mass fractions and isotope ratios of Rb and Sr in pyrite from the Daliuhang gold deposit are listed in Table 5. The Rb and Sr mass fractions were 0.1369–6.2452 ppm and 0.9369–19.8666 ppm, respectively. The $^{87}Rb/^{86}Sr$ ratios ranged from 0.0446–5.6666, and the $^{87}Sr/^{86}Sr$ ratios ranged from 0.71340 to 0.72280. The graphs of $^{87}Rb/^{86}Sr$ and $^{87}Sr/^{86}Sr$ for pyrite in the Daliuhang gold deposit show an excellent linear relationship. ISOPLOT software was used to calculate the isochron age, which was t = 117.60 ± 0.10 Ma, and the initial $^{87}Sr/^{86}Sr$ (Isr) value was 0.713319 ± 0.000028 (Figure 8).

Table 5. The mass fractions and isotope ratios of Rb and Sr in pyrite from the Daliuhang gold deposit.

Sample Serial Number	Ore Type	Rb (ppm)	Sr (ppm)	$^{87}Rb/^{86}Sr$		$^{87}Sr/^{86}Sr$	StdErr (%)
SD44B1-1	Coarse-grained pyrite	0.1369	8.8797	0.0446	0.5	0.71340	0.0006
SD44B1-2	Coarse-grained pyrite	6.2452	12.5817	1.4373	0.5	0.71571	0.0010
SD44B1-4	Coarse-grained pyrite	2.3568	19.8666	0.3435	0.5	0.71390	0.0012
SD46B1-1	Coarse-grained pyrite	1.8323	0.9369	5.6666	0.5	0.72280	0.0010
SD46B1-2	Coarse-grained pyrite	1.1963	0.9697	3.5734	0.5	0.71930	0.0012

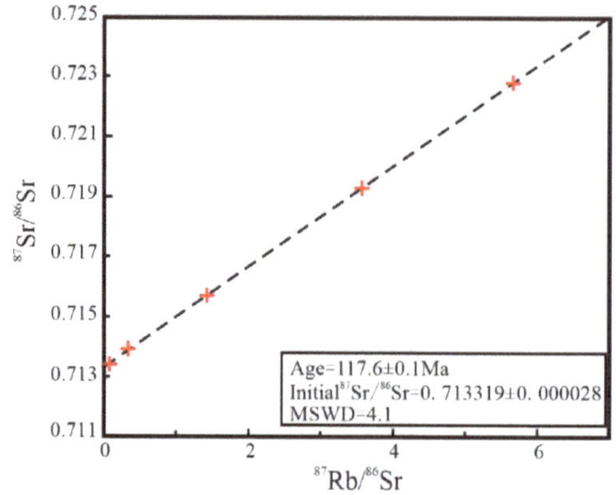

Figure 8. Rb-Sr isotope data and isochron ages of pyrite from the Daliuhang gold deposit.

6. Discussion

6.1. Evolution of Ore-Forming Fluid

The hydrogen and oxygen isotope compositions of the ore-forming fluids in the Daliuhang gold deposit in this study had a limited range of variation. The oxygen isotope composition $\delta^{18}O_{smow}$ values ranged from 15.80‰ to 18.30‰, and the $\delta^{18}O_{H2O}$ values ranged from 7.63‰ to 10.90‰. The hydrogen isotope composition δD_{smow} values ranged from −101.0‰ to −84.0‰. The magmatic water had typical hydrogen isotope compositions of −80.0‰ and −40.0‰ and oxygen isotopes of 5.0‰ to 10.0‰ [39–42]. The results of this study do not entirely fall into the category of magma water (Figure 5). According to the isotopic compositions of oxygen and hydrogen in the Daliuhang gold deposit (Figure 5), it can be seen that the projection points of each group of data were mainly concentrated in the lower part of the metamorphic water and magma water side, which was closer to the magma water, indicating that magma fluids dominated the fluids in the initial period of the gold deposits. With the advancement of mineralization, there is a mixing of meteoric water.

Generally, after the formation of the deposit, the fluid inclusions in the ore sulfides were susceptible to diffusive loss, which may have been affected by later geological processes. However, if the host minerals were sulfides, the inert gases inside them were generally not significantly lost after being captured [43,44], and the U and Th contents of sulfides such as pyrite are very low, which can rule out the adverse effects caused by late diffusion loss. In addition, noble gas isotopes will not produce obvious isotope fractionation during the capture and extraction process of inclusions [43,45,46]. Therefore, the He isotope analysis results of pyrite can represent the initial value when fluid inclusions were captured. The He and Ar isotopic compositions of the fluid inclusions in pyrite are shown in Figure 9. The proportions of mantle fluids (Rm) and crustal fluids (Rc) in the binary mixing model can be deduced by using the $^3He/^4He$ ratio with the following equation: Mantle ^4He (%) = [(R − Rc)/(Rm − Rc)] × 100, where Rc, Rm, and R represent the $^3He/^4He$ ratio of crustal fluid, mantle fluid, and the sample, respectively [47]. According to the calculations, the mantle fluid participation ratio in the Daliuhang gold deposit is only 10.39 to 21.55%, with an average value of 14.86% (Table 4). The above data indicate that the ore-forming fluids are mainly from crustal sources, with mantle fluids mixed in.

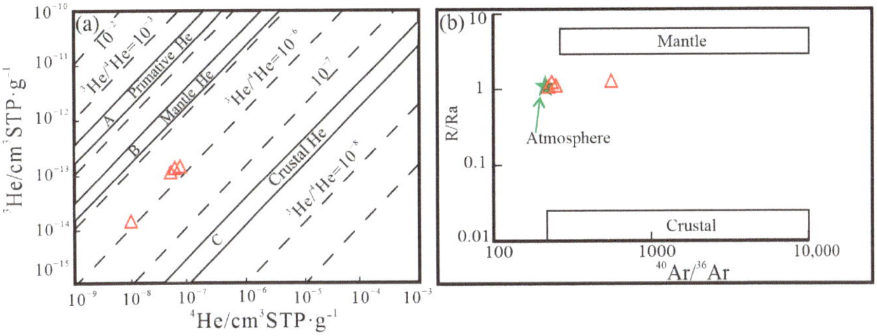

Figure 9. (**a**) Helium isotopes of pyrite from the Daliuhang gold deposit. A–C From Mamyin and Tolstikhin (1984) [48]; (**b**) He (R/Ra) vs. $^{40}Ar/^{36}Ar$ diagram for fluid in pyrite from the Daliuhang gold deposit. Values of crustal and mantle end-members [47,49–51].

The pyrite $^{40}Ar/^{36}Ar$ ratio of the Daliuhang gold deposit ranged from 446.30 to 753.80 (Table 4), which is higher than the $^{40}Ar/^{36}Ar$ ratio in the atmosphere of 295.5, but much lower than the $^{40}Ar/^{36}Ar$ ratio of the mantle (generally greater than 40,000), suggesting that the argon in the fluid is not from a single source. Atmospheric argon and radiotracer $^{40}Ar^*$ in the crust have both contributed to mineralization, but the mantle argon participation ratio is low and can be nearly negligible [38]. The radioactive origin $^{40}Ar^*$ (%) in the crust can be obtained according to the formula: $^{40}Ar^*$ (%) = $[(^{40}Ar/^{36}Ar)_{sample} - 295.5]/(^{40}Ar/^{36}Ar)_{sample} \times 100$ [49,52]. The calculated $^{40}Ar^*$ (%) is 33.79–60.80%, with a wide range, and atmospheric Ar also had a particular contribution. There was an addition of an atmospheric noble gas component to the mantle through subduction [53–56].

It is speculated that the ore-forming fluids were produced during the dehydration and decarbonization of the overlying sediments during the subduction of the Paleo-Pacific plate beneath the eastern North China Craton in the Early Cretaceous, and the atmospheric components were also mixed in during this process.

6.2. Sources of Ore-Forming Material

Earth has three main sources of sulfur: sulfur in the mantle, seawater, and continental crust. The average value of $\delta^{34}S$ in the mantle is close to zero, with a variation range of $0 \pm 3.0‰$, and it mainly exists in a reduced state. The average value of $\delta^{34}S$ in seawater is +20.0‰. In addition, the sulfur in crustal sources is affected by its original rock, and its $\delta^{34}S$ may be greater or less than the value of the above two sources. The in situ sulfur isotope results of pyrite in this study show that the $\delta^{34}S$ values were all greater than 4.82‰, and the test results generally show higher $\delta^{34}S$ values (Figure 6), indicating that the possibility of sulfur only coming from mantle-derived magma is relatively tiny. Sulfur in ore-forming fluids mainly comes from $\delta^{34}S$-rich magmatic rocks [57] or mantle degassing with high oxygen fugacity in the subduction zones [11]. The results of these two processes will have the characteristics of high $\delta^{34}S$ values. According to previous research results, the $\delta^{34}S$ value of pyrite in the Jiaodong Group ranges from 6.10‰ to 7.80‰ [58,59], in the Jingshan Group, it ranges from 5.60‰ to 9.80‰ [58,60], in the Late Jurassic granite, it ranges from 6.10‰ to 10.70‰ [59–61], in the early Early Cretaceous porphyritic granodiorite, it ranges from 2.7‰ to 11.8‰ [59,61], and in the intermediate–basic dike rocks, it ranges from 5.30‰ to 10.80‰ [62]. The $\delta^{34}S$ values of pyrite from the Daliuhang gold deposit overlap somewhat with the $\delta^{34}S$ values of all types of rocks, and in general, overlap more with the Jingshan Group, Early Cretaceous porphyritic granodiorite, and intermediate–basic dike rocks. The $\delta^{34}S$ value of pyrite (Py3) in Stage III is slightly higher than the pyrite (Py2) in Stage II, which indicates that new sulfur was added to the mineralized hydrothermal fluids as mineralization proceeded. A large number of isotopic exchanges were presumed to exist between the mineralized fluids and host rocks. It can be assumed that during the thinning and melting of the lithospheric mantle, the enriched volatile fraction promoted the release of $\delta^{34}S$. In contrast, the $\delta^{34}S$ in the surrounding rock was extracted when the fluid rose to the location of the enclosing rock, and the two effects resulted in the generation of high $\delta^{34}S$ values.

Pyrite, galena, and other ore minerals are generally free of U and Th, or their contents are low enough to be negligible. The composition of lead in the ore is mainly affected by the initial Pb, μ ($^{238}U/^{204}Pb$), ω ($^{232}Th/^{204}Pb$) in the source area and the initial formation time. It has little impact on the geochemical environment after its formation. Therefore, the lead isotope composition was analyzed to obtain information about the source of the minerals [63]. Previous studies have believed that lead with μ values of <9.74 [64] or <9.58 [65] is derived from the lower crust or upper mantle, respectively. Samples from the Daliuhang deposit have Pb μ values of 9.41–9.58 (average of 9.476), suggesting a lower crust or upper mantle source. It is generally believed that the $^{206}Pb/^{207}Pb$ ratio between 0.89 and 1.20 belongs to the range of normal lead, while a range between 1.20 and 1.95 belongs to the content of abnormal Pb (J–type Pb). It is known that the lead source of the Daliuhang gold deposit is normal lead according to the $^{206}Pb/^{207}Pb$ ratio. According to the lead isotopic tectonic environment model of the deposit (Figure 10), it can be seen that the injection points mainly fall within the lower crust and are close to the upper mantle, which indicates that the sources of the ore-forming materials have the characteristics of a mixed crust–mantle source.

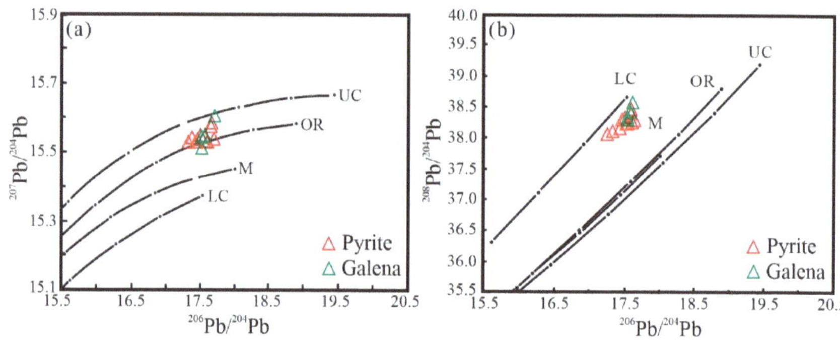

Figure 10. $^{207}Pb/^{204}Pb$ vs. $^{206}Pb/^{204}Pb$ diagram (**a**) and $^{208}Pb/^{204}Pb$ vs. $^{206}Pb/^{204}Pb$ diagram (**b**) of sulfides from the Daliuhang gold deposit (after [66]). Abbreviations: LC, lower crust; UC, upper crust; M, mantle; OR, orogenic belt.

The $^{87}Sr/^{86}Sr$ initial value is an essential indicator for determining the diagenetic source of ore-forming materials. The determination of the Sr isotope composition can be used in deposit geology research on sources to trace the deep origins or crust–mantle mixing of magmatic fluids and the source of ore-forming fluids. Generally, values of $^{87}Sr/^{86}Sr > 0.710$ are considered to reflect a crustal source, whereas values of $^{87}Sr/^{86}Sr < 0.705$ are considered to reflect a mantle source [13,36,67,68]. The initial value of pyrite in the Daliuhang gold deposit was lower than the continental crust, but significantly higher than the mantle. The Sr isotope composition also shows that the source of the minerals has the attributes of a mixed crust–mantle source.

6.3. Relationship between Gold Mineralization and Ore-Hosting Intrusions

The LA–ICP–MS zircon U–Pb age of the granodiorite was 130.35 ± 0.55 Ma, which is consistent with the results of previous studies [69–71]. The Rb–Sr isochron age of the pyrite from the main mineralization period was 117.60 ± 0.10 Ma, representing the ore formation age of the gold deposit belonging to the Early Cretaceous. This is consistent with the metallogenic age of gold deposits in the Jiaodong Peninsula (120 ± 5 Ma) [17,72–74], indicating that the gold mineralization in the Jiaodong Peninsula was accomplished in the same structure within a short period of time. Microscopic thermometry data of fluid inclusions in the Jiaodong gold deposits show that the maximum temperature of the mineralizing fluids was about 350 °C [75,76]. The granite of the Late Jurassic and Early Cretaceous cooled to 300 ± 50 °C at 123 Ma ± [77,78], indicating that the P–T conditions of the granite in this period were most conducive to the beginning of gold precipitation. The diagenetic time of the surrounding rocks (granodiorite) was more than 10 Ma earlier than the mineralization time. There is no direct association between magmatism and mineralization in terms of time [79]. However, more than 90% of the gold deposits in the Jiaodong Peninsula are hosted between the Late Jurassic and Early Cretaceous granite, and there is a particular connection between the magmatism and the mineralization in terms of spatial location. The geochemical composition of Guojialing granodiorite shows that its origin was crust–mantle mixing, which is related to the subduction of the North China Craton by the Paleo-Pacific Plate in the Mesozoic [80–82]. The lead and strontium isotopic compositions also show that the source of gold mineralization has the attribute of crust–mantle-mixed source, and the isotope composition is highly similar to the surrounding rocks. The tectonics of the Jiaodong Peninsula changed from compressional to extensional tectonics in the Jurassic–Cretaceous [83,84]. It is inferred that gold-rich initial mantle fluids containing volatile matter followed structural weak zones, such as faults. During the ascent and intrusion, a large amount of crustal sulfur was mixed, forming mixed sulfur dominated by crustal sulfur. The fluid carried the minerals and continued to rise. The fluid had the mineralized material and continued to rise, and in the crushed zone of the granite with increased tectonic space, fluid phase separation and sulfidation occurred, and the gold in the surrounding rock was extracted, thus forming the gold deposits.

7. Conclusions

The ore-forming fluids were the result of the dehydration and decarbonization of the overlying sediments during the subduction of the Paleo-Pacific plate beneath the eastern North China Craton in the Early Cretaceous. The magma fluids dominated the formation of the gold deposits in the early stages, with atmospheric components mixed in during later stages.

The LA–ICP–MS zircon U–Pb age of the granodiorite was 130.35 ± 0.55 Ma. The Rb–Sr isochron age of the pyrite from the main mineralization period was 117.60 ± 0.10 Ma, representing the ore formation age of gold deposits, belonging to the Early Cretaceous.

The mineralization of the gold deposits originated from the fluid of mantle magma mixed with a large amount of crustal material during their ascent. Fluid phase separation and sulfidation occurred under suitable temperature and pressure conditions, leading to the extraction of gold from the granite and the formation of gold deposits in the Early Cretaceous.

Author Contributions: J.T. conceived and designed the research ideas; J.L., J.T., C.F., Z.D., and P.Z. participated in the field investigation; J.T., X.W., P.Z., J.H., and R.T. performed the data processing; J.T., J.L., and W.T. reviewed and edited the draft. All the data were obtained from previous work performed by the project team. All authors have read and agreed to the published version of the manuscript.

Funding: This study was financially supported by the National Key Research and Development Program of China (grant 2016YFC0600107), the projects from the China Geological Survey (grants DD20190379-31, DD20221695-30 and 200110200038), and the Doctoral Research Foundation of Shandong Jianzhu University (grant X21005Z).

Data Availability Statement: All data are contained within the article.

Conflicts of Interest: The authors declare no conflict of interest.

References

1. Tian, R.C.; Li, D.P.; Zhang, W.; Tian, J.X.; Yu, X.W.; Geng, K.; Zhang, Y. The mixing of Mesozoic crust-mantle magma is the key to the source of large amounts of gold deposits in the Jiaobei uplift, China. *Acta Pet. Sin.* **2022**, *38*, 23–40. [CrossRef]
2. Li, J.J.; Luo, Z.K.; Liu, X.Y.; Xu, W.D.; Luo, H. Geodynamic setting for formation of large-superlarge gold deposits and Mesozoic granites in Jiaodong area. *Miner. Depos.* **2005**, *24*, 361–372. (In Chinese)
3. Deng, J.; Yang, L.Q.; Ge, L.S.; Wang, Q.F.; Zhang, J.; Gao, B.F.; Zhou, Y.F.; Jiang, S.Q. Research advances in the Mesozoic tectonic regimes during the formation of Jiaodong ore cluster area. *Prog. Nat. Sci.* **2006**, *16*, 777–784.
4. Deng, J.; Wang, C.M.; Bagas, L.; Carranza, E.J.M.; Lu, Y.J. Cretaceous-Cenozoic tectonic history of the Jiaojia Fault and gold mineralization in the Jiaodong Peninsula, China: Constraints from zircon U-Pb, illite K-Ar, and apatite fission track thermochronometry. *Min. Depos.* **2015**, *50*, 987–1006. [CrossRef]
5. Li, S.R.; Santosh, M. Metallogeny and craton destruction: Records from the North China Craton. *Ore Geol. Rev.* **2014**, *56*, 376–414. [CrossRef]
6. Yang, L.Q.; Deng, J.; Wang, Z.L.; Zhang, L.; Guo, L.N.; Song, M.C.; Zheng, X.L. Mesozoic gold metallogenic system of the Jiaodong gold province, eastern China. *Acta Pet. Sin.* **2014**, *30*, 2447–2467. (In Chinese)
7. Goldfarb, R.J.; Santosh, M. The dilemma of the Jiaodong gold deposits: Are they unique? *Geosci. Front.* **2014**, *5*, 139–153. [CrossRef]
8. Fan, H.R.; Feng, K.; Li, X.H.; Hu, F.F.; Yang, K.F. Mesozoic gold mineralization in the Jiaodong and Korean peninsulas. *Acta Pet. Sin.* **2016**, *32*, 3225–3238. (In Chinese)
9. Song, M.C.; Ding, Z.J.; Zhang, J.J.; Song, Y.X.; Bo, J.W.; Wang, Y.Q.; Liu, H.B.; Li, S.Y.; Li, J.J.; Li, R.X. Geology and mineralization of the Sanshandao supergiant gold deposit (1200 t) in the Jiaodong Peninsula, China: A review. *China Geol.* **2021**, *4*, 686–719. [CrossRef]
10. Song, M.C.; Song, Y.X.; Ding, Z.Z.; Li, S.Y. Jiaodong Gold Deposits:Essential Characteristics and Major Controversy. *Gold Sci. Technol.* **2018**, *26*, 406–422. (In Chinese)
11. Zhu, R.X.; Fan, H.R.; Li, J.W.; Meng, Q.R.; Li, S.R.; Zeng, Q.D. Decratonic gold deposits. *Sci. China Earth Sci.* **2015**, *58*, 1523–1537. (In Chinese) [CrossRef]
12. Zheng, Y.F.; Xu, Z.; Zhao, Z.F.; Dai, L.Q. Mesozoic mafic magmatism in North China: Implications for thinning and destruction of cratonic lithosphere. *Sci. China Earth Sci.* **2018**, *61*, 353–385. (In Chinese) [CrossRef]
13. Chen, Y.J.; Pirajno, F.; Lai, Y.; Li, C. Metallogenic time and tectonic setting of the Jiaodong gold province, eastern China. *Acta Pet. Sin.* **2004**, *20*, 907–922.
14. Wang, T.J.; Yan, F. Ore fluid evolution and ore prediction of magmatic hydrothermal gold deposits in Jiaodong area. *Conributions Geol. Miner. Resour. Res.* **2002**, *17*, 169–174. (In Chinese)
15. Goldfarb, R.J.; Hart, C.; Davis, G.; Groves, D. East Asian gold: Deciphering the anomaly of phanerozoic gold in precambrian cratons. *Econ. Geol.* **2007**, *102*, 341–345. [CrossRef]
16. Yang, M.Z. *The Geochemistry of Wallrock Alteration Zone of Gold Deposits—As Exemplified by Jiaodong Gold Deposits*; Geology Press: Beijing, China, 1998; pp. 109–122. (In Chinese)
17. Zhai, M.G.; Fan, H.R.; Yang, J.H.; Miao, L.C. Large-scale cluster of gold deposits in east shandong: Anorogenic metallogenes. *Earth Sci. Front.* **2004**, *11*, 85–98. (In Chinese)
18. Li, L.; Santosh, M.; Li, S.R. The 'Jiaodong type' gold deposits: Characteristics, origin and prospecting. *Ore Geol. Rev.* **2015**, *65*, 589–611. [CrossRef]
19. Deng, J.; Wang, Q.F. Gold mineralization in China: Metallogenic provinces, deposit types and tectonic framework. *Gondwana Res.* **2016**, *36*, 219–274. [CrossRef]
20. Song, M.C.; Song, Y.X.; Li, J.; Liu, H.B.; Li, J.; Dong, L.L.; He, C.Y.; Wang, R.S. Thermal doming-extension metallogenic system of Jiaodong type gold deposits. *Acta Pet. Sin.* **2023**, *39*, 1241–1260. (In Chinese) [CrossRef]
21. Miao, L.C.; Luo, Z.K.; Guan, K.; Huang, J.Z. The implication of the SHRIMP U-Pb age in zircon to the petrogenesis of the Linglong granite, East Shangdong Province. *Acta Pet. Sin.* **1998**, *14*, 198–206. (In Chinese)
22. Qiu, L.G.; Ren, F.L.; Cao, Z.X.; Zhang, Y.Q. Late mesozoic magmatic activities and their constraints on geotectonics of jiaodong region. *Geotecton. Metallog.* **2008**, *1*, 117–123. (In Chinese) [CrossRef]
23. Song, M.C.; Song, Y.X.; Li, J.; Li, S.Y. Metallogenic series of gold and nonferrous metal deposits related to cretaceous granites in eastern Shandong Peninsula, China. *Geotecton. Metallog.* **2015**, *39*, 828–843. (In Chinese) [CrossRef]
24. Tian, J.P. The Mesozoic Gold Polymetallic Regional Metallogeny in Qipengfu Ore Concentration Area, Jiaodong Peninsula. Ph.D. Thesis, China University of Geosciences, Beijing, China, 2020. (In Chinese)
25. Wan, D.F.; Fan, T.Y.; Tian, S.H. The chromium analytical technique for hydrogen isotopes. *Acta Geosci. Sin.* **2005**, *26*, 35–38. (In Chinese)
26. Liu, G.Q.; Zhao, K.D.; Jiang, S.Y.; Chen, W. In-situ sulfur isotope and trace element analysis of pyrite from the Xiwang uranium ore deposit in South China: Implication for ore genesis. *J. Geochem. Explor.* **2018**, *195*, 49–65. [CrossRef]
27. Fu, J.L.; Hu, Z.C.; Zhang, W.; Yang, L.; Liu, Y.S.; Li, M.; Zong, K.Q.; Gao, S.; Hu, S.H. In situ sulfur isotopes (δ34S and δ33S) analyses in sulfides and elemental sulfur using high sensitivity cones combined with the addition of nitrogen by laser ablation MC-ICP-MS. *Anal. Chim. Acta* **2016**, *911*, 14–26. [CrossRef] [PubMed]

28. Zhu, Z.Y.; Cook, N.J.; Yang, T.; Ciobanu, C.L.; Zhao, K.D.; Jiang, S.Y. Mapping of Sulfur Isotopes and Trace Elements in Sulfides by LA-(MC)-ICP-MS: Potential Analytical Problems, Improvements and Implications. *Minerals* **2016**, *6*, 110. [CrossRef]
29. Jackson, S.E.; Pearson, N.J.; Griffin, W.L.; Belousova, E.A. The application of laser ablation-inductively coupled plasma-mass spectrometry to in situ U-Pb zircon geochronology. *Chem. Geol.* **2004**, *211*, 47–69. [CrossRef]
30. Slama, J.; Kosler, J.; Condon, D.J.; Crowley, J.L.; Gerdes, A.; Hanchar, J.M.; Horstwood, M.S.A.; Morris, G.A.; Nasdala, L.; Norberg, N.; et al. Plesovice zircon—A new natural reference material for U-Pb and Hf isotopic microanalysis. *Chem. Geol.* **2008**, *249*, 1–35. [CrossRef]
31. Liu, Y.S.; Hu, Z.C.; Zong, K.Q.; Gao, C.G.; Gao, S.; Xu, J.A.; Chen, H.H. Reappraisement and refinement of zircon U-Pb isotope and trace element analyses by LA-ICP-MS. *Chin. Sci. Bull.* **2010**, *55*, 1535–1546. [CrossRef]
32. Andersen, T. Correction of common lead in U–Pb analyses that do not report 204 Pb. *Chem. Geol.* **2002**, *192*, 59–79. [CrossRef]
33. Ludwig, K.R. *User's Manual for Isoplot 3.75, A Geochronological Toolkit for Microsoft Excel*; Berkeley Geochronology Center Special Publication: Berkeley, CA, USA, 2012; Volume 5, pp. 1–75.
34. Liu, W.G.; Liu, H.B.; Li, G.Z.; Xiao, Z.B.; Tu, J.R.; Li, H.M. The Application of Ion Exchange Resins in Sr-Nd Isotopic Assay of Geological Samples. *Acta Geol. Sin.* **2017**, *91*, 2584–2592. (In Chinese)
35. Clayton, R.N.; O'Neil, J.R.; Mayeda, T.K. Oxygen isotope exchange between quartz and water. *J. Geophys. Res.* **1972**, *77*, 3057–3067. [CrossRef]
36. Hou, M.L.; Jiang, S.Y.; Jiang, Y.H.; Ling, H.F. S-Pb isotope geochemistry and Rb-Sr geochronology of the Penglai gold field in the eastern Shangdong province. *Acta Pet. Sin.* **2006**, *22*, 2525–2533. (In Chinese)
37. Lu, Y.F. GeoKit—A geochemical toolkit for Microsoft Excel. *Geochimica* **2004**, *33*, 459–464. (In Chinese)
38. Simmons, S.F.; Sawkins, F.J.; Schlutter, D.J. Mantle-derived helium in two Peruvian hydrothermal ore deposits. *Nature* **1987**, *329*, 429–432. [CrossRef]
39. Hugh, P.; Taylor, J.R. The Application of Oxygen and Hydrogen Isotope Studies to Problems of Hydrothermal Alteration and Ore Deposition. *Econ. Geol.* **1974**, *69*, 843–883.
40. Ohmoto, H. Stable isotope geochemistry of ore deposits. *Rev. Mineral.* **1986**, *16*, 491–559.
41. Sheppard, S.M.F. Characterization and isotopic variations in natural waters. Reviews of Mineralogy. *Rev. Mineral.* **1986**, *16*, 165–183.
42. Mao, J.W.; Li, H.M.; Wang, Y.T.; Zhang, C.Q.; Wang, R.T. The relationship between mantle-drived fluid gold ore-formation in the eastern Shandong peninsula: Evidences from D-O-C-S isotopes. *Acta Pet. Sin.* **2005**, *79*, 839–857. (In Chinese)
43. Baptiste, P.J.; Fouquet, Y. Abundance and isotopic composition of helium in hydrothermal sulfides from the East Pacific Rise at 13 °N. *Geochim. Cosmochim. Acta* **1996**, *60*, 87–93. [CrossRef]
44. Trull, T.W.; Kurz, M.D.; Jenkins, W.J. Diffusion of cosmogenic 3He in olivine and quartz: Implications for surface exposure dating. *Earth Planet. Sci. Lett.* **1991**, *103*, 241–251. [CrossRef]
45. Podosek, K.A.; Bernatowica, T.J.; Kramer, F.E. Absorption of xenon and krypton on shales. *Geochim. Cosmochim. Acta* **1980**, *45*, 2401–2415. [CrossRef]
46. Turner, G.; Stuart, F. Helium/heat ratios and deposition temperatures of sulphides from the ocean floor. *Nature* **1992**, *375*, 581–583. [CrossRef]
47. Tolstikhin, I.N. A Review—Some Recent Advances in Isotope Geochemistry of Light Rare Gases. *Adv. Earth Planet. Sci.* **1978**, *3*, 33–62.
48. Mamyrin, B.A.; Tolstikhin, I.N. *Helium Isotopes in Nature*; Elsevier: Amsterdam, The Netherlands, 1984; pp. 1–273.
49. Ballentine, C.J.; Burgess, R.; Marty, B. Tracing fluid origin, transport and interaction in the crust. *Rev. Mineral. Geochem.* **2002**, *47*, 539–614. [CrossRef]
50. Burnard, P.G.; Hu, R.Z.; Turner, G.; Bi, X.W. Mantle, crustal and atmospheric noble gases in Ailaoshan Gold deposits, Yunnan Province, China. *Geochim. Cosmochim. Acta J. Geochem. Soc. Meteorit. Soc.* **1999**, *63*, 1595–1604. [CrossRef]
51. Lai, Y.; Liu, Y.L.; Huang, B.L.; Chen, Y.J. The characteristics of noble gases in mantle-derived xenoliths in Wudalianchi and Kuandian, NE China: MORB-like mantle and metasomated mantle. *Acta Pet. Sin.* **2005**, *21*, 1373–1381. (In Chinese)
52. Kendrick, M.A.; Burgess, R.; Pattrick, R.A.D.; Turner, G. Fluid inclusion noble gas and halogen evidence on the origin of Cu-Porphyry mineralising fluids. *Geochim. Cosmochim. Acta* **2001**, *65*, 2651–2668. [CrossRef]
53. Matsumoto, T.; Chen, Y.l.; Matsuda, J.I. Concomitant occurence of primordial and recycled noble gasesin the Earth's mantle. *Earth Planet Sci. Lett.* **2001**, *185*, 35–47. [CrossRef]
54. Gautheron, C.; Moreira, M.; Allegre, C. He, Ne and Ar composition of the European lithospheric mantle. *Chem. Geol.* **2005**, *217*, 97–112. [CrossRef]
55. Hopp, J.; Ionov, D.A. Tracing partial melting and subduction-related metasomatism in the Kamchatkan mantle wedge using noble gas compositions. *Earth Planet Sci. Lett.* **2011**, *302*, 121–131. [CrossRef]
56. Broadley, M.W.; Ballentine, C.J.; Chavrit, D.; Dallai, L.; Burgess, R. Sedimentary halogens and noble gases within Western Antarctic xenoliths: Implications of extensive volatile recycling to the sub continental lithospheric mantle. *Geochim. Cosmochim. Acta* **2016**, *176*, 139–156. [CrossRef]
57. Hoefs, J. *Stable Isotope Geochemistry*, 6th ed.; Springer: Berlin/Heidelberg, Germany, 2009; Volume 72.

58. Gao, T.Z.; Zhao, L.S.; Yang, M.Z. Gold mineralization and its evolution in the Muping-Rushan gold deposits, Shandong province. *Geotecton. Metallog.* **2001**, *25*, 155–160. (In Chinese)
59. Wang, Y.W.; Zhu, F.S.; Gong, R.T. Tectonic isotope geochemistry--Further study on sulphur isotope of Jiaodong Gold Concentration Area. *Gold* **2002**, *23*, 1–16. (In Chinese)
60. Deng, J.; Liu, X.F.; Wang, Q.F.; Pan, R.G. Origin of the Jiaodong-type Xinli gold deposit, Jiaodong Peninsula, China: Constraints from fluid inclusion and C-D-O-S-Sr isotope compositions. *Ore Geol. Rev.* **2015**, *65*, 674–686. [CrossRef]
61. Li, Z.L.; Yang, M.Z. *The Geology-Geochemistry of Gold Deposits in Jiaodong Region*; Tianjin Science and Technology Press: Tianjin, China, 1993; pp. 1–300. (In Chinese)
62. Huang, D.Y. Sulfur isotope studies of the metallogenic series of gold deposits in Jiaodong area. *Miner. Depos.* **1994**, *13*, 75–86. (In Chinese)
63. Wu, K.X.; Hu, R.Z.; Bi, X.W.; Peng, J.T.; Tang, Q.L. Ore lead isotopes as a tracer for ore-forming material sources: A review. *Geol.-Geochem.* **2002**, *30*, 73–81.
64. Stacey, J.S.; Kramers, J.D. Approximation of terrestrial lead isotope evolution by a two stage model. *Earth Planet. Sci. Lett.* **1975**, *26*, 207–221. [CrossRef]
65. Doe, B.R.; Zartman, R.E. Plumbotectonics, the Phanerozoic. In *Geochemistry of Hydrothermal Ore Deposits*; Barnes, H.L., Ed.; Wiley: Hoboken, NJ, USA, 1979; pp. 22–70.
66. Zartman, R.E.; Doe, B.R. Plumbotectonics—The model. *Tectonophysics* **1981**, *75*, 135–162. [CrossRef]
67. Faure, M.; Caridroit, M.; Charvet, J. The Late Jurassic oblique collisional orogen of SW Japan: New structural data and synthesis. *Tectonics* **1986**, *5*, 1089–1114. [CrossRef]
68. Sun, X.L. Sr and O Isotopic Characteristics of Porphyries in the Qinling Molybdenum Deposit Belt and Their Implication to Genetic Mechanism and Type. Ph.D. Thesis, Chengdu University of Technology, Chengdu, China, 2001. (In Chinese)
69. Hou, M.L.; Jiang, Y.H.; Jiang, S.Y.; Ling, H.F.; Zhao, K.D. Contrasting origins of late Mesozoic adakitic granitoids from the northwestern Jiaodong Peninsula, east China: Implications for crustal thickening to delamination. *Geol. Mag.* **2007**, *144*, 619–631. [CrossRef]
70. Yang, K.F.; Fan, H.R.; Santosh, M.; Hu, F.F.; Wilde, S.A.; Lan, T.G.; Lu, L.N.; Liu, Y.S. Reactivation of the Archean lower crust: Implications for zircon geochronology, elemental and Sr-Nd-Hf isotopic geochemistry of late Mesozoic granitoids from northwestern Jiaodong Terrane, the North China Craton. *Lithos* **2012**, *146*, 112–127. [CrossRef]
71. Li, X.H.; Fan, H.R.; Zhang, Y.W.; Hu, F.F.; Yang, K.F.; Liu, X.; Cai, Y.C.; Zhao, K.D. Rapid exhumation of the northern Jiaobei Terrane, North China Craton in the Early Cretaceous: Insights from Al-in-hornblende barometry and U-Pb geochronology. *J. Asian Earth Sci.* **2018**, *160*, 365–379. [CrossRef]
72. Deng, J.; Qiu, K.F.; Wang, Q.F.; Goldfarb, R.; Yang, L.Q.; Zi, J.W.; Geng, J.Z.; Ma, Y. In Situ Dating of Hydrothermal Monazite and Implications for the Geodynamic Controls on Ore Formation in the Jiaodong Gold Province, Eastern China. *Econ. Geol.* **2020**, *115*, 671–685. [CrossRef]
73. Yang, K.F.; Jiang, P.; Fan, H.R.; Zuo, Y.B.; Yang, Y.H. Tectonic transition from a compressional to extensional metallogenic environment at similar to 120 Ma revealed in the Hushan gold deposit, Jiaodong, North China Craton. *J. Asian Earth Sci.* **2018**, *160*, 408–425. [CrossRef]
74. Zhang, L.; Weinberg, R.F.; Yang, L.Q.; Groves, D.I.; Sai, S.X.; Matchan, E.; Phillips, D.; Kohn, B.P.; Miggins, D.P.; Liu, Y.; et al. Mesozoic Orogenic Gold Mineralization in the Jiaodong Peninsula, China: A Focused Event at 120 +/− 2 Ma During Cooling of Pregold Granite Intrusions. *Econ. Geol.* **2020**, *115*, 415–441. [CrossRef]
75. Fan, H.R.; Zhai, M.G.; Xie, Y.H.; Yang, J.H. Ore-forming fluids associated with granite-hosted gold mineralization at the Sanshandao deposit, Jiaodong gold province, China. *Min. Depos.* **2003**, *38*, 739–750. [CrossRef]
76. Wen, B.J.; Fan, H.R.; Santosh, M.; Hu, F.F.; Pirajno, F.; Yang, K.F. Genesis of two different types of gold mineralization in the Linglong gold field, China: Constrains from geology, fluid inclusions and stable isotope. *Ore Geol. Rev.* **2015**, *65*, 643–658. [CrossRef]
77. Li, J.W.; Vasconcelos, P.M.; Zhang, J.; Zhou, M.F.; Zhang, X.J.; Yang, F.H. 40Ar/39Ar constraints on a temporal link between gold mineralization, magmatism, and continental margin transtension in the Jiaodong Gold Province, eastern China. *J. Geol.* **2003**, *111*, 741–751. [CrossRef]
78. Charles, N.; Augier, R.; Gumiaux, C.; Monie, P.; Chen, Y.; Faure, M.; Zhu, R.X. Timing, duration and role of magmatism in wide rift systems: Insights from the Jiaodong Peninsula (China, East Asia). *Gondwana Res.* **2013**, *24*, 412–428. [CrossRef]
79. Feng, K.; Fan, H.R.; Groves, D.I.; Yang, K.F.; Hu, F.F.; Liu, X.; Cai, Y.C. Geochronological and sulfur isotopic evidence for the genesis of the post-magmatic, deeply sourced, and anomalously gold-rich Daliuhang orogenic deposit, Jiaodong, China. *Min. Depos.* **2020**, *55*, 293–308. [CrossRef]
80. Qu, X.M.; Wang, H.M.; Rao, B. Partial melting experiments of Jiaodong Group and their implication for the origin of the granites. *Geochimica* **2000**, *2*, 153–161. (In Chinese)
81. Yang, J.H.; Zhou, X.H. The Rb-Sr isochron of ore and pyrite sub-samples from Linglong gold deposit, Jiaodong Peninsula, eastern China and their geological significance. *Chin. Sci. Bull.* **2000**, *45*, 2272–2277. (In Chinese) [CrossRef]
82. Yang, J.H.; Chu, M.F.; Liu, W.; Zhai, M.G. Geochemistry and petrogenesis of Guojialing granodiorites from the northwestern Jiaodong Peninsula, eastern China. *Acta Pet. Sin.* **2003**, *19*, 692–700. (In Chinese)

83. Maruyama, S.; Isozaki, Y.; Kimura, G.; Terabayashi, M. Paleogeographic maps of the Japanese Islands: Plate tectonic synthesis from 750 Ma to the present. *Isl. Arc.* **1997**, *6*, 121–142. [CrossRef]
84. Zhu, G.; Niu, M.L.; Xie, C.L.; Wang, Y.S. Sinistral to Normal Faulting along the Tan-Lu Fault Zone: Evidence for Geodynamic Switching of the East China Continental Margin. *J. Geol.* **2010**, *118*, 277–293. [CrossRef]

minerals

Article

Petrogenesis of the Laoshan Suite in the Jiaodong Peninsula (Eastern China): An Oxidized Low Ba–Sr A1-Type Granite

Jian Li [1,2,3], Changwei Wang [1,3], Mingchun Song [4,*], Changjiang Wang [1,3], Shiyong Li [5], Xiao Liu [1] and Qingyi Cui [1]

1 School of Resources and Environmental Engineering, Shandong University of Technology, Zibo 255049, China; lijian_sdut@163.com (J.L.); wangchangwei_sdut@163.com (C.W.); cj85520819@163.com (C.W.); liuxiaogis@163.com (X.L.); 15265520559@163.com (Q.C.)
2 Institute of Geology, Chinese Academy of Geological Sciences, Beijing 100037, China
3 Shandong Zhengyuan Geological Resources Exploration Co., Ltd., Zibo 255049, China
4 Hebei Province Collaborative Innovation Center for Strategic Critical Mineral Research, Hebei GEO University, Shijiazhuang 050031, China
5 Shandong Province Nuclear Industry Geological Group 273, Yantai 264006, China; dikechulsy@126.com
* Correspondence: mingchuns@163.com; Tel.: +86-137-0531-7191

Abstract: The Jiaodong Peninsula is closely related to Mesozoic granites in terms of spatial and temporal aspects. However, the specific association between the genesis of gold mineralization and these granites remains unclear. It is also ambiguous why Laoshan-type granites, which are similar to Mesozoic granites, are not gold mineralized. In this study, we analyzed the Laoshan granites and compiled Mesozoic magmatic rock data (Linglong, Guojialing, and Weideshan suites) of the Jiaodong Peninsula. We performed whole-rock major and trace elements, LA–ICP–MS zircon U–Pb geochronology and geochemistry analyses. Our zircon U–Pb data denote that the Laoshan granite was emplaced during the 118 ± 1 Ma. The Laoshan granite is characterized by high SiO_2 content (76.03–80.28 wt.%), high TFe_2O_3/MgO (11.1–27.1) and Ga/Al (3.0–3.5) ratios, high zircon saturation temperature (809–850 °C), and negative Eu (Eu/Eu* = 0.05–0.08) anomalies, showing A-type granite characteristics. Furthermore, the Laoshan granite is identified as an A1-type granite with low Ba (33.1–42.0 ppm) and Sr (14.1–21.0 ppm) contents. It was formed in an extensional tectonic environment induced by the subducting slab roll-back of the Paleo-Pacific Plate, mainly from the partial melting of lower crustal materials, mixed with a small amount of mantle components. Zircon trace elements indicate that the Laoshan suite had relatively high oxygen fugacity and temperature (ΔFMQ = + 2.43 to + 4.22, T-Ti in zircon (mean) = 796 °C) compared to the pre-mineralization and contemporaneous mineralization magma. We propose that oxidized adakitic magma (Weideshan suite) may contribute to the enrichment and mineralization of gold. Although Laoshan-type granite also has a high oxygen fugacity, its location is distant from the ore-controlling faults, which ultimately hinders the formation of gold mineralization. From the Jurassic to the Cretaceous, the tectonic and geochemical properties of magmatic rocks in the Jiaodong Peninsula have changed, and gold deposits are formed in such a transitional process (compression to extension, reduction to oxidation).

Keywords: A1-type granite; high oxygen fugacity; adakitic rocks; Laoshan suite; Jiaodong Peninsula

Citation: Li, J.; Wang, C.; Song, M.; Wang, C.; Li, S.; Liu, X.; Cui, Q. Petrogenesis of the Laoshan Suite in the Jiaodong Peninsula (Eastern China): An Oxidized Low Ba–Sr A1-Type Granite. *Minerals* **2023**, *13*, 1012. https://doi.org/10.3390/min13081012

Academic Editors: Fan Yang, Cun Zhang, Jian Chang and Jaroslav Dostal

Received: 3 July 2023
Revised: 26 July 2023
Accepted: 26 July 2023
Published: 29 July 2023

1. Introduction

The Jiaodong Peninsula is the world's third-largest gold province, surpassed only by the Witwatersrand Basin in South Africa and the Muruntau in Uzbekistan [1]. Located in the eastern part of the North China Craton (NCC) (Figure 1), the Jiaodong Peninsula hosts a nearly complete record of the complex NCC geological evolution from the Archean to the Cenozoic. Currently, the proven gold reserves in the Jiaodong Peninsula (only 0.3% of Chinese territory) are nearly 6000 t [2], accounting for more than 40% of China's total gold reserves [1,3,4]. Geochronological data from multiple methods (quartz/sericite $^{40}Ar/^{39}Ar$, pyrite Rb–Sr, and hydrothermal monazite U–Pb) show that the Jiaodong gold deposits were formed in the late Early Cretaceous (120 ± 5 Ma) [1,3,5–7], with explosive and transient characteristics [8–11]. According to geochronology and geochemical characteristics, the granites in Jiaodong can be divided into Linglong-, Guojialing-, Weideshan-, and Laoshan-type [4]. Previous studies have shown that 80% of gold deposits occur in granite (mainly Linglong and Guojialing) [7]. Although gold mineralization is temporally overlain with Weideshan- and Laoshan-type granites, gold mineralization does not exist in Laoshan-type granites, which are far from gold deposits. It is not yet clear whether the differences in magmatic physicochemical conditions or genetic types have caused the Laoshan-type granite to be unrelated to gold mineralization.

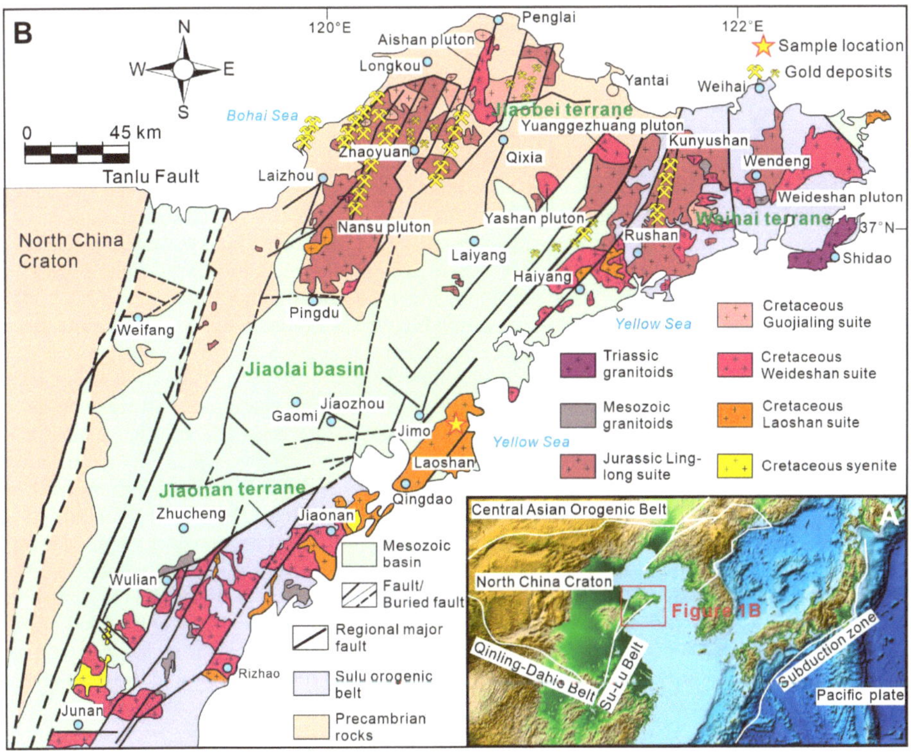

Figure 1. (**A**). Distribution map of the main tectonic zones in the NCC (after Yang et al. [12]); (**B**). Detailed geological map of the Jiaodong Peninsula (after Song et al. [4,7]).

To better understand the constraints of Mesozoic magmatic rocks on the formation of gold mineralization in the Jiaodong Peninsula, we present a comprehensive analysis that includes petrographic observations, whole-rock major and trace elements geochemistry data, and the results of laser ablation inductively coupled plasma mass spectrometry (LA–ICP–MS) U–Pb dating of zircon ages and trace elements for the Laoshan suite from the Jiaodong Peninsula. These data provide valuable insights into key parameters that may constrain gold mineralization, while also shedding light on the petrogenesis and physicochemical conditions of the Laoshan granite.

2. Geological Setting and Samples

The Jiaodong Peninsula is situated to the east of the Tanlu fault zone in the eastern part of the NCC as shown in Figure 1. The NCC, which is one of the largest cratons in East Asia, is made up of eastern and western blocks that were assembled at 1.85 Ga along the Trans-North China Orogen [13,14]. Subsequently, the NCC remained stable for a considerable period with minimal deformation, metamorphism, and magmatism until the Early Mesozoic. However, the deep continental subduction of the Yangtze Block relative to the NCC initiated the destruction of the craton [15,16], which climaxed in the Early Cretaceous under the subduction of the Paleo-Pacific plate [8,17–21]). During the Mesozoic, the NCC was significantly influenced by multiple tectonic processes, leading to the extensive thinning of the lithospheric mantle (>80 km) and strong interactions between the crust and mantle [20,22–25], resulting in the formation of a large number of polymetallic deposits. Consequently, numerous polymetallic deposits were formed, and the Jiaodong gold province is a typical example of this tectonic setting (Figure 1).

The Jiaodong Peninsula is comprised of three main geological regions: the Jiaobei terrane, the Jiaolai Basin, and the Sulu orogenic belt (Figure 1B). A Precambrian basement and Jurassic to Early Cretaceous granites constitute the main lithology of the Jiaobei terrane (Figure 1B). The Precambrian basement is made up of the Neoarchean Jiaodong Complex (inclu: metamorphosed volcanic-sedimentary rocks and Archean tonalite–trondhjemite–granodiorite), Paleoproterozoic Jingshan and Fenzishan groups, and the Neoproterozoic Penglai group (meta-sedimentary rock series) [4]. The Jiaolai continental basin was formed in the Cretaceous period and is situated at the boundary between the Sulu orogen and Jiaobei terrane. This basin is mainly composed of volcanic-sedimentary rocks (Figure 1B). The Sulu orogen can be further divided into the Jiaonan terrane in the south and Weihai terrane in the northeast (Figure 1B), which are mainly composed of ultrahigh-pressure metamorphic rock series [4]. Faults are mainly distributed in the NE-NNE-striking of the Jiaodong Peninsula, with NNE-striking faults playing a significant role as the main ore-controlling structures (Figure 1B).

The Mesozoic granites are mainly distributed in the Jiaobei terrane and the Sulu orogen (Figure 1B), panning two periods: the Middle Jurassic and Early Cretaceous [4]. The Middle Jurassic granites (~160 Ma) are mainly composed of Linglong, Luanjiahe, Kunyushan, and Wendeng plutons (Figure 1B) (hereafter collectively referred to as the Linglong suite), which have similar mineral compositions and geochemical characteristics [26]. The Early Cretaceous granites are classified into Guojialing (132–125 Ma), Weideshan (125–110 Ma), and Laoshan (125–109 Ma) suites based on their formation age, mineral assemblage, and geochemical information [26,27]. Over 200 gold deposits have been discovered in the Jiaodong Peninsula, with most of them hosted in granites (77% hosted in the Linglong type, 10% in the Guojialing type [4]), and the formation time of the Jiaodong gold deposits are concentrated around ~120 Ma [1,3–7]. These magmatic rocks are closely related to the Jiaodong gold deposit, both temporally and spatially. The gold deposits are mainly found in magmatic rocks and are distributed along NNE-striking faults (Figure 1B).

The Laoshan-type granites are mainly distributed in the eastern and southern Jiaodong Peninsula. There is no direct spatial distribution relationship with gold mineralization and non-ferrous metal mineralization. In this study, the Laoshan granite is classified as part of the Xiaopinglan unit, which is a medium- to fine-grained alkali feldspar granite (labelled as 21LS01 samples herein; location: N 36°15.3′; E 120°27.8′) (Figure 2B–D). The alkali feldspar granite primarily contains alkali feldspar (50–60%), quartz (25–30%), plagioclase (5–10%), and biotite (~1%), and zircon and apatite as the accessory phases (Figure 2B–D).

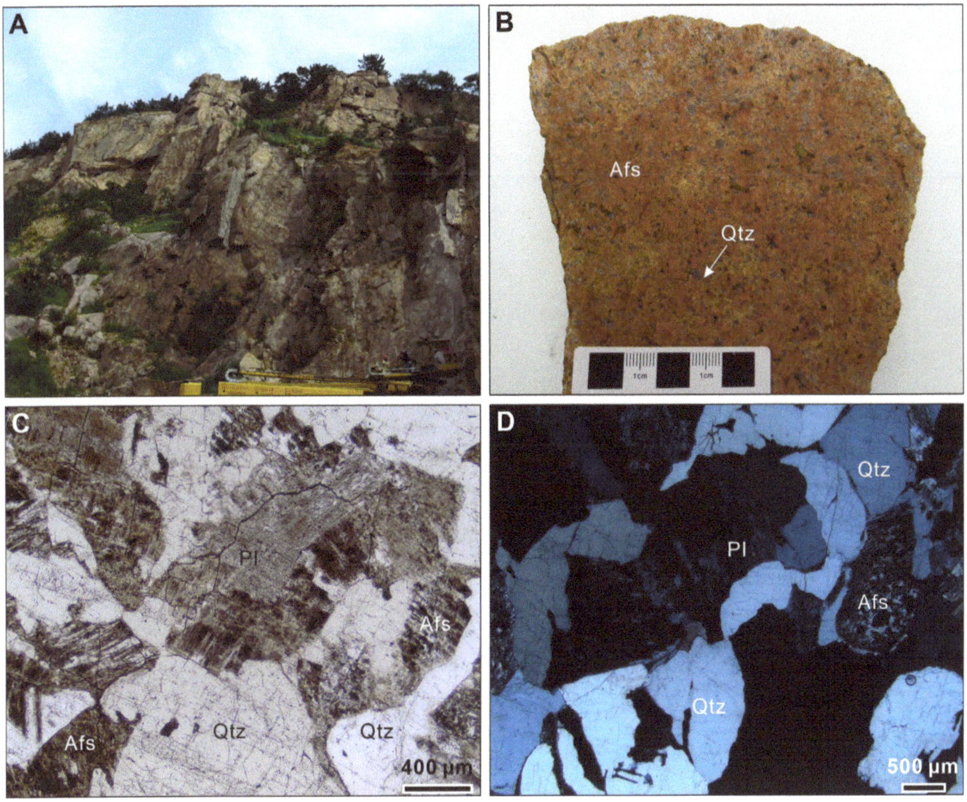

Figure 2. Representative photomicrographs of the Laoshan suite from the Jiaodong Peninsula ((**A**). Field photo; (**B**). Hand specimen photo; (**C**). Plane-polarized photo; (**D**). Cross-polarized photo). Afs = Alkali feldspar; Pl = Plagioclase; Qtz = Quartz.

3. Analytical Methods

3.1. Zircon U–Pb Dating and Trace Elements

The zircon grains of the Laoshan granites (fresh and unaltered) were separated through traditional magnetic and heavy liquid techniques and manually selected under a binocular microscope, which was completed by Guangzhou Tuoyan Analytical Technology Co., Ltd. (Guangzhou, China). The selected zircons were carefully examined using transmitted, reflected light, and cathodoluminescence (CL) images to analyze the internal texture of the zircons, avoiding cracks and inclusions during laser ablation. The U–Pb age and trace element analyses of zircon were completed using laser ablation inductively coupled plasma mass spectroscopy (LA–ICP–MS) (193 nm LA system and Agilent 7900 ICP–MS instrument) at the Institute of Geology, Chinese Academy of Geological Sciences (Beijing, China). The test parameters were as follows: the spot was 30 μm, the energy was 2 J/cm^2, and the repetition rate was 5 Hz. The selection of standard samples was to use 91500 (1062.4 ± 7.7 Ma; n = 34) as the external standard and NIST 610 glass and SA01 (533.7 ± 5.3 Ma; n = 17) were used to calibrate the age and trace elements (the analysis included two 91500 standard samples and one SA01 standard sample at every 10 sample points). Typically, a gas blank of 20 s was collected, and a signal interval of 35–40 s was used for data processing, followed by deep fractionation correction using an exponential equation [28]. Isoplot (Version 3.0 [29]) was used to obtain the concordia and weighted diagrams. The common Pb was corrected following the method described by T. Andersen [30].

3.2. Whole-Rock Major and Trace Element Analyses

We conducted 10 major element (21LS-1-1~5, 21LSS-1-1~5) and 5 trace element (21LS-1-1~5) analyses on Laoshan granite samples in this study, all of which were completed at Guangzhou Tuoyan Analytical Technology Co., Ltd. The major elements of the whole rock were determined with the Primus II X-ray fluorescence spectrometer (XRF), burned in a muffle furnace at 1000 °C for 2 h and the ignition loss (LOI) was calculated after cooling. The whole-rock trace elements analysis was completed using the Semferri CAP RQ system. During the test, the accuracy of data results was monitored with OU-6, BCR-1, and GBPG-1 as standard samples, and the RSD of data results was better than 5%.

4. Results

4.1. Zircon U–Pb Ages and Trace Elements Composition

Zircon grains of Laoshan granite exhibit clear oscillatory zones, with a length of 100–300 μm and a width of 40–100 μm, and contain a few mineral inclusions (Figure 3A). The REEs composition in the zircons reveal notable positive Ce anomalies (ranging from 3.49 to 1448.73) (Figure 3B), along with high Th/U ratios (0.54–2.77), indicating that these are magmatic zircons [31,32]. Twenty-nine analyses of zircon from the alkali feldspar granite 21LS01 gave $^{206}Pb/^{238}U$ ages of 121 ± 3 Ma to 113 ± 4 Ma (Supplementary Table S1), yielding a weighted mean $^{206}Pb/^{238}U$ age of 118 ± 1 Ma (MSWD = 0.31) (Figure 3C,D). This age is interpreted as the emplacement age of the alkali feldspar granite. Zircon trace element data are presented in Figure 4, showing significant negative correlations between zircon Ce/Yb, Sc/Yb, ΔFMQ, Th/U, and Gd/Yb and Hf elements, while U/Yb shows a positive correlation (Figure 4).

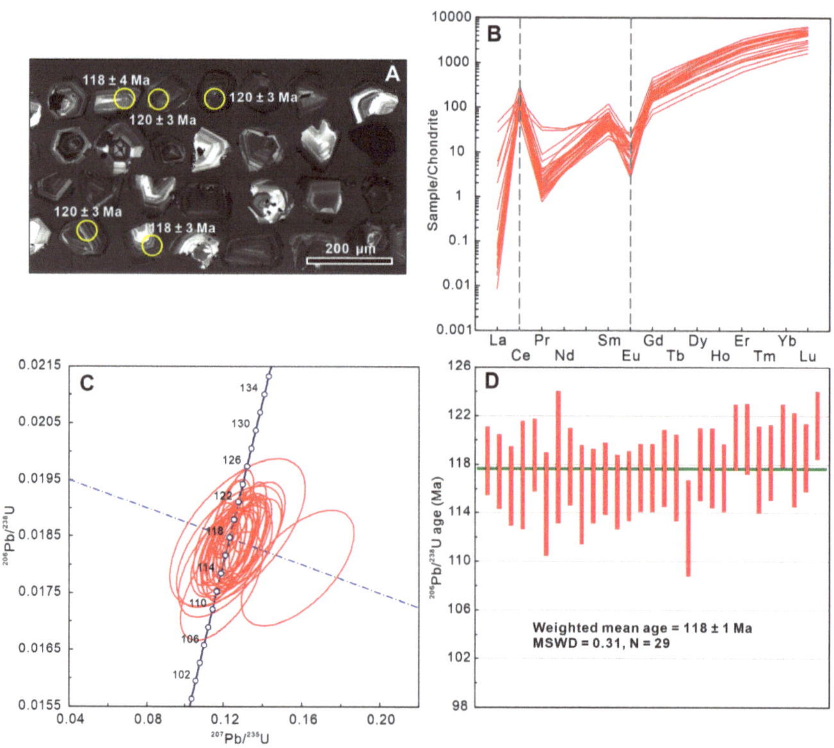

Figure 3. (**A**). Representative CL images; (**B**). Chondrite-normalized REE patterns; (**C**). Zircon U–Pb concordia diagram. (**D**). Weighted mean $^{206}Pb/^{238}U$ age diagram.

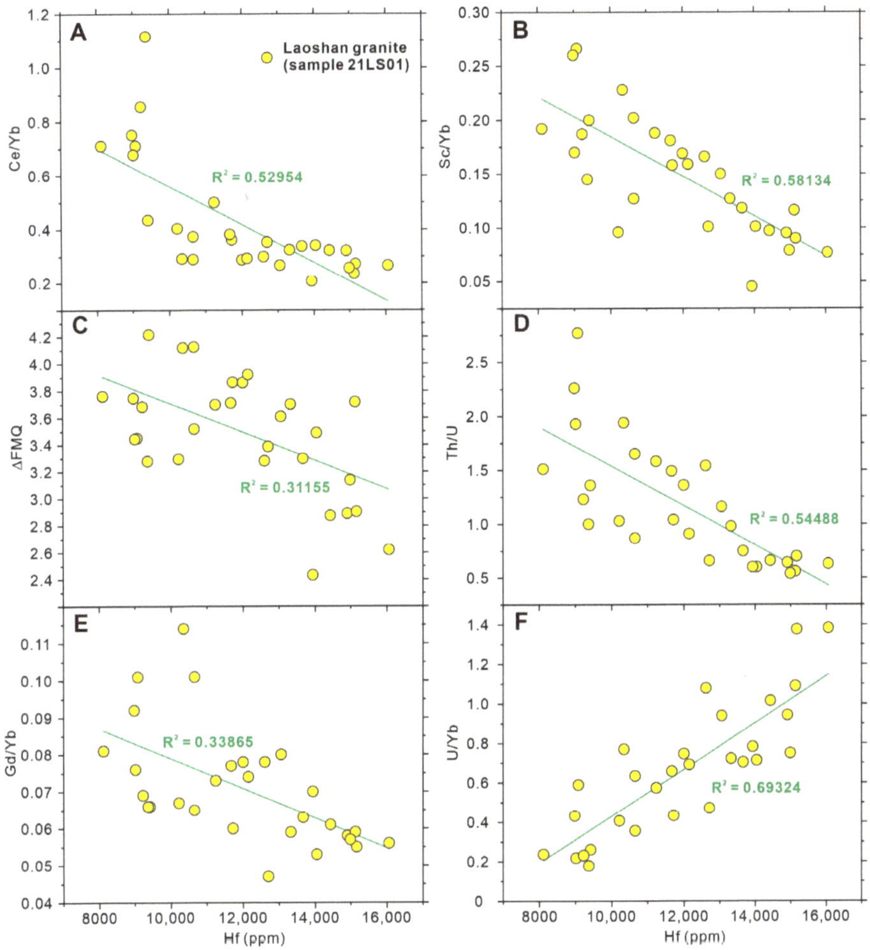

Figure 4. Ce/Yb (**A**), Sc/Yb (**B**), ΔFMQ (**C**), Th/U (**D**), Gd/Yb (**E**), and U/Yb (**F**) vs. Hf plots of zircons from the Laoshan suite.

4.2. Whole-Rock Major and Trace Element Geochemistry

The whole-rock major and trace element compositions of the Laoshan granite are listed in Supplementary Table S2. We also systematically collected the geochemical data of the Mesozoic igneous rock in the whole Jiaodong Peninsula, which are listed in Figures 5–7 together with the Laoshan granite. The Laoshan granite exhibits high contents of SiO_2 (76.03–80.28 wt.%), Na_2O (3.51–4.01 wt.%), and K_2O (3.94–4.61 wt.%), falling within the granite range and belonging to the subalkaline series in the SiO_2 vs. $Na_2O + K_2O$ diagram (Figure 5A). Some previous data fall within the quartz monzonite range of the alkaline series (Figure 5A). They belong to the high-K calc-alkaline series in the plot of SiO_2 vs. K_2O (Figure 5B). All samples in this study fall within the peraluminous range in the A/CNK-ANK diagram, while previous data fall within the metaluminous to peraluminous range (Figure 5C). In the Harker diagrams, the correlation between SiO_2 vs. Al_2O_3, CaO, MgO, P_2O_5, TFe_2O_3, and TiO_2 (Figure 6) is significant (especially adding published data), indicating the crystallization differentiation of minerals.

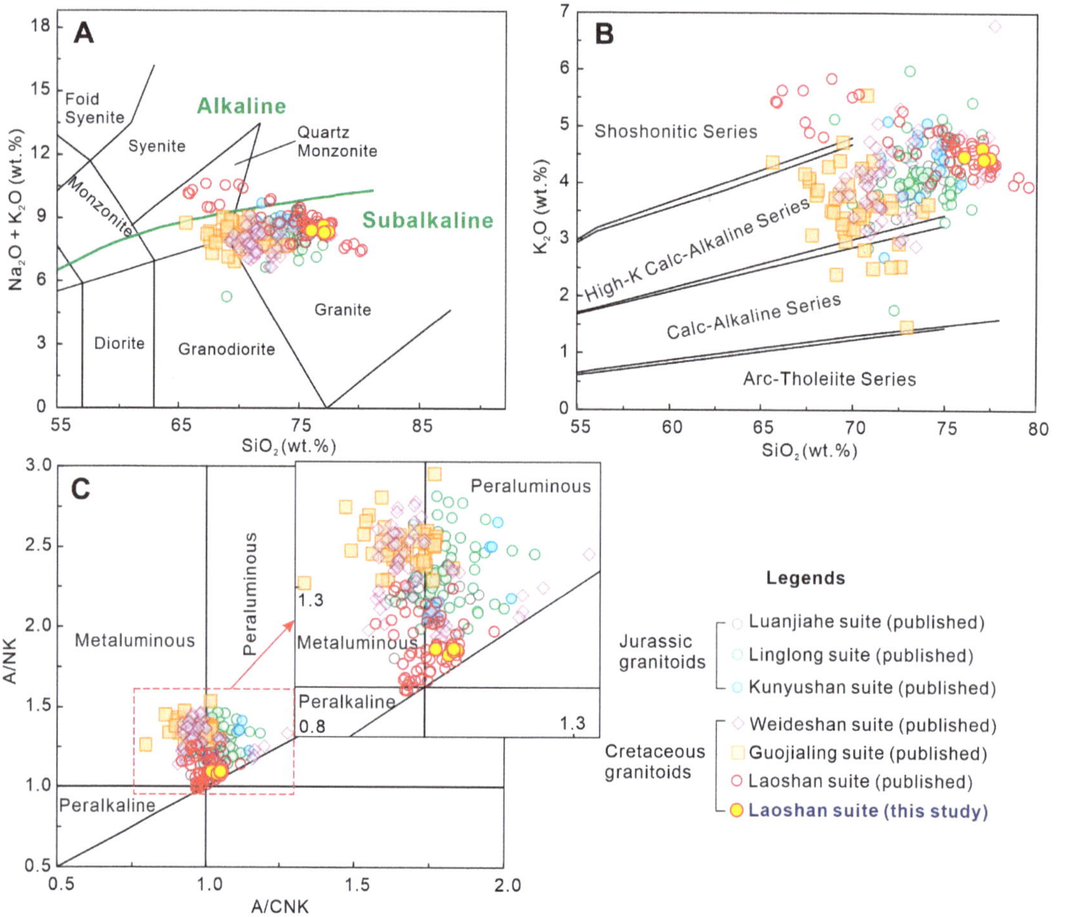

Figure 5. Whole-rock geochemistry plots for the Jiaodong Peninsula Mesozoic granitoids. (**A**). Total alkalis ($Na_2O + K_2O$) vs. silica (SiO_2) [33]; (**B**). K_2O vs. SiO_2 [34]; (**C**). A/NK vs. A/CNK [35]. Note: A/CNK = molar ratio of $Al_2O_3/(CaO + Na_2O + K_2O)$; A/NK = molar ratio of $Al_2O_3/(Na_2O + K_2O)$. Published data are from this study and [26,27,36–46].

In the REEs composition, the Laoshan granite has obvious characteristics of being enriched in LREE and depleted in HREE (La_N/Yb_N = 6.30–8.09; LREE/HREE = 5.94–6.90) (Figure 7G). The negative Eu anomaly is significant (Eu/Eu* = 0.06–0.08), with a clear V-shaped pattern, and the published data also exhibit this characteristic (Figure 7G). In the trace element spider diagrams, all samples are enriched in large-ion lithophile elements (LILEs, e.g., Rb and Pb) and depleted in high-field-strength elements (HFSEs, e.g., Nb, Ta, and Ti) and P (Figure 7H). The Linglong, Guojialing, and Weideshan suites also exhibit such characteristics (Figure 7B,D,F).

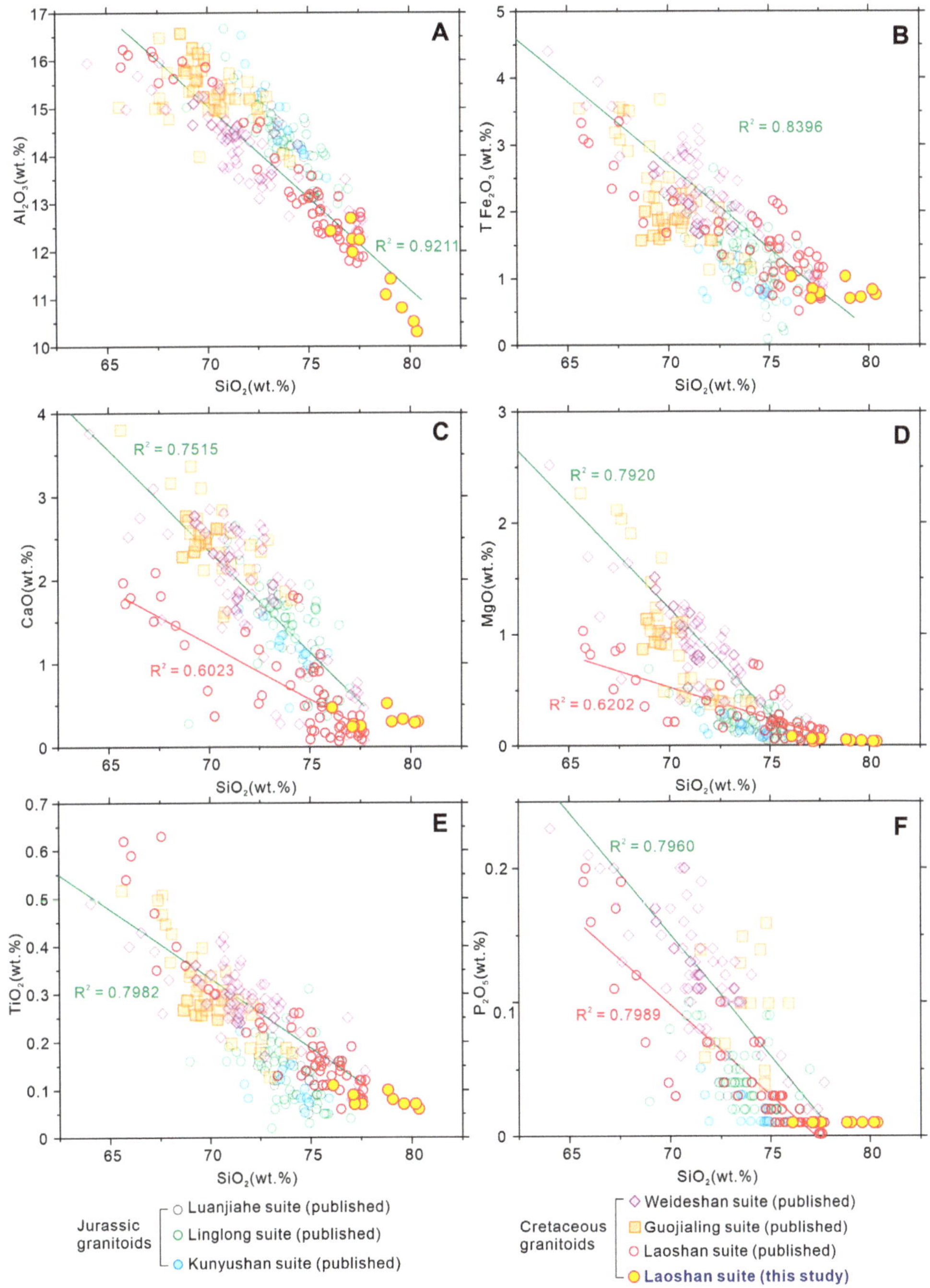

Figure 6. Al_2O_3 (**A**), TFe_2O_3 (**B**), CaO (**C**), MgO (**D**), TiO_2 (**E**), and P_2O_5 (**F**) vs. SiO_2 plots of Mesozoic granitoids from the Jiaodong Peninsula. The published data are the same as in Figure 5.

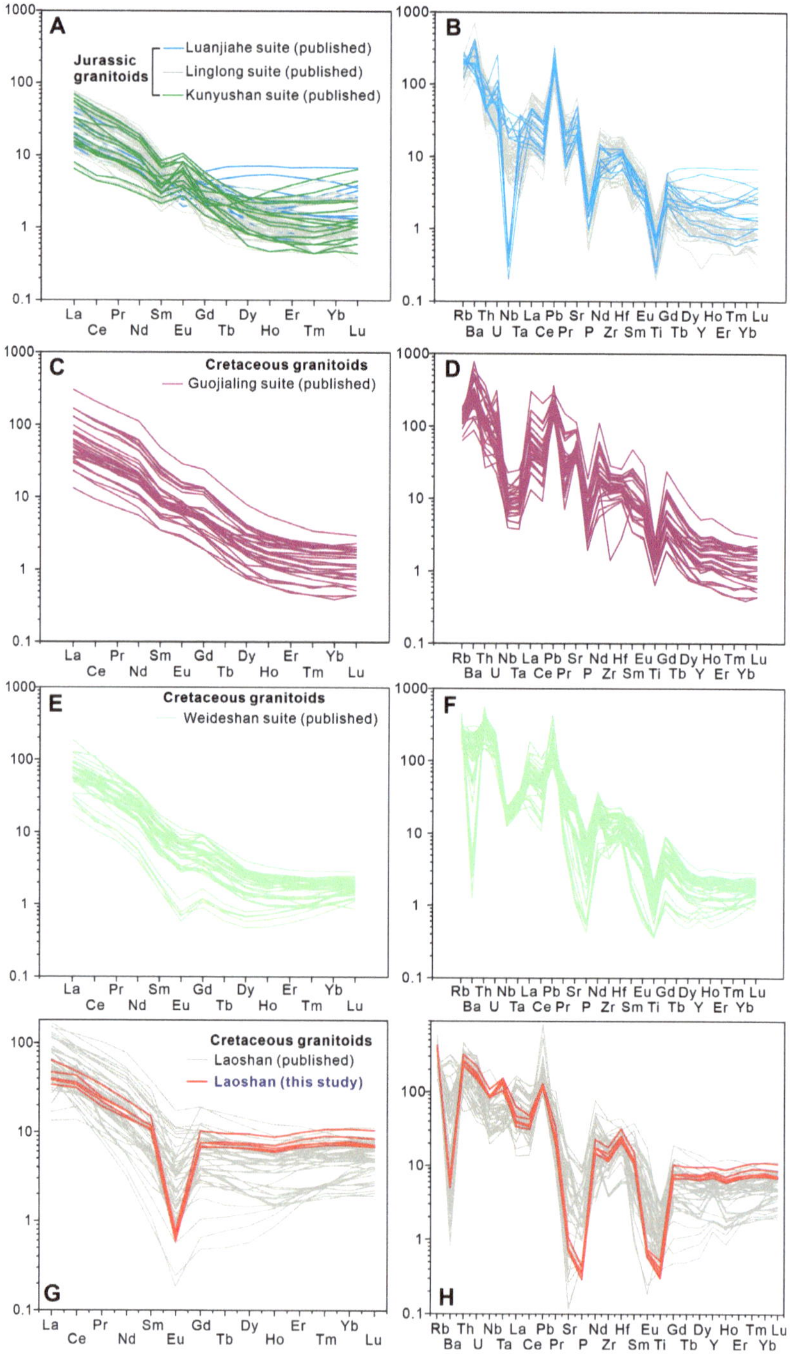

Figure 7. Chondrite-normalized REE patterns (**A**,**C**,**E**,**G**) and primitive mantle-normalized trace element variation diagrams (**B**,**D**,**F**,**H**) for the Jiaodong Peninsula Mesozoic granitoids. Normalizing values are from McDonough and Sun [47]. The published data are the same as in Figure 5.

5. Discussion

5.1. Magma Types, Sources, and Petrogenesis

Adakitic geochemical features are widely regarded as important for the formation of large deposits (e.g., porphyry Cu deposits [48–52]). The Linglong, Guojialing, and Weideshan suites have highly variable Sr/Y (35.54–136.50) and La/Yb (1.43–176.43) ratios, and low Y (3.26–11.20 ppm) and Yb (0.33–0.97 ppm) contents [26,36–39,42,43], indicating an adakitic affinity. Furthermore, in the Sr/Y vs. Y and $(La/Yb)_N$ vs. Yb_N discrimination diagrams, these samples fall in the adakite field (Figure 8).

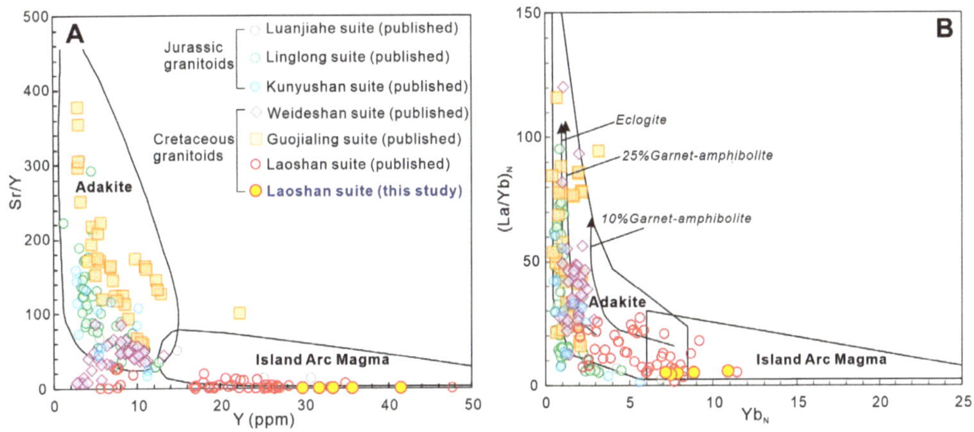

Figure 8. (**A**). Sr/Y vs. Y (after Martin [53]; Defant and Drummond [54]); (**B**). $(La/Yb)_N$ vs. Yb_N (after Martin [55]; Drummond et al. [56]; Martin et al. [57]). The published data are the same as in Figure 5.

In contrast, the Laoshan suite has low Sr/Y (0.09–27.78) and La/Yb (2.58–40.27) ratios, and low Y (5.56–41.50 ppm) and Yb (0.85–4.80 ppm) contents (this study and [40,41,44,46]), indicating a normal-arc geochemical character (Figure 8).

5.1.1. Adakite-like Rocks: Linglong, Guojialing, and Weideshan Suites

The Linglong, Guojialing, and Weideshan suites exhibit high SiO_2 (65.60–77.69 wt %), and low TFe_2O_3 (0.09–3.70 wt.%) and MgO (0.05–2.28 wt.%) contents [26,36,38,39,42,43]. These rocks are depleted in Nb, Ta, Ti, and P (Figure 7), indicating that they may have originated from the partial melting of crustal material [58,59]. The characteristic trace element ratios suggest that these rocks are primarily crustal-derived. Firstly, the average Zr/Hf (33.7) and Nb/Ta (13.0) ratios are significantly lower than those of the primitive mantle values (37 and 17.8, respectively [47]), but closer to the values of the average crustal values (33 and 11.4, respectively [60,61]). Secondly, the average Ce/Pb (3.3) and Nb/U (5.0) ratios reflect a continental crustal source (4 and 10, respectively [62]). Thirdly, the low compatible element (e.g., Co and Ni) contents indicate a lower crustal source [63]. In addition, the granites fall within the thickened lower crustal range in SiO_2 (wt%) versus MgO (wt%), Mg#, and Ni diagrams (Figure 9). Considering the regional geological setting, the Jiaodong Peninsula was in a subduction environment of the Paleo-Pacific Plate during the Middle Jurassic–Early Cretaceous. Therefore, the Linglong, Guojialing, and Weideshan adakitic magmatic rocks are products of the partial melting of the thickened lower crust. However, the variable Mg# value (10–59) suggests that there may also be mixing of mantle components in the magma source. The Linglong granite is characterized by a gneissic structure, no enclaves, and a Mg# < 40 [36,38,39,42]. It contains xenoliths of Precambrian strata, indicating that it originated from the remelting of ancient crustal materials. The Sr–Nd isotope diagram shows that the Linglong granite is closer to the crust and also supports crustal sources (Figure 10A). As for the Early Cretaceous Guojialing and Weideshan granites, a large number of mafic enclaves were developed [4], and their Sr–Nd isotopic composition was also close to the mantle evolution trend, indicating a significant involvement of mantle-derived materials (Figure 10A). The adakitic rocks formed by these granites have a lower zirconium saturation temperature than the Laoshan granite (Figure 10B).

The La vs. La/Yb and La/Sm diagrams indicate that these granites were formed by partial melting of crustal materials (Figure 11A,B). On the Sr/Nd vs. Th/Yb and Nb/Y vs. Ba diagrams (Figure 11C,D), the Weideshan suite shows a distinct trend from other granites, suggesting a melt-related enrichment feature, while the Linglong and Guojialing suites are fluid-related. This indicates that the mafic magma during the formation of the Weideshan granite may be attributed to the partial melting of a mantle wedge that had been metasomatized by subduction slab-derived melts rather than slab-derived hydrous fluids (e.g., Ji et al. [64]).

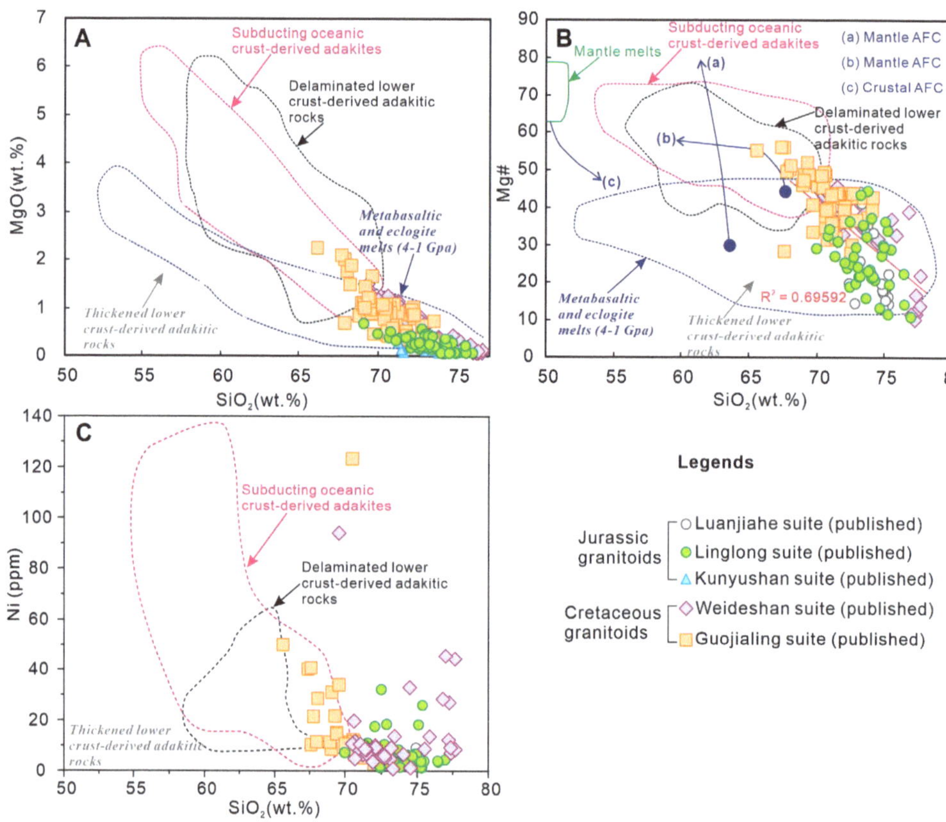

Figure 9. Plots of MgO (**A**), Mg# (**B**), and Ni (**C**) vs. SiO_2 for the Mesozoic adakitic rocks in the Jiaodong Peninsula. The published data are the same as in Figure 5. Note: the subducting oceanic crust-derived adakites are from Defant and Drummond [54], Stern and Kilian [65], Martin [66], Smithies [67], and Defant et al. [68]; the thickened lower crust-derived adakitic rocks are from Atherton and Petford [69], Petford and Atherton [70], Muir et al. [71], and Smithies [67]; the delaminated crust-derived adakitic rocks are from Xu et al. [72].

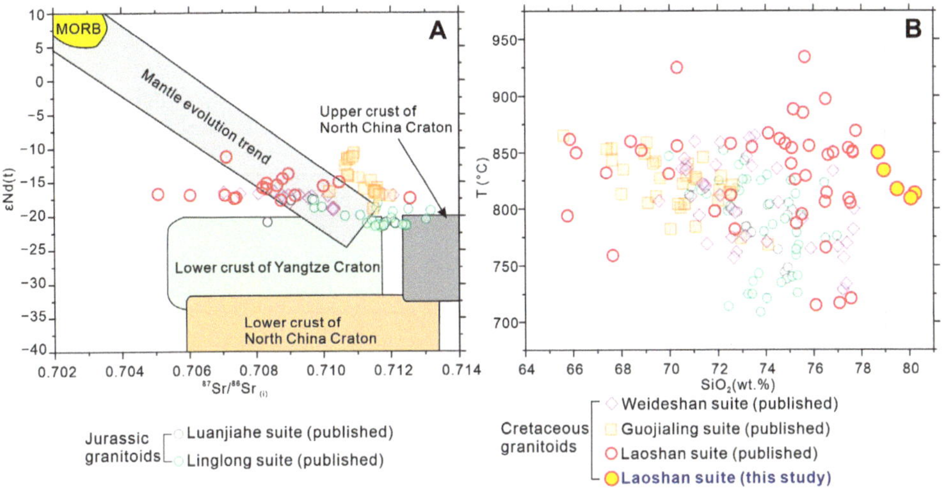

Figure 10. Plots of εNd(t) vs. $(^{87}Sr/^{86}Sr)_i$ ratio (**A**) and SiO_2 vs. T-Zr (**B**) diagrams for the granites from the Jiaodong Peninsula. The published data are the same as in Figure 5.

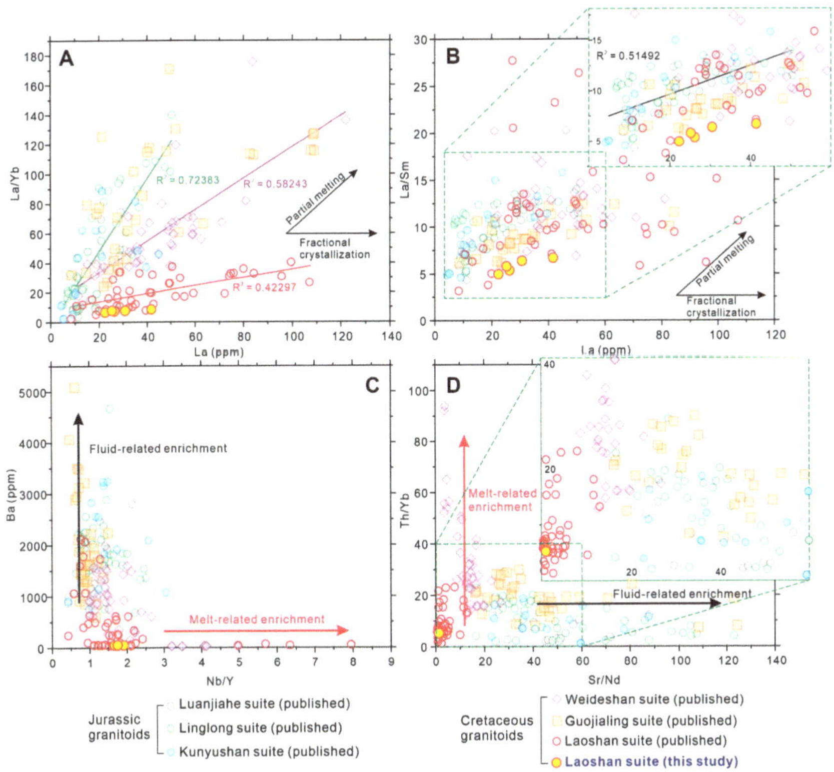

Figure 11. Plots of (**A**) La vs. La/Yb, (**B**) La vs. La/Sm, (**C**) Sr/Nd vs. Th/Yb [73], and (**D**) Nb/Y vs. Ba [74] for the Mesozoic granites from Jiaodong Peninsula. The published data are the same as in Figure 5.

5.1.2. A-Type Granite: Laoshan Suites

As previously mentioned, the Laoshan suite exhibits low Sr/Y and La/Yb ratios, as well as low Y and Yb contents, which are indicative of a typical normal-arc magma. These igneous rocks have a high SiO_2 content (65.76–80.28 wt.%), low Mg# value (6–44) and compatible element (e.g., Co and Ni) content, and an aluminum deficiency (A/CNK = 0.92–1.12, lack of Al-rich minerals) (this study and [40,41,44,46]). This set of characteristics rules out the possibility of M- and S-type granites [75]. Additionally, these granites have high total alkalis (7.45–10.63 wt%), Zr (39.3–437.5 ppm), Ce (23.12–209.28 ppm), Nb (12.8–132.0 ppm), and Y (5.56–41.50 ppm) contents, and low CaO (0.07–2.09 wt%) and MgO (0.03–1.03 wt%) contents. They are also characterized by high TFe_2O_3/MgO (3.2–33.5) and Ga/Al (0.6–4.0) ratios, high varying zircon saturation temperatures (715–934 °C [76]), and remarkable negative Eu (Eu/Eu* = 0.05–0.80) anomalies (this study and [40,41,44,46]). These features strongly suggest that the Laoshan suite has the characteristics of A-type granite (e.g., Li et al. [77]). The data points in the 10000 Ga/Al vs. K_2O/MgO, Nb, (Na_2O + K_2O)/CaO, and TFe_2O_3/MgO plots also fall within the range of A-type granite (Figure 12A–D) and belong to A1-type granite (Figure 12E,F). Furthermore, these granites all have lower Ba and Sr contents, which are typical of low Ba–Sr granites (Figure 13).

The A-type granites discussed in this study are characterized by high SiO_2 and low MgO contents, and are depleted in Nb, Ta, P, and Ti, while enriched in Rb and Pb. These features suggest that their formation can be attributed to partial melting of the crust [58,59]. These rocks also exhibit low average ratios of Nb/Ta (15.8), Zr/Hf (28.6), Ce/Pb (3.5), Nb/U (13.8), Rb/Sr (13.7), Ti/Zr (8.6), and Ti/Y (67.1) (this study and [40,41,44,46]), further supporting a lower crustal source [47,60–62,78–80]. However, the Laoshan suite appears to be located in or near the mantle channel on the $^{87}Sr/^{86}Sr$ vs. εNd(t) diagram (Figure 10A), suggesting possible mixing with mantle material. Moreover, similar to the Weideshan suite, the Laoshan granite exhibits a trend of melt-related enrichment in Sr/Nd vs. Th/Yb and Nb/Y vs. Ba diagrams (Figure 11C,D), indicating that it is also a result of partial melting of mantle wedges derived from subduction plate-derived melt metasomatism.

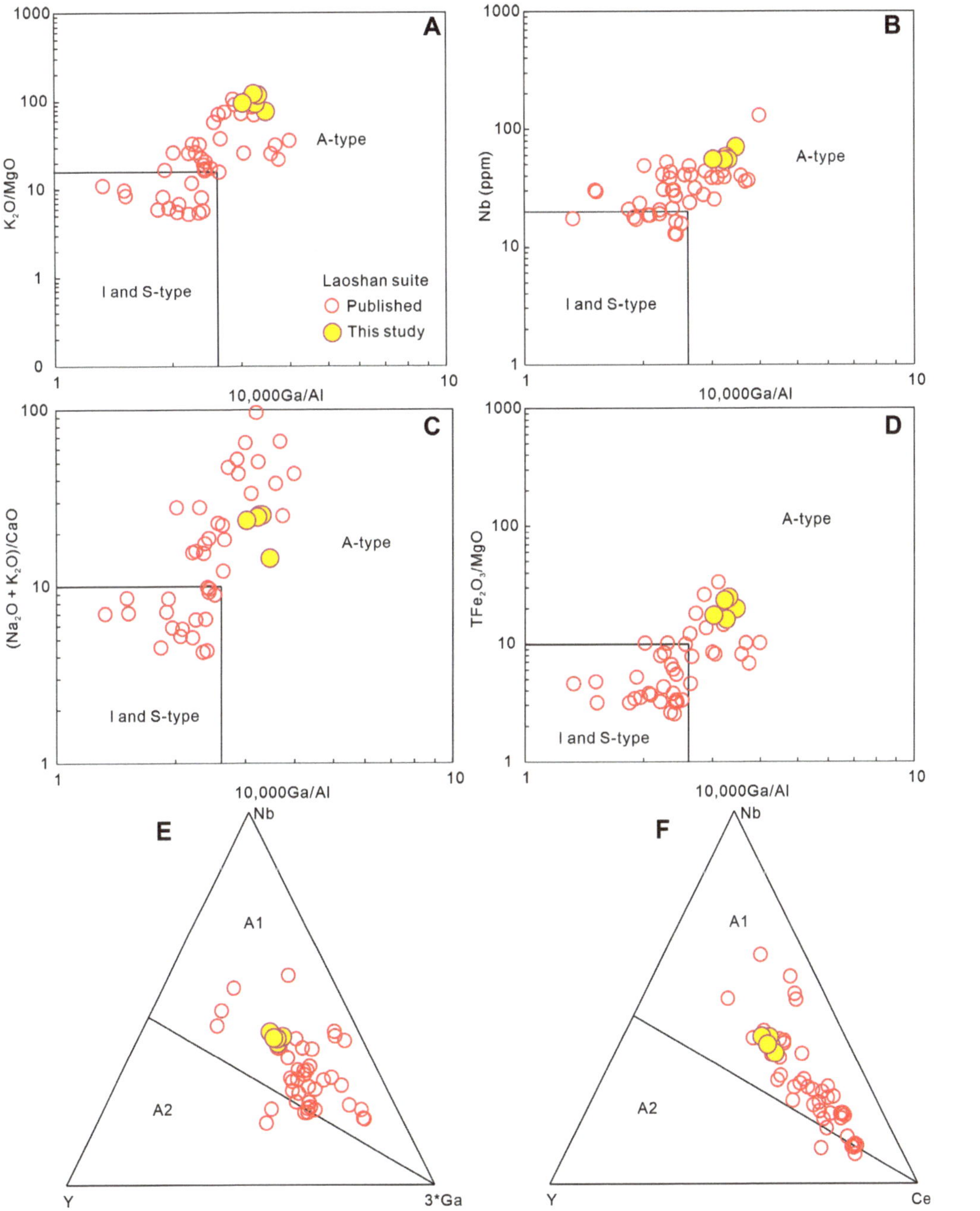

Figure 12. Plots of K_2O/MgO (**A**), Nb (**B**), $(Na_2O + K_2O)/CaO$ (**C**), and TFe_2O_3/MgO (**D**) vs. 10000 Ga/Al (Whalen et al. [81]), Nb–Y–3*Ga (**E**), and Nb–Y–Ce (**F**) (Eby [82]) discrimination diagrams for the Laoshan suite.

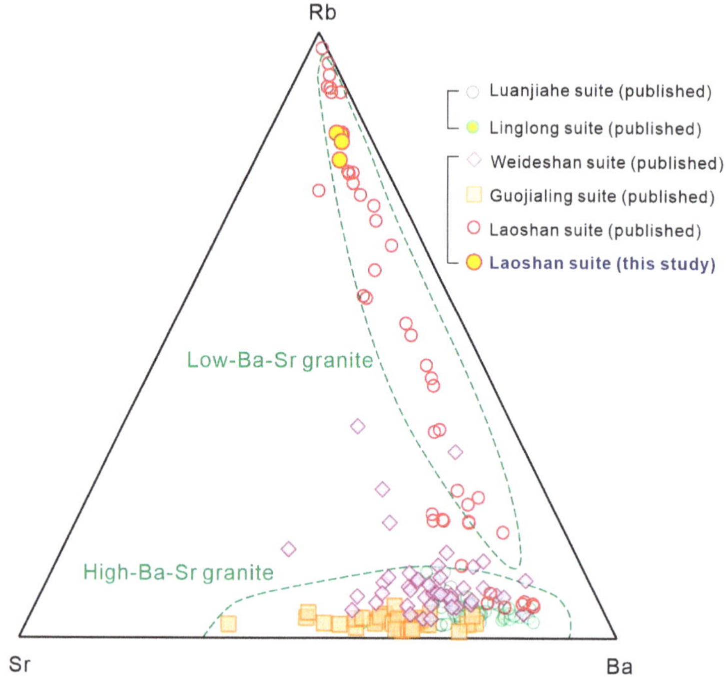

Figure 13. Sr–Rb–Ba plot for the Laoshan suite in the Jiaodong Peninsula (Qian et al. [83]).

5.2. An oxidized Low Ba–Sr Granite

Previous studies on porphyry mineralization systems, particularly Cu–Mo mineralization, have indicated that the oxidation state of magma is closely related to mineralization [52,84]. Under high oxygen fugacity conditions, sulfur exists in the form of sulfates in magma (S^{6+}: HSO_4^-/SO_4^{2-}), which can significantly inhibit the early saturation and decomposition of sulfides (S^{2-}), as well as the consequent loss of chalcophile elements in the deeper crust [85,86]. Additionally, the migration of Au in hydrothermal fluids is mainly related to sulfur complexes [87], meaning that high oxygen fugacity in the magma source area also inhibits the early precipitation of Au. To calculate the Laoshan granite, we used the latest method provided by Loucks et al. [88], which avoids the disadvantage of the prevailing zircon oxybarometries. The analysis uncertainty of La is large in these methods, because minerals such as titanate, plagioclase, or apatite are fractionated before zircon [89,90]. The pressure data used in the specific calculation process are based on the average of EPMA data of amphibole and biotite from Dong et al. [91]. The results show that the ΔFMQ values of the Laoshan suite are 2.43–4.22, which are significantly higher than those of the Weideshan and pre-mineralization Linglong suite (Figure 14) [11,91]. Furthermore, the Ti-saturation temperature is also significantly higher than that of the Weideshan and Linglong suites, which is also consistent with the basic characteristics of A-type granite (Figure 14A) [82]. This studied Laoshan granite was formed at ~118 Ma, and the entire Laoshan suite was formed over a large time span of 125–108 Ma [44,92,93]. The location is distant from the known gold deposits (Figure 1), and it is considered to be a post-mineralization granite. Therefore, the Laoshan suite belongs to a low Ba–Sr granite with high oxygen fugacity after mineralization.

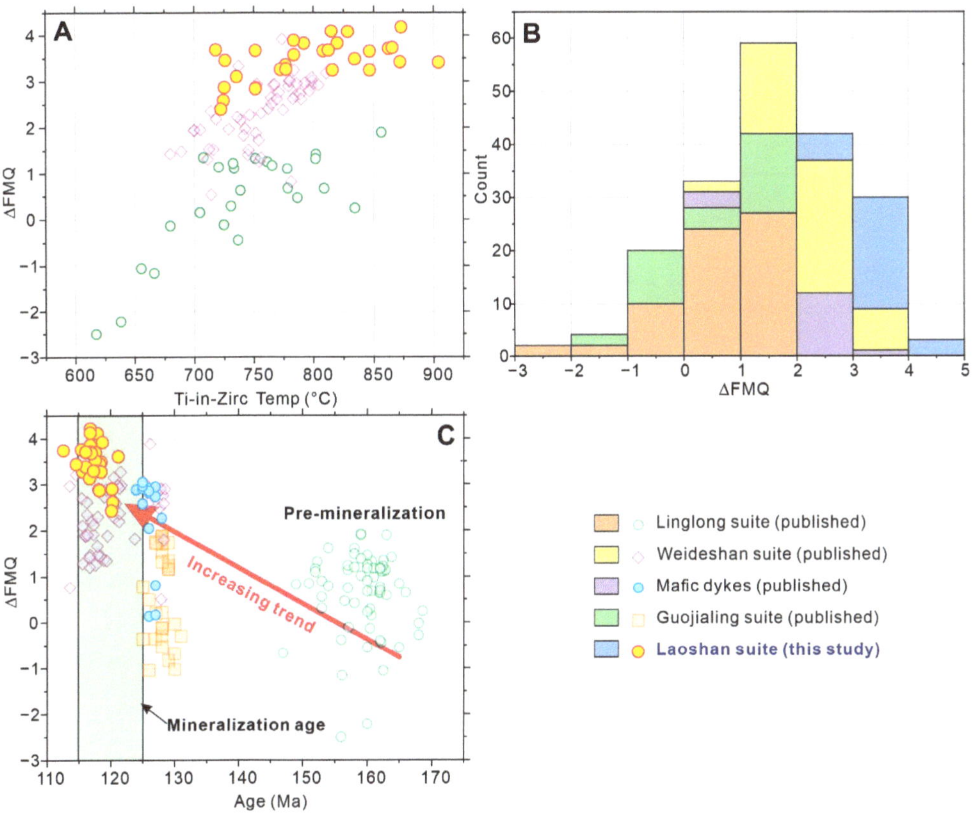

Figure 14. (**A**). T-Ti (°C) (Ti-saturation temperature) vs. ΔFMQ; (**B**). ΔFMQ histogram; (**C**). Age (Ma) vs. ΔFMQ. Ti-saturation temperature and ΔFMQ values are calculated from zircon trace elements (Table S3).

5.3. Tectonic Setting and Gold Minerlization Implication

The tectonic setting of the eastern NCC has been extensively studied, and a consensus has been reached that it is mainly influenced by the subduction of the Paleo-Pacific Plate [52,94–96]. The studied suites of Linglong, Guojialing, and Weideshan, which are of Jurassic to Early Cretaceous age, all fall within the category of volcanic arc granites (VAG) in the tectonic discrimination diagrams (Figure 15). This suggests that they were formed in a subduction environment. On the other hand, the Laoshan granites in this study and published data fall within the range of post-collision and syn-collision (Figure 15), indicating that they were formed in an extensional environment. Therefore, the tectonic nature of the Jiaodong Peninsula shifted from compression to extension from the Jurassic to the Early Cretaceous, and gold mineralization occurred during this transition. Additionally, the geochemical characteristics of granitoids in Jiaodong have changed significantly from the Jurassic to Early Cretaceous: (i) The Linglong, Guojialing, and Weideshan suites exhibit geochemical properties of adakitic rocks (Figure 8), whereas the Laoshan suite is a typical A-type granite (Figure 12); (ii) The positive Eu anomaly of the Jurassic Linglong granite transformed into the negative Eu anomaly of the Early Cretaceous Laoshan granite in terms of REE composition (Figure 7); (iii) The Mesozoic granites of the Jiaodong Peninsula are all related to partial melting of thickened lower crustal material associated with oceanic crustal subduction. The Linglong suite has no mantle material admixture, while the Guojialing, Weideshan, and Laoshan suites have a large amount of mantle material mixing

(Figures 9 and 10) and petrographic development of a mafic microgranular enclave (MME) (e.g., Song et al., [2]); (iv) From the Jurassic to Early Cretaceous, the oxidation state of the magma source area changes and becomes gradually oxidized, with the highest ΔFMQ value for Laoshan granite (Figure 14); (v) From the Jurassic to the Cretaceous, the lithospheric mantle properties changed from enrichment to depletion [2].

The concentrated mineralization in the Jiaodong region occurred during the Early Cretaceous (120 $\pm$ 5 Ma; [3,5]), which corresponds to the period of transformation of tectonic setting and magma geochemistry mentioned above. The adakitic magma with high oxygen fugacity, medium negative Eu anomaly, extensional tectonic system, and mixed mantle material may have facilitated the formation of gold deposits. In the Jurassic Linglong granite, there are no MMEs or amphiboles, while in the Early Cretaceous granites, there are a large number of MMEs, amphiboles, and contemporaneous mafic dykes. This suggests the mixing of mantle materials may have contributed to magma oxidation and water enrichment (e.g., Dong et al. [91]). Therefore, we have established such a tectonic evolution model from the Jurassic to the Cretaceous: (a) During the Jurassic period, the ancient crustal materials were partially melted under the subduction of the Paleo-Pacific Plate, forming Linglong granite. In the region, it is common to find xenoliths containing strata in Linglong granite and displaying gneissic structures. Moreover, the Jurassic mafic dykes are almost not developed, which also suggests that the Linglong granite with low oxygen fugacity and low water content was formed by partial melting of the lower crust. This process may have promoted the saturation and accumulation of sulfides (which may also include gold) [91], causing the pre-enrichment of some key elements (e.g., sulfur) related to mineralization. (b) During the Cretaceous period, the subduction slab of the Paleo-Pacific Plate roll-back resulted in a significant upwelling of asthenosphere materials, making the tectonic nature change into extensional. The upwelling of mantle material and the mixing with a partially melted lower crust formed Early Cretaceous MME-rich granites. These crust–mantle magmas have high oxygen fugacity, carry a large amount of metal sulfides, and enrich at fault locations to form gold deposits. This suggests that the injection of mafic magma contributes to the increase in oxygen fugacity, which in turn leads to the enrichment of metal sulfides at fault locations, resulting in gold deposits. The current model provides valuable insights into the explosive and instantaneous mineralization of Jiaodong gold deposits, and also suggests that Jiaodong gold deposits may differ from traditional orogenic gold deposit categories [97]. In contrast, the formation of low Ba–Sr granites with higher oxidation levels, such as those in the Laoshan suite, was facilitated by the continued subduction of the Paleo-Pacific Plate. However, the lack of fault development in and around the Laoshan suite has prevented the formation of gold deposits. Additionally, the low Ba–Sr content may be attributed to less mantle material injection, which is one of the reasons for non-mineralization.

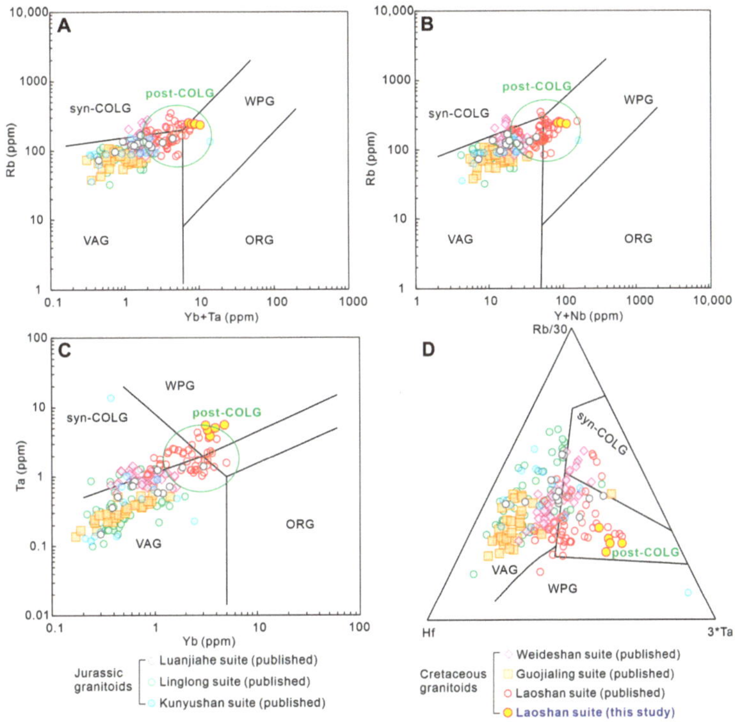

Figure 15. Tectonic discrimination diagrams for granitoids from the Jiaodong Peninsula. (**A**). Rb vs. (Yb + Ta) [98]; (**B**). Rb vs. (Y + Nb); (**C**). Ta vs. Yb [99]; (**D**). Rb/30-Hf-3*Ta [100]. Abbreviations: VAG = volcanic arc granites; ORG = ocean ridge granites; WPG = within-plate granites; syn-COLG = syn-collision granites; post-COLG = post-collision granites.

6. Conclusions

Through this study and the data compiled, we recognize that the Mesozoic magmatic rocks in Jiaodong Peninsula are adakitic rocks (inclu. Linglong, Guojialing, and Weideshan suites) and A-type granite (Laoshan suite). The adakitic rocks were mainly formed through partial melting of the thickened lower crust due to the subduction of the Paleo-Pacific Plate, and in the Early Cretaceous it was also mixed with a large amount of mantle material. The A1-type granite, represented by the Laoshan suite, was formed through partial melting of the crust with a small amount of mantle material. From the Jurassic to the Early Cretaceous, the tectonic and geochemical properties of magmatic rocks in Jiaodong Peninsula underwent significant changes. This transitional process, from compression to extension and reduction to oxidation, led to the formation of gold deposits.

Supplementary Materials: The following supporting information can be downloaded at: https://www.mdpi.com/article/10.3390/min13081012/s1, Table S1: LA–ICP–MS zircon U–Pb data for Laoshan granities from the Jiaodong Peninsula; Table S2: Major (wt.%) and trace element (ppm) data for Laoshan granitic intrusive rocks from the Jiaodong Peninsula; Table S3: Zircon trace element data for Laoshan granitic intrusive rocks from the Jiaodong Peninsula.

Author Contributions: Conceptualization, J.L. and M.S.; methodology, J.L., X.L. and C.W. (Changwei Wang); software, C.W. (Changjiang Wang) and S.L.; data curation, J.L., X.L. and Q.C.; investigation, J.L. and M.S.; project administration, M.S. All authors have read and agreed to the published version of the manuscript.

Funding: This research was funded by the "Shandong Natural Science Foundation, grant number ZR2021QD056" and the "National Natural Science Foundation of China-Shandong Joint Fund, grant number No. U2006201".

Data Availability Statement: Data are contained within the Supplementary Materials.

Acknowledgments: We gratefully acknowledge the assistance of four anonymous reviewers for their insightful comments, which greatly improved the quality of this manuscript. We thank Fan Yang for his handling of the manuscript and editorial input. This work was financially supported by the Shandong Natural Science Foundation (No. ZR2021QD056) and the National Natural Science Foundation of China-Shandong Joint Fund (No. U2006201).

Conflicts of Interest: The authors declare no conflict of interest.

References

1. Song, M.C.; Li, J.; Yu, X.F.; Ding, Z.J.; Li, S.Y. Metallogenic characteristics and tectonic setting of the Jiaodong gold deposit, China. *Solid Earth Sci.* **2021**, *6*, 385–405. [CrossRef]
2. Song, M.C.; Song, Y.X.; Li, J.; Liu, H.B.; Li, J.; Dong, L.L.; He, C.Y.; Wang, R.S. Thermal doming-extension metallogenicsystem of Jiaodong type gold deposits. *Acta Petrol. Sin.* **2023**, *39*, 1241–1260, (In Chinese with English abstract). [CrossRef]
3. Deng, J.; Yang, L.Q.; Groves, D.I.; Zhang, L.; Qiu, K.F.; Wang, Q.F. An integrated mineral system model for the gold deposits of the giant Jiaodong province, eastern China. *Earth-Sci. Rev.* **2020**, *208*, 103274. [CrossRef]
4. Song, M.C.; Lin, S.Y.; Yang, L.Q.; Song, Y.X.; Ding, Z.J.; Li, J.; Li, S.Y.; Zhou, M.L. Metallogenic model of Jiaodong Peninsula gold deposits. *Miner. Depos.* **2020**, *39*, 215–236. (In Chinese with English abstract).
5. Yang, L.Q.; Deng, J.; Wang, Z.L.; Zhang, L.; Guo, L.N.; Song, M.C.; Zheng, X.L. Mesozoic gold metallogenic system of the Jiaodong gold province, eastern China. *Acta Petrol. Sin.* **2014**, *30*, 2447–2467. (In Chinese with English abstract).
6. Fan, H.R.; Feng, K.; Li, X.H.; Hu, F.F.; Yang, K.F. Mesozoic gold mineralization in the Jiaodong and Korean peninsulas. *Acta Petrol. Sin.* **2016**, *32*, 3225–3238. (In Chinese with English abstract).
7. Song, M.C.; Song, Y.X.; Ding, Z.J.; Li, S.Y. Jiaodong Gold Deposits: Essential Characteristics and Major Controversy. *Gold Sci. Technol.* **2018**, *26*, 406–422. (In Chinese with English abstract).
8. Zhu, R.X.; Fan, H.R.; Li, J.W.; Meng, Q.R.; Li, S.R.; Zeng, Q.D. Decratonic gold deposits. *Sci. China-Earth Sci.* **2015**, *58*, 1523–1537. [CrossRef]
9. Li, J.; Song, M.C.; Yu, J.T.; Bo, J.W.; Zhang, Z.L.; Liu, X. Genesis of Jinqingding Gold Deposit in eastern Jiaodong Peninsula, China: Constrain from trace elements of sulfide ore and wall-rock. *Geol. Bull. China* **2022**, *41*, 1010–1022. (In Chinese with English abstract).
10. Li, J.; Yang, Z.M.; Song, M.C.; Dong, L.L.; Li, S.Y.; Wang, R.S.; Liu, X.; Li, Z.S.; Song, Y.X.; Lai, C.K. Gold remobilization of the Sanshandao gold deposit, Jiaodong Peninsula, Eastern China: Perspective from in-situ sulfide trace elements and sulfur isotopes. *Ore Geol. Rev.* **2023**, *158*, 105505.
11. Li., J.; Dai, C.G.; Wang, C.W.; Song, M.C.; Wang, C.J.; Li, S.Y.; Wang, R.S.; Shi, H.J.; Xu, K.L.; Wang, P. Possible genetic relationship between Mesozoic magmatic rocks and gold mineralization in the Jiaodong Peninsula (Eastern China): Constraints of magmatic evolution and physicochemical conditions. *Front. Earth Sci.* 2023; *under review*.
12. Yang, J.H.; Xu, L.; Sun, J.F.; Zeng, Q.D.; Zhao, Y.N.; Wang, H.; Zhu, Y.S. Geodynamics of decratonization and related magmatism and mineralization in the North China Craton. *Sci. China-Earth Sci.* **2021**, *64*, 1409–1427. [CrossRef]
13. Zhao, G.C.; Wilde, S.A.; Cawood, P.A.; Sun, M. Archean blocks and their boundaries in the North China Craton: Lithological, geochemical, structural and P-T, path constraints and tectonic evolution. *Precambrian Res.* **2001**, *107*, 45–73.
14. Zhao, G.C.; Sun, M.; Wilde, S.A.; Li, S.Z. Late Archean to Paleoproterozoic evolution of the North China Craton: Key issues revisited. *Precambrian Res.* **2005**, *136*, 177–202.
15. Yang, J.H.; Wu, F.Y. Triassic magmatism and its relation to decratonization in the eastern North China Craton. *Sci. China Ser. D-Earth Sci.* **2009**, *52*, 1319–1330.
16. Li, J.; Cai, W.Y.; Wang, K.Y.; Kim, N.H.; Liu, H.L.; Lee, J.G.; Yoo, B.C. Initial decratonization of the eastern North China Craton: New constraints from geochronology, geochemistry, and Hf isotopic compositions of Mesozoic igneous rocks in the Qingchengzi district. *Geol. J.* **2020**, *55*, 3796–3820.
17. Wu, F.Y.; Yang, J.H.; Xu, Y.G.; Wilde, S.A.; Walker, R.J. Destruction of the North China Craton in the Mesozoic. *Annu. Rev. Earth Planet. Sci.* **2019**, *47*, 173–195. [CrossRef]
18. Jahn, B.M.; Auvray, B.; Cornichet, J.; Bai, Y.L.; Shen, Q.H.; Liu, D.Y. 3.5 Ga old amphibolites from eastern Hebei province, China: Field occurrence, petrology, Sm–Nd isochron age and REE geochemistry. *Precambrian Res.* **1987**, *34*, 311–346.
19. Griffin, W.L.; O'Reilly, S.Y.; Ryan, C.G. Composition and thermal structure of the lithosphere beneath South Africa, Siberia and China: Proton microprobe studies. In Proceedings of the International Symposium on Cenozoic Volcanic Rocks and Deep-Seated Xenoliths of China and Its Environs, Beijing, China, 8–10 September 1992; pp. 65–66.

20. Griffin, W.L.; Zhang, A.; O'Reilly, S.Y.; Ryan, C.G. Phanerozoic evolution of the lithosphere beneath the Sino-Korean Craton. In *Mantle Dynamics and Plate Interactions in East Asia;* Geodynamics Series; Flower, M.F.J., Chung, S.-L., Lo, C.-H., Lee, T.-Y., Eds.; American Geophysical Union: Washington, DC, USA, 1998; Volume 27, pp. 107–126. [CrossRef]
21. Gao, S.; Rudnick, R.L.; Yuan, H.L.; Liu, X.M.; Liu, Y.S.; Xu, W.L.; Ling, W.L.; Ayers, J.; Wang, X.C.; Wang, Q.H. Recycling lower continental crust in the North China Craton. *Nature* **2004**, *432*, 892–897. [CrossRef]
22. Basu, A.R.; Wang, J.W.; Huang, W.K.; Xie, G.; Tatsumoto, M. Major element, REE, and Pb and Sr isotopic geochemistry of Cenozoic volcanic rocks of eastern China: Implications for their origin from suboceanic type mantle reservoirs. *Earth Planet. Sci. Lett.* **1991**, *105*, 149–169.
23. Menzies, M.A.; Fan, W.; Zhang, M. Paleozoic and Cenozoic.lithoprobes and loss of N120 km of Archean lithosphere, Sino-Korean/Craton. *Magmat. Process. Plate Tecton.* **1993**, *76*, 71–81.
24. Fan, W.M.; Zhang, H.F.; Baker, J.; Jarvis, K.E.; Mason, P.R.D.; Menzies, M.A. On and off the North China Craton: Where is the Archaean keel? *J. Petrol.* **2000**, *41*, 933–950. [CrossRef]
25. Yang, J.H.; Wu, F.Y.; Wilde, S.A. A review of the geodynamic setting of large-scale Late Mesozoic gold mineralization in the North China Craton: An association with lithospheric thinning. *Ore Geol. Rev.* **2003**, *23*, 125–152.
26. Dong, L.L.; Yang, Z.M.; Song, M.C.; Bai, X. Petrogenesis of Mesozoic Magmatic Suites in the Jiaodong Peninsula: Implications for Crust-Mantle Interactions and Decratonization. *Lithosphere* **2023**, *2023*, 6226908.
27. Song, M.C.; Zhou, J.B.; Song, Y.X.; Wang, B.; Li, S.Y.; Li, J.; Wang, S.S. Mesozoic Weideshan granitoid suite and its relationship to large-scale gold mineralization in the Jiaodong Peninsula, China. *Geol. J.* **2020**, *55*, 5703–5724. [CrossRef]
28. Paton, C.; Woodhead, J.D.; Hellstrom, J.C.; Hergt, J.M.; Greig, A.; Maas, R. Improved laser ablation U-Pb zircon geochronology through robust downhole fractionation correction. *Geochem. Geophys. Geosyst.* **2010**, *11*, Q0AA06. [CrossRef]
29. Ludwig, K.R. User's manual for isoplot 3.0: A geochronological toolkit for Microsoft Excel. *Berkeley Geochronol. Cent. Spec. Publ.* **2003**, *4*, 25–32.
30. Andersen, T. Correction of common lead in U–Pb analyses that do not report ^{204}Pb. *Chem. Geol.* **2002**, *192*, 59–79. [CrossRef]
31. Hoskin, P.W.O.; Black, L.P. Metamorphic zircon formation by solid-state recrystallization of protolith igneous zircon. *J. Metamorph. Geol.* **2000**, *18*, 423–439. [CrossRef]
32. Hoskin, P.W.O. Trace-element composition of hydrothermal zircon and the alteration of hadean zircon from the Jack Hills, Australia. *Geochim. Cosmochim. Acta* **2005**, *69*, 637–648. [CrossRef]
33. Irvine, T.H.; Baragar, W.R.A. A guide to the chemical classification of the common volcanic rocks. *Can. J. Earth Sci.* **1971**, *8*, 523–548. [CrossRef]
34. Peccerillo, A.; Taylor, D.R. Geochemistry of Eocene calc-alkaline volcanic rocks from the Kastamonu area, northern Turkey. *Contrib. Mineral. Petrol.* **1976**, *58*, 63–81. [CrossRef]
35. Maniar, P.D.; Piccoli, P.M. Tectonic discrimination of granitoids. *Geol. Soc. Am. Bull.* **1989**, *101*, 635–643. [CrossRef]
36. Hou, M.L.; Jiang, Y.H.; Jiang, S.Y.; Ling, H.F.; Zhao, K.D. Contrasting origins of late mesozoic adakitic granitoids from the northwestern, Jiaodong Peninsula, east China: Implications for crustal thickening to delamination. *Geol. Mag.* **2007**, *144*, 619–631.
37. Li, X.C.; Fan, H.R.; Santosh, M.; Hu, F.F.; Yang, K.F.; Lan, T.G.; Liu, Y.; Yang, Y.H. An evolving magma chamber within extending lithosphere: An integrated geochemical, isotopic and zircon U-Pb geochronological study of the Gushan granite, eastern North China Craton. *J. Asian Earth Sci.* **2012**, *50*, 27–43.
38. Yang, K.F.; Fan, H.R.; Santosh, M.; Hu, F.F.; Wilde, S.; Lan, T.C.; Lu, L.N.; Liu, Y.S. Reactivation of the Archean lower crust: Implications for zircon geochronology, elemental and Sr–Nd–Hf isotopic geochemistry of late Mesozoic granitoids from northwestern Jiaodong Terrane, the North China Craton. *Lithos* **2012**, *146*, 112–127.
39. Ma, L.; Jiang, S.Y.; Dai, B.Z.; Jiang, Y.H.; Hou, M.L.; Pu, W.; Bin, X. Multiple sources for the origin of Late Jurassic Linglong adakitic granite in the Shandong Peninsula, eastern China: Zircon U-Pb geochronological geochemical and Sr-Nd-Hf isotopic evidence. *Lithos* **2013**, *162*, 175–194.
40. Yan, Q.S.; Shi, X.F. Geochemistry and petrogenesis of the Cretaceous A-type granites in the Laoshan granitic complex, eastern China. *Isl. Arc* **2014**, *23*, 221–235. [CrossRef]
41. Gao, Y.J.; Niu, Y.L.; Duan, M.; Xue, Q.Q.; Sun, P.; Chen, S.; Xiao, Y.Y.; Guo, P.Y.; Wang, X.H.; Chen, Y.H. The petrogenesis and tectonic significance of the Early Cretaceous intraplate granites in eastern China: The Laoshan granite as an example. *Lithos* **2019**, *328–329*, 200–211.
42. Li, X.H.; Fan, H.R.; Hu, F.F.; Hollings, P.; Yang, K.F.; Liu, X. Linking lithospheric thinning and magmatic evolution of Late Jurassic to Early Cretaceous granitoids in the Jiaobei Terrane, southeastern North China Craton. *Lithos* **2019**, *324–325*, 280–296. [CrossRef]
43. He, J.T. Gold Mineralization and Post Ore Enrollment in the Muping Rushan Gold Belt, Jiaodong Peninsula, Eastern China. Ph.D. Thesis, China University of Geosciences, Beiijing, China, 2021; pp. 1–171. (In Chinese with English abstract).
44. Wang, B.; Song, M.C.; Huo, G.; Zhou, M.L.; Xu, Z.H.; Jiang, L.; Song, Y.X.; Li, J. Source characteristics and tectonic evolution of Late mesozoic granites in Jiaodong and their implications for gold mineralization. *Acta Petrol. Mineral.* **2021**, *40*, 288–320. (In Chinese with English abstract).
45. Dong, L.L.; Yang, Z.M.; Bai, X.; Deng, C. Generation of the Early Cretaceous granitoid in the Dazeshan region, Jiaodong Peninsula: Implications for the crustal reworking in the North China Craton. *Front. Earth Sci.* **2023**, *10*, 1083608. [CrossRef]
46. Wang, D.; Wang, T.Q.; Li, H.Y. Petrogenesis of Early Cretaceous Laoshan A-type granites and the implications for the tectonic evolution of Jiaodong Peninsula. *Acta Petrol. Sin.* **2023**, *39*, 317–339. (In Chinese with English abstract). [CrossRef]

47. McDonough, W.F.; Sun, S.S. The composition of the Earth. *Chem. Geol.* **1995**, *120*, 223–253. [CrossRef]
48. Richards, J.P. Tectono-magmatic precursors for porphyry Cu–(Mo–Au) deposit formation. *Econ. Geol.* **2003**, *98*, 1515–1533.
49. Richards, J.P. High Sr/Y arc magmas and porphyry Cu Mo Au deposits: Just add water. *Econ. Geol.* **2011**, *106*, 1075–1081.
50. Richards, J.P.; Kerrich, R. Special paper: Adakite-like rocks: Their diverse origins and questionable role in metallogenesis. *Econ. Geol.* **2007**, *102*, 537–576.
51. Moyen, J.F. High Sr/Y and La/Yb ratios: The meaning of the "adakitic signature". *Lithos* **2009**, *112*, 556–574. [CrossRef]
52. Cai, W.Y.; Wang, K.Y.; Li, J.; Fu, L.J.; Lai, C.K.; Liu, H.L. Geology, geochronology and geochemistry of large Duobaoshan Cu–Mo–Au orefield in NE China: Magma genesis and regional tectonic implications. *Geosci. Front.* **2021**, *12*, 265–292.
53. Martin, H. Effect of steeper Archean geochemical gradient on geochemistry of subduction-zone magmas. *Geology* **1986**, *14*, 753–756. [CrossRef]
54. Defant, M.J.; Drummond, M.S. Derivation of some modern arc magmas by melting of young subducted lithosphere. *Nature* **1990**, *347*, 662–665. [CrossRef]
55. Martin, H. Petrogenesis of Archaean trondhjemites, tonalities and granodioritesfrom eastern Finland: Major and trace element geochemistry. *J. Petrol.* **1987**, *28*, 921–953. [CrossRef]
56. Drummond, M.S.; Defant, M.J.; Kepezhinskas, P.K. Petrogenesis of slab-derived trondhjemite–tonalite–dacite/adakite magmas. In *Third Hutton Symposium on the Origin of Granites and Related Rocks*; Special Paper; Brown, M., Candela, P.A., Pecket, D.L., Eds.; Geological Society of America: Boulder, CO, USA, 1996; Volume 315, pp. 205–215.
57. Martin, H.; Smithies, R.H.; Rapp, R.; Moyen, J.F.; Champion, D.C. An overview of adakite, tonalite–trondhjemite–granodiorite (TTG), and sanukitoid: Relationships and some implications for crustal evolution. *Lithos* **2005**, *79*, 1–24.
58. Gao, S.; Luo, T.C.; Zhang, B.R.; Zhang, H.F.; Han, Y.W.; Zhao, Z.D.; Hu, Y.K. Chemical composition of the continental crust as revealed by studies in East China. *Geochim. Cosmochim. Acta* **1998**, *62*, 1959–1975. [CrossRef]
59. Chappell, B.W.; White, A.J.R. Two contrasting granite types: 25 years later. *Aust. J. Earth Sci.* **2001**, *48*, 489–499. [CrossRef]
60. Taylor, S.R.; McLennan, S.M. *The Continental Crust: Its Composition and Evolution*; Oxford Press: Blackwell, UK, 1985; p. 312.
61. Rudnick, R.L.; Gao, S.; Ling, W.L.; Liu, Y.S.; McDonough, W.F. Petrology and geochemistry of spinel peridotite xenoliths from Hannuoba and Qixia, North China Craton. *Lithos* **2004**, *77*, 609–637. [CrossRef]
62. Hofmann, A.W.; Jochum, K.P.; Seufert, M.; White, W.M. Nb and Pb in oceanicbasalts: New constraints on mantle evolution. *Earth Planet Sci. Lett.* **1986**, *79*, 33–45. [CrossRef]
63. Rapp, R.P.; Shimizu, N.; Norman, M.D.; Applegate, G.S. Reaction between slab derived melts and peridotite in the mantle wedge: Experimental constraints at 3.8 GPa. *Chem. Geol.* **1999**, *160*, 335–356.
64. Ji, Z.; Ge, W.C.; He, Y.; Bi, J.H.; Dong, Y.; Yang, H.; Hao, Y.J. Mixing of cognate magmas as a process for producing high-silica granites: Insights from Guanmenshan Complex in Liaodong Peninsula, China. *Lithos* **2021**, *406–407*, 106495.
65. Stern, C.R.; Kilian, R. Role of the subducted slab, mantle wedge and continental crust in the generation of adakites from the Austral Volcanic Zone. *Contrib. Mineral. Petrol.* **1996**, *123*, 263–281.
66. Martin, H. Adakitic magmas: Modern analogues of Archean granitoids. *Lithos* **1999**, *46*, 411–429. [CrossRef]
67. Smithies, R.H. The Archean tonalite–trondhjemite–granodiorite (TTG) series is not an analogue of Cenozoic adakite. *Earth Planet. Sci. Lett.* **2000**, *182*, 115–125. [CrossRef]
68. Defant, M.J.; Xu, J.F.; Kepezhinskas, P.; Wang, Q.; Zhang, Q.; Xiao, L. Adakites: Some variations on a theme. *Acta Petrol. Sin.* **2002**, *18*, 129–142.
69. Atherton, M.P.; Petford, N. Generation of sodium-rich magmas from newly underplated basaltic crust. *Nature* **1993**, *362*, 144–146. [CrossRef]
70. Petford, N.; Atherton, M. Na-rich partial melts from newly underplated basaltic crust: The Cordillera Blanca Batholith, Peru. *J. Petrol.* **1996**, *37*, 1491–1521.
71. Muir, R.J.; Weaver, S.D.; Bradshaw, J.D.; Eby, G.N.; Evans, J.A. Geochemistry of the Cretaceous Separation Point Batholith, New Zealand: Granitoid magmas formed by melting of mafic lithosphere. *J. Geol. Soc.* **1995**, *152*, 689–701.
72. Xu, J.F.; Shinjo, R.; Defant, M.J.; Wang, Q.; Papp, R.P. Origin of Mesozoic adakitic intrusive rock in the Ningzhen area of east China: Partial melting of delaminated lower continental crust? *Geology* **2002**, *30*, 1111–1114. [CrossRef]
73. Woodhead, J.D.; Eggins, S.M.; Johnson, R.W. Magma genesis in the New Britain island arc: Further insights into melting and mass transfer processes. *J. Petrol.* **1998**, *39*, 1641–1668.
74. Kepezhinskas, P.; McDermott, F.; Defant, M.J.; Hochstaedter, A.; Drummond, M.S.; Hawkesworth, C.J.; Koloskov, A.; Maury, R.C.; Bellon, H. Trace element and SrNd-Pb isotopic constraints on a three-component model of Kamchatka Arc petrogenesis. *Geochim. Cosmochim. Acta* **1997**, *61*, 577–600. [CrossRef]
75. Whalen, J.B. Geochemistry of an island-arc plutonic suite: The Uasilau–Yau Yau intrusive complex, New Britain P.N.G. *J. Petrol.* **1985**, *26*, 603–632. [CrossRef]
76. Watson, E.B.; Harrison, T.M. Zircon saturation revisited: Temperature and composition effects in a variety of crustal magma types. *Earth Planet. Sci. Lett.* **1983**, *64*, 295–304. [CrossRef]
77. Li, J.; Wang, K.Y.; Fu, L.J.; Zhang, M.; Liu, H.L.; Liu, Q.Z.; Tang, W.H.; Wang, C.H. Adakitic rocks and A-type felsic dykes in the Changlingzi area, NE China: Constraints on multistage tectonism in the southern Great Xing'an Range. *Geol. J.* **2020**, *55*, 5451–5478. [CrossRef]

78. Pearce, J.A. Role of the sub-continental lithosphere in magma genesis at active continental margins. In *Continental Basalts and Mantle Xenoliths*; Hawkesworth, C.J., Norry, M.J., Eds.; Shiva Press: Nantwich, UK, 1983; pp. 230–249.
79. Tischendorf, G.; Paelchen, W. On the classification of granitoids. *Z. Fuer Geol. Wiss.* **1985**, *13*, 615–627.
80. Wilson, M. *Igneous Petrogenesis a Global Tectonic Approach*; Unwin Hyman: London, UK, 1989; p. 466.
81. Whalen, J.B.; Currie, K.L.; Chappell, B.W. A-type granites: Geochemical characteristics, discrimination and petrogenesis. *Contrib. Mineral. Petrol.* **1987**, *95*, 407–419.
82. Eby, G.N. Chemical subdivision of the A-type granitoids: Petrogenetic and tectonic implications. *Geology* **1992**, *20*, 641–644. [CrossRef]
83. Qian, Q.; Zhong, S.L.; Li, T.Y.; Wen, D.R. Geochemical characteristics and petrogenesis of the Badaling high Ba–Sr granitoids: A comparison of igneous rocks from North China and the Dabie–Sulu Orogen. *Acta Petrol. Sin.* **2002**, *18*, 275–292. (In Chinese with English abstract).
84. Shu, Q.H.; Chang, Z.S.; Lai, Y.; Hu, X.L.; Wu, H.Y.; Zhang, Y.; Wang, P.; Zhai, D.G.; Zhang, C. Zircon trace elements and magma fertility: Insights from porphyry (-skarn) Mo deposits in NE China. *Miner. Depos.* **2019**, *54*, 645–656.
85. Li, Y.; Audétat, A. Partitioning of V, Mn Co, Ni, Cu, Zn, As, Mo, Ag, Sn, Sb, W, Au, Pb, and Bi between sulfide phases and hydrous basanite melt at upper mantle conditions. *Earth Planet Sci. Lett.* **2012**, *355–356*, 327–340. [CrossRef]
86. Sun, W.D.; Liang, H.Y.; Ling, M.X.; Zhan, M.Z.; Ding, X.; Zhang, H.; Yang, X.Y.; Li, Y.L.; Ireland, T.R.; Wei, Q.R.; et al. The link between reduced porphyry copper deposits and oxidized magmas. *Geochem. Cosmochim. Acta* **2013**, *103*, 263–275. [CrossRef]
87. Williams-Jones, A.E.; Bowell, R.J.; Migdisov, A.A. Gold in solution. *Elements* **2009**, *5*, 281–287. [CrossRef]
88. Loucks, R.R.; Fiorentini, M.L.; Henríquez, G.J. New magmatic oxybarometer using trace elements in zircon. *J. Petrol.* **2020**, *61*, egaa034. [CrossRef]
89. Trail, D.; Bruce Watson, E.; Tailby, N.D. Ce and Eu anomalies in zircon as proxies for the oxidation state of magmas. *Geochim. Cosmochim. Acta* **2012**, *97*, 70–87. [CrossRef]
90. Rezeau, H.; Moritz, R.; Wotzlaw, J.F.; Hovakimyan, S.; Tayan, R. Zircon Petrochronology of the Meghri-Ordubad Pluton, Lesser Caucasus: Fingerprinting Igneous Processes and Implications for the Exploration of Porphyry Cu-Mo Deposits. *Econ. Geol.* **2019**, *114*, 1365–1388.
91. Dong, L.L.; Yang, Z.M.; Liu, Y.H.; Song, M.C. Possible source of Au in the Jiaodong area from lower crustal sulfide cumulates: Evidence from oxygen states and chalcophile elements contents of Mesozoic magmatic suites. *Ore Geol. Rev.* **2023**, *153*, 105268.
92. Goss, C.S.; Wilde, S.A.; Wu, F.Y.; Yang, J.H. The age, isotopicsignature and significance of the youngest Mesozoic granitoids in the Jiaodong Terrane. Shandong Province. North China Craton. *Lithos* **2010**, *120*, 309–326. [CrossRef]
93. Yan, Q.S.; Metcalfe, I.; Shi, X.F.; Zhang, P.Y.; Li, F.C. Early Cretaceous granitic rocks from the southern Jiaodong Peninsula eastern China: Implications for lithospheric extension. *Int. Geol. Rev.* **2019**, *61*, 821–838.
94. Cai, W.Y.; Wang, Z.G.; Li, J.; Fu, L.J.; Wang, K.Y.; Yasssa, K.; Li, S.D. Zircon U–Pb and molybdenite Re–Os geochronology and geochemistry of Jinchang porphyry gold–copper deposit, NE China: Two-phase mineralization and the tectonic setting. *Ore Geol. Rev.* **2019**, *107*, 735–753.
95. Yang, F.; Santosh, M.; Kim, S.W. Mesozoic magmatism in the eastern North China Craton: Insights on tectonic cycles associated with progressive craton destruction. *Gondwana Res.* **2018**, *60*, 153–178.
96. Yang, F.; Santosh, M.; Glorie, S.; Jepson, G.; Xue, F.; Kim, S.W. Meso-Cenozoic multiple exhumation in the Shandong Peninsula, eastern North China Craton: Implications for lithospheric destruction. *Lithos* **2020**, *370–371*, 105597.
97. Groves, D.I.; Goldfarb, R.J.; Gebre-Mariam, M.; Hagemann, S.G.; Robert, F. Orogenic gold deposits: A proposed classification in the context of their crustal distribution and relationship to other gold deposit types. *Ore Geol. Rev.* **1998**, *13*, 7–27.
98. Pearce, J.A.; Harris, N.B.W.; Tindle, A.G. Trace element discrimination diagrams for the tectonic interpretation of granitic rocks. *J. Petrol.* **1984**, *25*, 956–983. [CrossRef]
99. Pearce, J.A.; Peate, D.W. Tectonic implications of the composition of volcanic arc magmas. *Annu. Rev. Earth Planet Sci.* **1995**, *23*, 251–285. [CrossRef]
100. Harris, R.A.; Stone, D.B.; Turner, D.L. Tectonic implications of paleomagnetic and geochronologic data from theYukon-Koyukuk province, Alaska. *Geol. Soc. Am. Bull.* **1987**, *99*, 362–375. [CrossRef]

MDPI AG
Grosspeteranlage 5
4052 Basel
Switzerland
Tel.: +41 61 683 77 34

Minerals Editorial Office
E-mail: minerals@mdpi.com
www.mdpi.com/journal/minerals

www.ingramcontent.com/pod-product-compliance
Lightning Source LLC
Chambersburg PA
CBHW041350210926
51850CB00013B/3036
* 9 7 8 3 7 2 5 8 3 4 0 6 8 *